# Riemann Surfaces
by Way of Complex
Analytic Geometry

# Riemann Surfaces by Way of Complex Analytic Geometry

Dror Varolin

Graduate Studies
in Mathematics
Volume 125

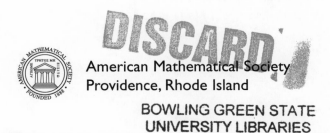

American Mathematical Society
Providence, Rhode Island

## EDITORIAL COMMITTEE
David Cox (Chair)
Rafe Mazzeo
Martin Scharlemann
Gigliola Staffilani

2010 *Mathematics Subject Classification.* Primary 30F10, 30F15, 30F30, 30F45, 30F99, 30G99, 31A05, 31A99, 32W05.

For additional information and updates on this book, visit
**www.ams.org/bookpages/gsm-125**

**Library of Congress Cataloging-in-Publication Data**
Varolin, Dror, 1970–
    Riemann surfaces by way of complex analytic geometry / Dror Varolin.
        p. cm. — (Graduate studies in mathematics ; v. 125)
    Includes bibliographical references and index.
    ISBN 978-0-8218-5369-6 (alk. paper)
    1. Riemann surfaces. 2. Functions of complex variables. 3. Geometry, Analytic. I. Title. II. Series.

QA333.V37   2011
515'.93—dc23
                                                                    2011014621

**Copying and reprinting.** Individual readers of this publication, and nonprofit libraries acting for them, are permitted to make fair use of the material, such as to copy a chapter for use in teaching or research. Permission is granted to quote brief passages from this publication in reviews, provided the customary acknowledgment of the source is given.

Republication, systematic copying, or multiple reproduction of any material in this publication is permitted only under license from the American Mathematical Society. Requests for such permission should be addressed to the Acquisitions Department, American Mathematical Society, 201 Charles Street, Providence, Rhode Island 02904-2294 USA. Requests can also be made by e-mail to reprint-permission@ams.org.

© 2011 by Dror Varolin. All rights reserved.
Printed in the United States of America.

∞ The paper used in this book is acid-free and falls within the guidelines
established to ensure permanence and durability.
Visit the AMS home page at http://www.ams.org/

10 9 8 7 6 5 4 3 2 1     16 15 14 13 12 11

Dedicated to Erin

# Contents

| | | |
|---|---|---|
| Preface | | xi |
| Chapter 1. Complex Analysis | | 1 |
| §1.1. | Green's Theorem and the Cauchy-Green Formula | 1 |
| §1.2. | Holomorphic functions and Cauchy Formulas | 2 |
| §1.3. | Power series | 3 |
| §1.4. | Isolated singularities of holomorphic functions | 4 |
| §1.5. | The Maximum Principle | 8 |
| §1.6. | Compactness theorems | 9 |
| §1.7. | Harmonic functions | 11 |
| §1.8. | Subharmonic functions | 14 |
| §1.9. | Exercises | 19 |
| Chapter 2. Riemann Surfaces | | 21 |
| §2.1. | Definition of a Riemann surface | 21 |
| §2.2. | Riemann surfaces as smooth 2-manifolds | 23 |
| §2.3. | Examples of Riemann surfaces | 25 |
| §2.4. | Exercises | 36 |
| Chapter 3. Functions on Riemann Surfaces | | 37 |
| §3.1. | Holomorphic and meromorphic functions | 37 |
| §3.2. | Global aspects of meromorphic functions | 42 |
| §3.3. | Holomorphic maps between Riemann surfaces | 45 |
| §3.4. | An example: Hyperelliptic surfaces | 54 |

| | | |
|---|---|---|
| §3.5. | Harmonic and subharmonic functions | 57 |
| §3.6. | Exercises | 59 |
| **Chapter 4.** | **Complex Line Bundles** | **61** |
| §4.1. | Complex line bundles | 61 |
| §4.2. | Holomorphic line bundles | 65 |
| §4.3. | Two canonically defined holomorphic line bundles | 66 |
| §4.4. | Holomorphic vector fields on a Riemann surface | 70 |
| §4.5. | Divisors and line bundles | 74 |
| §4.6. | Line bundles over $\mathbb{P}_n$ | 79 |
| §4.7. | Holomorphic sections and projective maps | 81 |
| §4.8. | A finiteness theorem | 84 |
| §4.9. | Exercises | 85 |
| **Chapter 5.** | **Complex Differential Forms** | **87** |
| §5.1. | Differential $(1,0)$-forms | 87 |
| §5.2. | $T_X^{*0,1}$ and $(0,1)$-forms | 89 |
| §5.3. | $T_X^*$ and 1-forms | 89 |
| §5.4. | $\Lambda_X^{1,1}$ and $(1,1)$-forms | 90 |
| §5.5. | Exterior algebra and calculus | 90 |
| §5.6. | Integration of forms | 92 |
| §5.7. | Residues | 95 |
| §5.8. | Homotopy and homology | 96 |
| §5.9. | Poincaré and Dolbeault Lemmas | 98 |
| §5.10. | Dolbeault cohomology | 99 |
| §5.11. | Exercises | 100 |
| **Chapter 6.** | **Calculus on Line Bundles** | **101** |
| §6.1. | Connections on line bundles | 101 |
| §6.2. | Hermitian metrics and connections | 104 |
| §6.3. | $(1,0)$-connections on holomorphic line bundles | 105 |
| §6.4. | The Chern connection | 106 |
| §6.5. | Curvature of the Chern connection | 107 |
| §6.6. | Chern numbers | 109 |
| §6.7. | Example: The holomorphic line bundle $T_X^{1,0}$ | 111 |
| §6.8. | Exercises | 112 |
| **Chapter 7.** | **Potential Theory** | **115** |

| §7.1. | The Dirichlet Problem and Perron's Method | 115 |
| §7.2. | Approximation on open Riemann surfaces | 126 |
| §7.3. | Exercises | 130 |

Chapter 8.  Solving $\bar{\partial}$ for Smooth Data — 133

| §8.1. | The basic result | 133 |
| §8.2. | Triviality of holomorphic line bundles | 134 |
| §8.3. | The Weierstrass Product Theorem | 135 |
| §8.4. | Meromorphic functions as quotients | 135 |
| §8.5. | The Mittag-Leffler Problem | 136 |
| §8.6. | The Poisson Equation on open Riemann surfaces | 140 |
| §8.7. | Exercises | 143 |

Chapter 9.  Harmonic Forms — 145

| §9.1. | The definition and basic properties of harmonic forms | 145 |
| §9.2. | Harmonic forms and cohomology | 149 |
| §9.3. | The Hodge decomposition of $\mathscr{E}(X)$ | 151 |
| §9.4. | Existence of positive line bundles | 157 |
| §9.5. | Proof of the Dolbeault-Serre isomorphism | 161 |
| §9.6. | Exercises | 161 |

Chapter 10.  Uniformization — 165

| §10.1. | Automorphisms of the complex plane, projective line, and unit disk | 165 |
| §10.2. | A review of covering spaces | 166 |
| §10.3. | The Uniformization Theorem | 168 |
| §10.4. | Proof of the Uniformization Theorem | 174 |
| §10.5. | Exercises | 175 |

Chapter 11.  Hörmander's Theorem — 177

| §11.1. | Hilbert spaces of sections | 177 |
| §11.2. | The Basic Identity | 180 |
| §11.3. | Hörmander's Theorem | 183 |
| §11.4. | Proof of the Korn-Lichtenstein Theorem | 184 |
| §11.5. | Exercises | 195 |

Chapter 12.  Embedding Riemann Surfaces — 197

| §12.1. | Controlling the derivatives of sections | 198 |
| §12.2. | Meromorphic sections of line bundles | 201 |

| | | |
|---|---|---|
| §12.3. | Plenitude of meromorphic functions | 202 |
| §12.4. | Kodaira's Embedding Theorem | 202 |
| §12.5. | Narasimhan's Embedding Theorem | 204 |
| §12.6. | Exercises | 210 |

Chapter 13.  The Riemann-Roch Theorem — 211
- §13.1. The Riemann-Roch Theorem — 211
- §13.2. Some corollaries — 217

Chapter 14.  Abel's Theorem — 223
- §14.1. Indefinite integration of holomorphic forms — 223
- §14.2. Riemann's Bilinear Relations — 225
- §14.3. The Reciprocity Theorem — 228
- §14.4. Proof of Abel's Theorem — 229
- §14.5. A discussion of Jacobi's Inversion Theorem — 231

Bibliography — 233

Index — 235

# Preface

**What**

This book presents certain parts of the basic theory of Riemann surfaces through methods of complex analytic geometry, many of which were developed at one time or another in the past 50 years or so.

The first chapter of the book presents a rapid review of the elementary part of basic complex analysis, introducing holomorphic, meromorphic, harmonic, and subharmonic functions and establishing some of the basic local properties of these functions. In the second chapter we define Riemann surfaces and give some examples, both of a concrete and abstract nature. The most abstract example is that any oriented 2-dimensional real manifold has a complex structure. In fact, we state the classical theorem of Korn-Lichtenstein: given any metric on an oriented surface, there is a complex structure that makes that metric locally conformal to the Euclidean metric in any holomorphic chart. The proof of this theorem is given only in Chapter 11.

Riemann surfaces are non-linear spaces on which, roughly speaking, local complex analysis makes sense, and so we can extend the inherently local notions of holomorphic, meromorphic, harmonic, and subharmonic functions to such surfaces. This point of view starts the third chapter, but by the end we prove our first main result, namely the constancy of the degree of a holomorphic map between two compact Riemann surfaces. With the recollection of some basic topology of surfaces, we are then able to establish the Riemann-Hurwitz Formula, which puts constraints on holomorphic maps between compact Riemann surfaces. The chapter ends with a proof of the Harnack Principle. Up to this point, the presentation is mostly classical, in the sense that it does not differ markedly from other presentations of the same material (though perhaps the particular blend of the ideas is unusual).

The novelty of the book begins in the fourth chapter, where we first define and give examples of complex and holomorphic line bundles. There are a number of important, albeit elementary, ideas that are presented in this chapter, and among these the correspondence between divisors and holomorphic line bundles with sections is first established. Another key idea is the correspondence between finite-dimensional vector spaces of sections of a holomorphic line bundle and projective maps. The chapter ends with an elementary proof of the finite-dimensionality of the space of all global holomorphic sections of a holomorphic line bundle over a compact Riemann surface.

The fifth chapter begins by defining complex differential forms. These forms play a central role in the rest of the book, but in this particular chapter we focus on their formal aspects—how to integrate them and how to describe the topology of a Riemann surface with them.

In the sixth chapter we develop the theory of differentiation for sections of complex line bundles (which is known as *defining a connection*). A complex line bundle admits many connections and, unlike the trivial bundle, there is no canonical choice of a connection for a general complex line bundle. The first goal of the chapter is to describe geometric structures that isolate a subset of the possible ways to differentiate sections. If a geometry is symmetric enough, there will be a canonical way to define differentiation of sections. The first instance of such a scenario is the classical Levi-Čivita Theorem for vector fields on a Riemannian manifold. The main fact we establish in Chapter 6, due to Chern, is that for each holomorphic line bundle with Hermitian metric there is a unique connection that is compatible with these structures in the appropriate sense. We also define the curvature of a connection, though we postpone a thorough demonstration of how curvature got its name. We demonstrate Chern's crucial observation that average curvature captures topological information.

We begin the seventh chapter with a discussion of potential theory. We use Perron's Method to solve the Dirichlet Problem and, what is almost the same thing, define a Green's Function on any Riemann surface that admits bounded non-constant subharmonic functions. Our use of the Green's Function is, to our knowledge, somewhat novel; we obtain a Cauchy-Green-type representation formula and use it to solve $\bar{\partial}$ on large (so topologically non-trivial) relatively compact domains with smooth boundary in a Riemann surface. In the second half we use the method of Runge-Behnke-Stein, together with our solution of $\bar{\partial}$, to approximate any local holomorphic function by global ones on an open Riemann surface, provided the complement of domain of the function to be approximated has no relatively compact components.

What was done in Chapter 7 immediately applies to obtain global solutions of the $\bar{\partial}$-equation on any open Riemann surface, and this is the first main result of Chapter 8. We then use this result to establish a number of classical facts about holomorphic and meromorphic functions on that Riemann surface. The chapter ends with an adaptation of the method of solving $\bar\partial$ to the problem of solving $\partial\bar\partial$.

If it were obvious that all Riemann surfaces admit holomorphic line bundles with metrics of positive curvature, our next chapter would be Chapter 11. However, the problem of finding a positive line bundle, especially on a compact Riemann surface, is a little bit subtle. We have three ways to find positive line bundles, and the first of these is linked to the intimate relationship between the Poisson Equation and curvature. We therefore establish in Chapter 9 the celebrated Hodge Theorem, in the case of Riemann surfaces. The Hodge Theorem is fundamental in its own right, and the higher-dimensional analog (whose proof is not much different) plays a significant role in algebraic geometry. After establishing the Hodge Theorem, we use it to complete the proofs of two theorems on cohomology of compact Riemann surfaces and to obtain positive line bundles.

Another approach to the existence of positive line bundles is through the Uniformization Theorem of Riemann-Köbe. We therefore establish this theorem in Chapter 10. From a direct observation, one then obtains the positivity of either the tangent or cotangent bundle of any Riemann surface, compact or open, that is not a complex torus. The case of a complex torus is treated in an *ad hoc* manner using Theta functions, which were defined in an example in Chapter 3.

In Chapter 11 we prove Hörmander's Theorem. By this point, we have all the geometric machinery we need, and we set up the Hilbert space method, which is a twisted version of part of the method in the proof of the Hodge Theorem. We then adapt the method to prove the Korn-Lichtenstein Theorem that we stated in Chapter 2.

Chapter 12 concerns the embedding problem for Riemann surfaces. If a Riemann surface is compact without boundary, it cannot be embedded in Euclidean space. (Indeed, if a Riemann surface is embedded in Euclidean space, the coordinate functions of the ambient space restrict to holomorphic functions on the surface, and since the surface is compact without boundary, the real parts (as well as the moduli) of these restricted coordinate functions have interior local maxima. By the Maximum Principle, the restrictions of the coordinate functions are thus locally constant, which contradicts the assumption that the surface is embedded.) The classical remedy for this problem was to puncture a few holes in the surface and embed the punctured surface using meromorphic functions with 'poles in the holes'. There is an interpretation of the resulting embedded surface as lying in projective space, but we take a more direct geometric approach and work directly on projective spaces. It is here that one sees the need for holomorphic line bundles most clearly.

In Chapter 13 we establish the Riemann-Roch Theorem. For an advanced book, our approach is elementary and admittedly less illuminating than a more sheaf-theoretic approach might be. Though the approach looks different from other proofs, it is simply a combination of two known facts: the usual sheaf-theoretic approach to the proof together with the realization of Serre duality (a result that is stated in the language of sheaf-theory) via residues.

The final chapter of the book states and proves Abel's Theorem, which characterizes linearly trivial divisors, i.e., divisors of meromorphic functions, on a compact Riemann surface. It is elementary to see—as we do on several occasions—that such a divisor must have degree zero and that if the surface in question has positive genus, this necessary condition is not sufficient. Based on computing the periods of global holomorphic forms, Abel constructed a map from the surface into a complex torus whose dimension is the genus of the surface. Abel's map extends to divisors in the obvious way because the torus is an additive group, and the map is a group homomorphism. Abel showed that a necessary and sufficient condition for a divisor to be linearly trivial is that this divisor is mapped to 0 in the torus under Abel's map. The torus is called the Jacobian of the Riemann surface, and Jacobi showed that the set of divisors on the surface maps surjectively onto the Jacobian. We do not prove Jacobi's Theorem, but we do give a sketch of the idea. The chapter, and thus the book, ends with an interpretation of the Abel-Jacobi Theorem as a classification of all holomorphic line bundles on a compact Riemann surface.

## Why

The present book arose from the need to bridge what I perceived as a substantial gap between what graduate students at Stony Brook know after they have passed their qualifying exams and higher-dimensional complex analytic geometry in its present state. At present, the generic post-qual student at Stony Brook is relatively well-prepared in algebraic topology and differential geometry but far from so in real and complex analysis or in partial differential equations (though two years after the first draft of this book, the number of students at Stony Brook interested either in algebraic geometry or in partial differential equations has increased significantly; those interested in both still form a very small, nearly empty set).

Courses in complex analysis do not typically emphasize the points most important in the study of Riemann surfaces, focusing instead on functions. For example, the courses focus on direct consequences of Cauchy's Theorem, the Schwarz Lemma, and the Riemann Mapping Theorem.

In this book, the Riemann Mapping Theorem makes way for results based on solving the inhomogeneous Cauchy-Riemann equations (and sometimes Poisson's equation). We present a number of methods for solving these equations, introducing and discussing Green's Functions and Runge-type approximation theorems for this purpose and giving a proof of the Hodge Theorem using basic Hilbert and Sobolev space theory. Perhaps the centerpiece is the Andreotti-Vesentini-Hörmander Theorem on the solution of $\bar{\partial}$ with $L^2$ estimates. (The theorem has come to be called *Hörmander's Theorem*, though the history is well known and the presentation we use here is closer in geometric spirit to Andreotti and Vesentini's version of the theorem.) The proof of Hörmander's Theorem in one complex dimension simplifies greatly, most naively because the $\bar{\partial}$-equation does not come with a compatibility condition, and more technically because a certain boundary condition that arises in the functional analytic formulation of the $\bar{\partial}$-problem on Hilbert spaces is a Dirichlet boundary condition, as opposed to its higher-dimensional and more temperamental relative, the $\bar{\partial}$-Neumann boundary condition of Spencer-Kohn-Morrey, which requires the introduction of the notion of pseudo-convexity of the boundary of a domain.

Line bundles play a major role in the book, providing the backdrop and geometric motivation for much of what is done. The 1-dimensional aspect makes the Hermitian geometry rather easy to deal with, and gives the novice a gentle introduction to the higher-dimensional differential geometry of Hermitian line bundles and vector bundles. For example the Kähler condition, which plays such an important role in the higher-dimensional theory, is automatic. Moreover, the meaning of curvature of line bundles, which is beautifully demonstrated in establishing what has come to be known as the Bochner-Kodaira Identity, is greatly simplified in complex dimension 1. (The Bochner-Kodaira Identity is used to estimate from below the smallest eigenvalue of the Laplace-Beltrami operator on sections of a Hermitian holomorphic line bundle.) The identity is obtained through integration by parts. At a certain point one must interchange the order of some exterior differential operators. When a non-flat geometry is present, the commutator of these operators is non-trivial and is the usual definition of curvature. If this commutator, which is a multiplier, is positive, then we obtain a positive lower bound for the smallest eigenvalue of a certain geometric Laplacian, and this lower bound is precisely what is needed to apply the functional analytic method to prove Hörmander's Theorem.

The main application of Hörmander's Theorem is to the existence and plenitude of holomorphic sections of sufficiently positive line bundles. Using these sections, we prove the existence of non-trivial meromorphic functions on Riemann surfaces, non-trivial meromorphic sections of any holomorphic line bundle, and the existence of a projective embedding for any compact Riemann surface—the Kodaira Embedding Theorem. Almost simultaneously, we prove that any open Riemann surface embeds in $\mathbb{C}^3$.

Despite embracing line bundles, I made the choice to avoid both vector bundles and sheaves, two natural extensions of line bundles. This choice shows simultaneously that (i) sheaves and higher-rank vector bundles are not needed in the basic theory of Riemann surfaces and (ii) at times the absence of vector bundles and sheaves makes the presentation cumbersome in certain places. A good example of (i) is the proof of Kodaira embedding without the artifice of sheaf cohomology, but rather through a simple-minded direct construction of certain sections using a beautiful idea first introduced by Bombieri. (Steve Zelditch has coined the perfect name: *designer sections*.) A reasonable example of (ii) is provided by a number of the results proved in Chapter 8, such as the Mittag-Leffler Theorem. A second example of (ii) is seen in our proof of the Riemann-Roch Theorem without the use of sheaves.

The book ends with two classical results, namely the theorems of Riemann-Roch and Abel. These results constitute an anti-climax of sorts, since they make little use of what was done in the book up to that point. In this regard, we do not add much to what is already in the literature, especially in the case of Abel's Theorem. The inclusion of the two theorems is motivated by seeing them as concluding remarks: the Riemann-Roch Theorem allows one to sharpen the Kodaira Embedding Theorem (or in the language of modern analytic geometry, give an effective embedding result), while Abel's Theorem and its complement, Jacobi's Inversion Theorem, are included because they provide a kind of classification for perhaps the most central group of characters in the book, namely, holomorphic line bundles.

At the urging of many people, I have included exercises for the first 12 chapters of the book. I chose to omit exercises for the last two chapters because, as I have suggested, they do not fall in line with the main pedagogical point of the book, namely, the use of the $\bar{\partial}$- and $\partial\bar{\partial}$-equations.

Of course, there are many glaring omissions that would appear in a standard treatise on Riemann surfaces. We do not discuss Weierstrass points and the finiteness of the automorphism group of a compact Riemann surface of genus at least two. Riemann's Theta Functions are not studied in any great detail. We bring them up only on the torus as a demonstrative tool. As a consequence, we do not discuss Torelli's Theorem. We also omit any serious discussion of basic algebraic geometry of curves (except a few brief remarks mostly scattered around the beginning and end of the book) or of monodromy. There are certainly other omissions, some of which I am not even aware of. Psychologically, the most difficult omission for me was that of a discussion of interpolation and sampling on so-called finite open Riemann surfaces. The theory of interpolation and sampling provides a natural setting (in fact, the only non-trivial natural setting I know in one complex dimension) in which to introduce the twisted $\bar{\partial}$-technique of Ohsawa-Takegoshi. This technique has had incredibly powerful applications in both several complex variables and algebraic geometry, and at the time of the writing of this book there remain many avenues of research to pursue. I chose to omit this topic because, by comparison with the rest of the book, it is disproportionately technical in nature.

## Who

The ideal audience for this book consists of students who are interested in analysis and geometry and have had basic first courses in real and complex analysis, differentiable manifolds, and topology. My greatest motivation in writing the book was to help such students in the transition from complex analysis to complex analytic geometry in higher dimensions, but I hope that the book will find a much wider audience.

In order to get through this book and emerge with a reasonable feeling for the subject, the reader must be at least somewhat prepared in the following sense.

I assume that the reader is well versed in advanced calculus and has seen basic differential topology. For example, the reader should be at ease with the definition of a manifold and the basics of integration of differential forms.

The reader should certainly have taken a first serious course in complex analysis. We state and basically prove all that we need from the early parts of such a course in the first chapter, but the presentation, though fairly complete, is terse and would not be the ideal place to learn the material.

Some minimal amount of real analysis is required in the book. For example, the reader should have seen the most elementary parts of the Riemann and Lebesgue theories of integration. On a couple of occasions we make use of the Hahn-Banach Theorem, the Banach-Alaoglu Theorem, and the Spectral Theorem for Compact, Self-adjoint Operators, but a deep understanding of the proofs of these theorems is not essential.

The topology of a Riemann surface certainly plays a role in much of the book, and in the later chapters the reader encounters a little bit of algebraic topology. Though the notions of homotopy, covering space, and fundamental group are introduced, there is not much detail for the uninitiated, and the reader truly interested in that part of the book should have preparation in those subjects and is moreover probably reading the wrong book on Riemann surfaces.

Very little basic familiarity with linear and, just barely, multilinear algebra is required on the part of the reader. So little is assumed that if the reader is not familiar with some of it but has the mathematical maturity of the aforementioned requirements, there should be no problem in filling the gaps during reading.

## How

In my days as a pizza delivery guy for Pizza Pizza in Toronto, I had a colleague named Vlad who used to say in a thick Russian accent: "No money, no funny!" I am grateful to the NSF for its generous financial support.

Much of what is presented in this book is motivated by the work of Jean-Pierre Demailly and Yum-Tong Siu, and I am grateful to both of them for all that they have taught me, both in their writings and in person. John D'Angelo and Jeff McNeal were very encouraging in the early parts of the project and gave me the inspiration I needed to start the project. Andy Raich and Colleen Robles read a preliminary version of these notes at Texas A&M, and Colleen communicated corrections and suggestions that were extremely useful. I am grateful to both of them. Steve Zelditch has used the notes for part of his course on Riemann surfaces and has sent back comments for which I owe him a debt of gratitude, as does the future reader of this book. The final blow was delivered at the University of Cincinnati, where I was fortunate enough to be the Taft Fellow for the spring quarter of 2010, for which I am grateful to the Charles Phelps Taft Foundation and all those involved in administering various parts of my fellowship. During the tenure of my Taft Fellowship I gave a mini-course, covering some of the material in the text, and most importantly I came up with exercises for the text. I am grateful to David Herron and David Minda for arranging the lectures and my visit. They attended my mini-course, as did Anders and Jana Bjorn, Robbie Buckingham, Andy Lorent, Diego Mejia, Mihaela Poplicher, and Nages Shanmugalingam, and I am grateful to all of them for their patience with my lectures and their generous hospitality.

There were a number of anonymous referees who communicated a number of valuable suggestions and corrections. I would like to thank them all for their service and help and apologize to them if I did not agree with all of their suggestions, perhaps wrongly.

The book would certainly not exist were it not for the efforts of Ina Mette, to whom I warmly express my heartfelt gratitude. Two other people at the AMS, namely Marcia Almeida and Arlene O'Sean, were instrumental in helping me to get the book finished, and I thank them for their help and their kind dealings.

Most of all, I am indebted to Mohan Ramachandran. Our frequent conversations, beyond giving me great pleasure, led to the inclusion of many important topics and to the correction of many errors, and Mohan's passion for the old literature taught me an enormous amount about the history and development of the subject.

**Where and when**

Dror Varolin
Brooklyn, NY
2010

*Chapter 1*

# Complex Analysis

## 1.1. Green's Theorem and the Cauchy-Green Formula

### 1.1.1. Green's Theorem.

THEOREM 1.1.1 (Green's Theorem). *Let $D \subset\subset \mathbb{R}^2$ be an open connected set[1] whose boundary $\partial D$ is piecewise smooth and positively oriented.[2] Let $P$ and $Q$ be smooth, complex-valued functions on some neighborhood of the closure $\overline{D}$ of $D$. Then*

$$\oint_{\partial D} P dx + Q dy = \iint_D \left( \frac{\partial Q}{\partial x} - \frac{\partial P}{\partial y} \right) dA.$$

A proof of Green's Theorem can be found in almost any calculus text.

### 1.1.2. The Cauchy-Green Formula.

Green's Theorem can be used to prove the Cauchy-Green Integral Formula. To state the latter, we recall Wirtinger's complex partial derivatives, defined as

$$\frac{\partial f}{\partial z} := \frac{1}{2}\left( \frac{\partial f}{\partial x} - \sqrt{-1}\frac{\partial f}{\partial y}\right) \quad \text{and} \quad \frac{\partial f}{\partial \bar{z}} := \frac{1}{2}\left( \frac{\partial f}{\partial x} + \sqrt{-1}\frac{\partial f}{\partial y}\right).$$

(The notation $A := B$ means that item $A$ is defined to be the known item $B$.)

THEOREM 1.1.2. *Let $D \subset\subset \mathbb{C}$ be open and connected with piecewise smooth boundary, and let $f : \overline{D} \to \mathbb{C}$ be a $\mathscr{C}^1$-function. Then for all $z \in D$,*

$$f(z) = \frac{1}{2\pi\sqrt{-1}} \int_{\partial D} \frac{f(\zeta)}{\zeta - z} d\zeta - \frac{1}{\pi} \iint_D \frac{\partial f}{\partial \bar{\zeta}} \frac{dA(\zeta)}{\zeta - z}.$$

---

[1] The notation $A \subset\subset B$ means '$A$ is relatively compact in $B$'.
[2] The boundary $\partial D$ is oriented positively if, when one moves *forward* along $\partial D$, one finds $D$ to one's left.

**Proof.** Fix $z \in D$. Fix $\varepsilon > 0$ such that $\mathbb{D}(z,\varepsilon) := \{x \in \mathbb{C} \,;\, |x-z| < \varepsilon\} \subset\subset D$. Applying Green's Theorem to the functions

$$P = \frac{1}{2\pi\sqrt{-1}} \frac{f(\zeta)}{\zeta - z} \quad \text{and} \quad Q = \frac{\sqrt{-1}}{2\pi\sqrt{-1}} \frac{f(\zeta)}{\zeta - z}$$

and the domain

$$D_\varepsilon := D - \mathbb{D}(z,\varepsilon),$$

we obtain, using the definition of the complex partial derivatives, the formula

$$(1.1) \quad \frac{1}{2\pi\sqrt{-1}} \int_{\partial D} \frac{f(\zeta)}{\zeta - z} d\zeta - \frac{1}{2\pi\sqrt{-1}} \int_{\partial \mathbb{D}(z,\varepsilon)} \frac{f(\zeta)}{\zeta - z} d\zeta = \frac{1}{\pi} \iint_{D_\varepsilon} \frac{\partial f}{\partial \bar\zeta} \frac{dA(\zeta)}{\zeta - z}.$$

In polar coordinates centered at $z$, the measure $dA$ is $r\,dr\,d\theta$, and since $\zeta - z = re^{i\theta}$, the integrand of the right integral is locally bounded. It follows that, as $\varepsilon \to 0$, the right side converges to

$$\frac{1}{\pi} \iint_{D} \frac{\partial f}{\partial \bar\zeta} \frac{dA(\zeta)}{\zeta - z}.$$

On the other hand, since $f$ is differentiable, we have

$$\int_{\partial \mathbb{D}(z,\varepsilon)} \frac{f(\zeta)}{\zeta - z} d\zeta = \int_{\partial \mathbb{D}(z,\varepsilon)} \frac{f(\zeta) - f(z)}{\zeta - z} d\zeta + 2\pi\sqrt{-1} f(z) \to 2\pi\sqrt{-1} f(z)$$

as $\varepsilon \to 0$. □

## 1.2. Holomorphic functions and Cauchy Formulas

**1.2.1. The homogeneous Cauchy-Riemann equations.** Recall the definition of a holomorphic function.

**DEFINITION 1.2.1.** A $\mathscr{C}^1$ function $f$ is *holomorphic* in a domain $D$ if and only if it satisfies the homogeneous Cauchy-Riemann equations

$$\frac{\partial f}{\partial \bar z} \equiv 0$$

on $D$. ◇

**REMARK.** As is well known, one can relax the condition that $f$ is $\mathscr{C}^1$ to assume only that $f$ is a distribution.[3] We will not worry so much about regularity issues in the definition of holomorphic functions, since such issues are often thoroughly treated in a first serious course in complex analysis. Later in this chapter we will establish regularity for harmonic functions, which can be used to prove the regularity of a holomorphic distribution. ◇

---

[3]It is assumed that the reader is familiar with distributions, though the lack of familiarity with this notion does not pose a significant problem to following this book. References for distributions abound, but perhaps the most thorough reference relevant to the present book is [**Hörmander-2003**].

## 1.3. Power series

REMARK. Some readers may wonder why the author refers to this equation in the plural form. The reason is classical; the equation is a complex one and therefore has two components. Moreover, from the point of view of partial differential equations, the pair of equations obtained by taking real and imaginary parts form an elliptic system, and ultimately this ellipticity is responsible for the regularity of the functions in the kernel, though in this setting one can give a direct proof without referring to ellipticity. In any case, we have chosen to keep the plural. ⋄

**1.2.2. Cauchy's Theorem and the Cauchy Integral Formula.** As a corollary of the Cauchy-Green Formula, we obtain the Cauchy Theorem and the Cauchy Integral Formula for $\mathscr{C}^1$-functions.

THEOREM 1.2.2. *Let $D \subset\subset \mathbb{C}$ be open with piecewise smooth boundary, and let $f$ be holomorphic on a neighborhood of the closure of $D$. Then the following hold.*

(1) *(Cauchy's Theorem)*
$$\int_{\partial D} f(z)dz = 0.$$

(2) *(Cauchy Integral Formula)*
$$f(z) = \frac{1}{2\pi\sqrt{-1}} \int_{\partial D} \frac{f(\zeta)}{\zeta - z} d\zeta.$$

**Proof.** The second result is obvious from the Cauchy-Green Formula. For the first result, apply the Cauchy-Green Formula to the function $z \mapsto (z - \zeta)f(z)$. □

### 1.3. Power series

**1.3.1. Series representation.** Suppose $f$ is holomorphic in a neighborhood of the closure of the disk $\mathbb{D}(0, r)$. For $z \in \mathbb{D}(0, r)$ and $|\zeta| = r$ we have the series
$$\frac{1}{\zeta - z} = \frac{1}{\zeta(1 - \frac{z}{\zeta})} = \sum_{j=0}^{\infty} \frac{z^j}{\zeta^{j+1}}.$$

Applying the Cauchy Integral Formula, we find that
$$f(z) = \sum_{j=0}^{\infty} a_j z^j, \quad \text{where} \quad a_j = a_j(f, 0) := \frac{1}{2\pi\sqrt{-1}} \int_{|\zeta|=r} \frac{f(\zeta)d\zeta}{\zeta^{n+1}}.$$

In particular, $f$ is given locally by a convergent power series. A similar result, properly scaled and shifted, holds at any point $z_o$ in place of the origin:
$$f(z) = \sum_{j=1}^{\infty} a_j(z - z_o)^j,$$

where
$$a_j = a_j(f, z_o) = \frac{1}{2\pi\sqrt{-1}} \int_{|\zeta - z_o|=r} \frac{f(\zeta)d\zeta}{(\zeta - z_o)^{n+1}}.$$

It follows that $f$ is infinitely complex-differentiable and that

$$(1.2) \qquad \left.\frac{\partial^n f}{\partial z^n}\right|_{z=z_o} = \frac{n!}{2\pi\sqrt{-1}} \int_{|\zeta-z_o|=r} \frac{f(\zeta)d\zeta}{(\zeta-z_o)^{n+1}}.$$

**1.3.2. Corollaries.** The fact that $f$ is given by a convergent power series implies the following theorems.

THEOREM 1.3.1. *The uniform limit of holomorphic functions is holomorphic.*

THEOREM 1.3.2 (Identity Theorem). *If two holomorphic functions are defined on an open connected set $D \subset \mathbb{C}$ and they agree on an open subset of $D$, then they agree on all of $D$.*

We also have the following fundamental definition.

DEFINITION 1.3.3 (Order). Let $f$ be holomorphic in a neighborhood of a point $x \in \mathbb{C}$. We say that $f$ *has a zero of order $n$ at $x$* if $n$ is the smallest integer such that

$$\left.\frac{\partial^n f}{\partial z^n}\right|_{z=z_o} \neq 0.$$

We write $n = \mathrm{Ord}_x f$. ◇

**1.3.3. Cauchy Estimates.** Estimating (1.2) gives us the following result.

PROPOSITION 1.3.4 (Cauchy Estimates). *Let $f : \overline{\mathbb{D}(0,R)} \to \mathbb{C}$ be a holomorphic function. Then*

$$\left|\frac{\partial^n f}{\partial z^n}\right|_{z=0} \leq \frac{n!}{R^n} \sup_{\mathbb{D}(0,R)} |f|.$$

As a corollary, we have Liouville's Theorem.

COROLLARY 1.3.5 (Liouville). *Any bounded holomorphic function $f : \mathbb{C} \to \mathbb{C}$ is constant.*

**Proof.** Let $C := \sup_{\mathbb{C}} |f|$. For any $p \in \mathbb{C}$ and $R > 0$, $|f'(p)| \leq CR^{-1}$. Thus $f' \equiv 0$. □

## 1.4. Isolated singularities of holomorphic functions

**1.4.1. Laurent series.** Let $A$ be an annulus centered at 0 with radii $R > r \geq 0$, and let $f$ be holomorphic on $A$. Choose $R'$ and $r'$ such that $R > R' > r' > r$. Then by the Cauchy Integral Formula,

$$f(z) = \frac{1}{2\pi\sqrt{-1}} \int_{|\zeta|=R'} \frac{f(\zeta)}{\zeta-z}d\zeta - \frac{1}{2\pi\sqrt{-1}} \int_{|\zeta|=r'} \frac{f(\zeta)}{\zeta-z}d\zeta, \qquad r' < |z| < R'.$$

## 1.4. Isolated singularities of holomorphic functions

In the first integral, $|z| < |\zeta| = R'$, while in the second $r' = |\zeta| < |z|$. If we write

$$\zeta - z = \zeta\left(1 - \frac{z}{\zeta}\right) \qquad \text{and} \qquad \zeta - z = -z\left(1 - \frac{\zeta}{z}\right)$$

in the first integral and in the second integral, respectively, then using the fact that that $(1-r)^{-1} = 1 + r + r^2 + \ldots$ we find

(1.3) $$f(z) = \sum_{n \in \mathbb{Z}} a_n z^n,$$

where

(1.4) $$a_n = \frac{1}{2\pi\sqrt{-1}} \int_{|z|=\rho} \frac{f(z)}{z^{n+1}} dz,$$

for some $\rho \in (r, R)$. Since $f(z)/z^n$ is analytic in $A$, the latter integral is independent of $\rho$.

DEFINITION 1.4.1. The series (1.3) is called the *Laurent series* of $f$. The number

$$\text{Res}_0(f) := a_{-1}$$

is called the *residue* of $f$ at 0. ◊

More generally, if $c \in \mathbb{C}$ and $g$ is holomorphic in a punctured neighborhood of $c$ (i.e., a neighborhood of $c$ from which $c$ is then removed), we define

$$\text{Res}_c(g) := \text{Res}_0(f), \qquad \text{where } f(z) = g(z+c).$$

Using the Laurent series of $f$ at each singularity of $f$, one obtains the following theorem.

THEOREM 1.4.2 (Residue Theorem). *If $z_1, \ldots, z_k \in D$ are distinct and $f : \overline{D} - \{z_1, \ldots, z_k\} \to \mathbb{C}$ is holomorphic, then*

$$\frac{1}{2\pi\sqrt{-1}} \int_{\partial D} f(z) dz = \text{Res}_{z_1}(f) + \ldots + \text{Res}_{z_k}(f).$$

### 1.4.2. Singularities of holomorphic functions at a puncture.

DEFINITION 1.4.3. Let $f$ be holomorphic in a punctured neighborhood of a point $z \in \mathbb{C}$.

(1) We say that $f$ has *a pole of order* $n \geq 1$ if the Laurent series $\sum a_j (\zeta - z)^j$ of $f$ at $z$ has the property that $a_{-n} \neq 0$ but $a_{-k} = 0$ for $k > n$.

(2) If $a_j = 0$ for all $j < 0$ at $z$, i.e., the Laurent series of $f$ at $z$ is holomorphic across $z$, we say that $z$ is a *removable singularity* of $f$.

(3) If the Laurent series of $f$ has infinitely many negative coefficients, we say that $f$ has an *essential singularity* at $z$.

(4) If $f$ does not have an essential singularity at $z$, we define
$$\operatorname{Ord}_z(f) := k$$
where $k$ is the largest integer such that $a_\ell = 0$ for each $\ell < k$. ◇

REMARK. Note that if $f$ has a pole of order $k > 0$ then $\operatorname{Ord}_x(f) = -k$. ◇

DEFINITION 1.4.4 (Meromorphic function). Let $D \subset \mathbb{C}$.

(1) We say that $f$ is *meromorphic at a point* $p \in D$ if there is a punctured neighborhood $U$ of $p$ in $D$ such that $f$ is holomorphic on $U$ and $p$ is not an essential singularity for $f$.

(2) We say that $f$ is *meromorphic on* $D$ if $f$ is meromorphic at all points of $D$. ◇

We have the following characterization of singularities.

THEOREM 1.4.5. *Let $U$ be a neighborhood of $p \in \mathbb{C}$ and let $f : U - \{p\} \to \mathbb{C}$ be holomorphic.*

(1) *(Riemann Removable Singularities Theorem) If $f$ is bounded in a neighborhood of $p$ then $f$ extends holomorphically to $U$. In particular, the limit $\lim_{z \to p} f(z)$ exists.*

(2) *If $\lim_{z \to p} |f(z)| = +\infty$, then $f$ has a pole at $z$.*

**Proof.** We may assume $p = 0$.

(1) Suppose $|f| \leq M$ in a small disk centered at the origin and let $\sum_{n \in \mathbb{Z}} a_n z$ be the Laurent series of $f$ at $0$. By (1.4), $|a_n| \leq 2\pi M \varepsilon^{-n}$ for all sufficiently small $\varepsilon$. For $n < 0$ this means that $a_n = 0$.

(2) Choose a disk $D$ centered at the origin such that $f$ does not vanish on $\overline{D}$. Then the function $g(z) = 1/f(z)$ is holomorphic and bounded on $D - \{0\}$. By part (1), $g$ extends to a holomorphic function on $D$, and since it has limit zero at $0$, the power series expansion on $g$ shows there is an integer $k \geq 1$ such that $g(z) = z^k h(z)$ for some holomorphic function $h$ satisfying $h(0) \neq 0$. Thus near the origin, $f(z) = z^{-k} \varphi(z)$ for some holomorphic function $\varphi$ satisfying $\varphi(0) \neq 0$. □

REMARK (Weierstrass-Casorati Theorem). If $f$ is unbounded on $U - \{p\}$ and $|f(q)|$ is not arbitrarily large for all points $q$ sufficiently near $p$, then according to Theorem 1.4.5, $f$ has an essential singularity at $p$. In this case we can say more: the image of $U$ is dense in $\mathbb{C}$. Indeed, if not, then there is a disk $D \subset \mathbb{C} \cup \{\infty\}$ such that $f(U - \{p\}) \subset D$. We can then choose constants $a, b, c, e \in \mathbb{C}$ such that
$$g(z) = \frac{a + bf(z)}{c + ef(z)} \in \mathbb{D}(0, 1) \qquad \text{for all } z \in U.$$
Then $g$ extends to $p$, contradicting that $f$ has an essential singularity. ◇

## 1.4. Isolated singularities of holomorphic functions

PROPOSITION 1.4.6. *If $D$ is a disk and $g : D \to \mathbb{C} - \{0\}$ is holomorphic then $g = e^h$ for some holomorphic function $h$ on $D$.*

**Proof.** Fix $p \in D$. Since $g'(z)/g(z)$ is holomorphic on $D$, we can define
$$H(z) = \int_p^z \frac{g'(\zeta)d\zeta}{g(\zeta)},$$
where the integral is over some path connecting $p$ to $z$. By Cauchy's Theorem, the integral is independent of the path connecting $p$ to $z$. We therefore find that $H'(z) = g'(z)/g(z)$, and hence
$$\frac{d}{dz}\left(\frac{g(z)}{e^{H(z)}}\right) = \frac{g'(z)}{e^{H(z)}} - \frac{g(z)H'(z)}{e^{H(z)}} = 0.$$
Thus $g = ce^H = e^h$. □

COROLLARY 1.4.7 (Normal Form Theorem). *If $f$ is holomorphic near $p$ and $k = \mathrm{Ord}_p(f)$ then there is a neighborhood $U$ of $p$ and an injective holomorphic function $g : U \to \mathbb{C}$ such that*
$$f(g^{-1}(w)) = w^k.$$

**Proof.** We may assume $p = 0$. By power series, we can write $f(z) = z^k G(z)$ with $G(0) \neq 0$. According to Proposition 1.4.6, $f(z) = z^k e^{h(z)}$ for some holomorphic $h$. Then $f(z) = (g(z))^k$ with $g(z) = ze^{k^{-1}h(z)}$. Since $g'(0) \neq 0$, $g$ is injective by the Inverse Function Theorem. □

REMARK. If $|a|$ is sufficiently small but non-zero, then the equation $f(z) = a$ has exactly $k$ distinct solutions $p_1, ..., p_k$ in $U$, namely $p_j = g^{-1}(|a|^{1/k}\omega_k^j)$, where $\omega_k$ is a $k^{\text{th}}$ root of unity. For this reason, we sometimes call the number $k$ the multiplicity of $f$ at $p$. ◇

The next theorem is an easy and important consequence of Corollary 1.4.7.

THEOREM 1.4.8. *Let $U \subset \mathbb{C}$ be open.*

(1) *If $f : U \to V$ is injective holomorphic then so is $f^{-1} : f(U) \to U$.*

(2) *If $f : U \to \mathbb{C}$ is a non-constant holomorphic function, then $f(U)$ is open.*

(3) *The zero set of a non-constant holomorphic function is discrete.*

Let $f$ be holomorphic in a punctured neighborhood of $0$ and assume $0$ is not an essential singularity. In this case, there is an integer $k$ and a holomorphic function $g$ such that $f(z) = z^k g(z)$ and $g(0) \neq 0$. Therefore
$$\frac{f'(z)}{f(z)} = \frac{k}{z} + \frac{g'(z)}{g(z)}.$$
Modification of this argument to points other than the origin yields the following theorem.

THEOREM 1.4.9 (Argument Principle). *Let $D \subset \mathbb{C}$ be an open connected set with piecewise smooth boundary, and let $f$ be a meromorphic function on a neighborhood of $D$ having no zeros or poles on $\partial D$. Then*

$$\frac{1}{2\pi\sqrt{-1}} \int_{\partial D} \frac{f'(z)}{f(z)} dz = \sum_{z \in D} \mathrm{Ord}_z(f).$$

## 1.5. The Maximum Principle

THEOREM 1.5.1 (Maximum Principle). *If $f : \{|z| < r\} \to \mathbb{C}$ is holomorphic and there is a local maximum for $|f|$ at some $z_0$ with $|z_0| < r$, then $f$ is constant.*

**Proof.** We can assume $f(z_0) \neq 0$. For $\rho > 0$ small, we find from the Cauchy Formula that

$$f(z_0) = \frac{1}{2\pi} \int_0^{2\pi} f(z_0 + \rho e^{\sqrt{-1}\theta}) d\theta$$

for any $\rho < \mathrm{dist}(z_0, \{|z| \geq r\})$. Rearranging and taking real parts, we have

$$\int_0^{2\pi} \mathrm{Re}\left(1 - \frac{f(z_0 + \rho e^{\sqrt{-1}\theta})}{f(z_0)}\right) d\theta = 0.$$

By hypothesis, the integrand is continuous and non-negative, and therefore $\mathrm{Re}\, f$ is constant. By the Cauchy-Riemann equations, $f$ is constant. $\square$

From the Maximum Principle we obtain the well-known Schwarz Lemma.

THEOREM 1.5.2 (Schwarz Lemma). *Let $f : \mathbb{D} \to \mathbb{D}$ be holomorphic and suppose $f(0) = 0$. Then for all $z \in \mathbb{D}$, $|f(z)| \leq |z|$. In particular, $|f'(0)| \leq 1$. Equality holds in the first estimate for some $z \in \mathbb{D} - \{0\}$ or in the second estimate if and only if $f(z) = e^{\sqrt{-1}\theta} z$ for some constant $\theta \in \mathbb{R}$.*

**Proof.** Fix $r \in (0,1)$. Let $g_r(z) := \frac{f(rz)}{rz}$ for $z \neq 0$ and $g_r(0) := f'(0)$. Then $g_r$ is holomorphic on $\{|z| < 1/r\}$, and thus we have

$$|g_r(z)| \leq \max_{|\zeta|=1} |g_r(\zeta)| = \max_{|\zeta|=1} |f(r\zeta)|/r \leq 1/r.$$

Thus $|f(rz)| \leq |z|$. Letting $r \to 1$ gives the estimates.

If equality holds at any point $z_o \in \mathbb{D}$, i.e., $g_1(z_o) = 1$, then for some $r \in (|z_o|, 1)$, $g_r(z_o/r) = 1$. By the Maximum Principle $g_r$ is constant. Thus $|f(rz)| = |rz|$ holds for $|z| < 1/r$. It follows that $f(z) = e^{\sqrt{-1}\theta} z$ for some constant $\theta \in \mathbb{R}$. The proof is complete. $\square$

## 1.6. Compactness theorems

### 1.6.1. Montel's Theorem.

THEOREM 1.6.1 (Montel). *Let $U \subset \mathbb{C}$ be an open set. Let $K_1 \subset\subset K_2 \subset\subset \ldots \subset U$ be a sequence of compact sets such that any compact subset $K \subset\subset U$ is contained in some $K_j$. Let $M_1 \leq M_2 \leq \ldots$ . Then the set of holomorphic functions*

$$\mathscr{B}_{\mathfrak{K},\mathfrak{M}} := \{f : U \to \mathbb{C} \, ; \, \sup_{K_j} |f| \leq M_j, \, j = 1, 2, \ldots\}$$

*is compact in $\mathcal{O}(U)$ in the compact-open topology.*

**Proof.** By the Cauchy Estimates, the family $\mathscr{B}_{\mathfrak{K},\mathfrak{M}}$ is uniformly bounded and equicontinuous on each open subset $V \subset\subset U$. By the Arzela-Ascoli Theorem, every sequence in $\mathscr{B}_{\mathfrak{K},\mathfrak{M}}|_{K_j}$ has a uniformly convergent subsequence. By a diagonal argument, every sequence in $\mathscr{B}_{\mathfrak{K},\mathfrak{M}}$ has a uniformly convergent subsequence. An application of Theorem 1.3.1 completes the proof. $\square$

COROLLARY 1.6.2. *Let $U \subset \mathbb{C}$ be an open set such that $\mathbb{C} - U$ has interior points. Then the set*

$$\mathscr{F} := \{f \in \mathcal{O}(\mathbb{D}) \, ; \, f(\mathbb{D}) \subset U\}$$

*is compact in $\mathcal{O}(D)$ in the compact-open topology.*

**Proof.** Let $p \in \mathbb{C} - U$ be an interior point. The transformation $f \mapsto (f - p)^{-1}$ maps $\mathscr{F}$ to a family of uniformly bounded holomorphic functions. An application of Montel's Theorem completes the proof. $\square$

### 1.6.2. Köbe's Compactness Theorem.
A fundamental result in complex analysis is the following theorem of P. Köbe.

THEOREM 1.6.3 (Köbe). *The collection $\mathscr{S}$ of all injective holomorphic functions $f : \mathbb{D} \to \mathbb{C}$ such that $f(0) = 0$ and $f'(0) = 1$ is compact in $\mathcal{O}(\mathbb{D})$ in the compact-open topology.*

**Proof.** Fix a sequence $\{f_n\} \subset \mathscr{S}$. Let

$$R_n := \sup\{R > 0 \, ; \, \mathbb{D}(0, R) \subset f_n(\mathbb{D})\}.$$

Since $f_n^{-1}|_{\mathbb{D}(0,R)} : \mathbb{D}(0, R) \to \mathbb{D}$ is holomorphic for any $\mathbb{D}(0, R) \subset f(\mathbb{D})$, the Schwarz Lemma implies that $R_n \leq 1$. Choose a point $x_n \in \partial\mathbb{D}(0, R_n) - f_n(\mathbb{D})$ and let $g_n := f_n/x_n$. Then $\mathbb{D} \subset g_n(\mathbb{D}) \not\ni 1$.

Now, $g_n(\mathbb{D})$ is simply connected, so there is a holomorphic branch $\psi$ of $\sqrt{z-1}$ such that $\psi(0) = \sqrt{-1}$. Then $h_n := \psi \circ g_n$ satisfies $h_n^2 = g_n - 1$.

We claim that $h_n(\mathbb{D}) \cap (-h_n(\mathbb{D})) = \emptyset$. Indeed, if $w = h_n(z)$ and $-w = h_n(z')$, then $g_n(-w) = g_n(w)$ and by injectivity $w = -w = 0$. But then $g_n(z) = 1$, which is impossible.

Since $\mathbb{D} \subset g_n(\mathbb{D})$, we have $U := \psi(\mathbb{D}) \subset h_n(\mathbb{D})$, and thus $(-U) \cap h_n(\mathbb{D}) = \emptyset$. By Corollary 1.6.2, $h_n$ has a convergent subsequence. Thus, since $|x_n| = R_n \leq 1$, $f_n = x_n(1 + h_n^2)$ has a convergent subsequence. Let $f$ be the limit of this subsequence.

Let $a \in f(\mathbb{D})$. By the argument principle, for any $z$ with $f(z) \neq a$,
$$\ell := \frac{1}{2\pi\sqrt{-1}} \int_{|z|=r} \frac{f'(z)dz}{f(z) - a}$$
is an integer, which evidently is $\geq 1$ for $r$ sufficiently close to 1. Since $f_{n_k}$ is arbitrarily close to $f$ on $|z| \leq r$, the injectivity of $f_{n_k}$ implies that
$$1 = \frac{1}{2\pi\sqrt{-1}} \int_{|z|=r} \frac{f'_{n_k}(z)dz}{f_{n_k}(z) - a} = \frac{1}{2\pi\sqrt{-1}} \int_{|z|=r} \frac{f'(z)dz}{f(z) - a} = \ell.$$
Thus $f$ is injective, and the proof is complete. □

REMARK. Clearly for each $f \in \mathscr{S}$, $f(\mathbb{D})$ contains a disk centered at the origin. Since $\mathscr{S}$ is compact, the radius of this disk is bounded below. For many applications, the fact that the radius is bounded below suffices. Nevertheless, Köbe conjectured that the smallest possible radius is $1/4$. Köbe knew that $1/4$ was sharp, realized for the following function:
$$f(z) = \frac{1}{4}\left(1 - \left(\frac{z-1}{z+1}\right)^2\right).$$
Köbe's conjecture was proved by Bieberbach, and though we will not need it, we will give a proof here. To this end, let $f \in \mathscr{S}$. Suppose $w \in \mathbb{C} - f(\mathbb{D})$. Then, with
$$g(z) := \frac{wf(z)}{w - f(z)},$$
$g \in \mathscr{S}$. Note that $g(0) = 0$, $g'(0) = 1$, and $g''(0) = f''(0) + 2w^{-1}$. Now, since $\mathscr{S}$ is compact,
$$D_2 := \sup\{|f''(0)| \,;\, f \in \mathscr{S}\}$$
is finite. We therefore have the estimate
$$D_2 \geq |g''(0)| \geq 2|w|^{-1} - |f''(0)| \geq 2|w|^{-1} - D_2,$$
so that $|w| \geq D_2^{-1}$.

Observe that if $f \in \mathscr{S}$ then the functions
$$\hat{f}(z) = (f(z^{-1}))^{-1} \quad \text{and} \quad \tilde{f}(z) := \frac{1}{\sqrt{f(z^{-2})}},$$
where we take the branch of $\sqrt{\phantom{x}}$ sending $-1$ to $\sqrt{-1}$, are both injective on the complement of the unit disk and have a Laurent series of the form
(1.5) $$z + b_1 z^{-1} + b_2 z^{-2} + \cdots.$$

Let us denote by $\Sigma$ the set of univalent maps of $\mathbb{C} - \mathbb{D}$ whose Laurent series has the form (1.5). For any $g \in \Sigma$ and any $r > 1$,

$$0 \leq \text{Area inside}(g(\partial \mathbb{D}(0,r))) = \frac{1}{2\sqrt{-1}} \int_{g(\partial \mathbb{D}(0,r))} \bar{w} dw$$

$$= \frac{1}{2\sqrt{-1}} \int_{|z|=r} \overline{g(z)} g'(z) dz.$$

As the reader can check,

$$\lim_{r \to 1} \frac{1}{2\sqrt{-1}} \int_{|z|=r} \overline{g(z)} g'(z) dz = \pi \left( 1 - \sum_{n=1}^{\infty} n |b_n|^2 \right).$$

We therefore see that $|b_1| \leq 1$.

Finally, the reader can check that $\tilde{f}(z) = z - \frac{f''(0)}{4} z^{-1} + \cdots$. Our bound for $|b_1|$ on $\Sigma$ thus shows $D_2 \leq 4$. We have therefore proved the following result.

THEOREM 1.6.4 (Köbe $\frac{1}{4}$-Theorem). *For all $f \in \mathscr{S}$, $f(\mathbb{D}) \supset \mathbb{D}(0, \frac{1}{4})$, and the number $\frac{1}{4}$ is sharp.* ◇

## 1.7. Harmonic functions

**1.7.1. Laplacian and harmonic.** The Laplace operator (or Laplacian) $\Delta$ is defined in a plane domain by

$$\Delta u = u_{xx} + u_{yy} = 4 \frac{\partial^2 u}{\partial z \partial \bar{z}}.$$

DEFINITION 1.7.1. A function $u \in \mathscr{C}^2(\Omega)$ is said to be harmonic if $\Delta u \equiv 0$, in which case we write $u \in H(\Omega)$. ◇

REMARK. As we will see, the requirement that $H(\Omega) \subset \mathscr{C}^2(\Omega)$ can be substantially reduced. ◇

REMARK. Since $\Delta$ is a real operator, $f = u + \sqrt{-1} v \in H(\Omega)$ if and only if $u, v \in H(\Omega)$. ◇

**1.7.2. Harmonic conjugate.** From the formula $\Delta = 4 \frac{\partial^2}{\partial z \partial \bar{z}}$, one sees that every holomorphic function is harmonic, as is the complex conjugate of a holomorphic function. On a simply connected domain, the converse holds, i.e., every harmonic function is the real part of a holomorphic function, which may be seen as follows. Let

$$d^c := \frac{1}{2\sqrt{-1}} \left( \partial - \bar{\partial} \right) = \frac{1}{2}(-dx \otimes \partial_y + dy \otimes \partial_x).$$

Then for a function $h$ one has

$$dd^c h = \sqrt{-1} \partial \bar{\partial} h = 2 (\Delta h) \, dx \wedge dy.$$

(A review of exterior calculus, in the setting of Riemann surfaces, is presented in Chapter 5.) Suppose $u$ is a real-valued harmonic function on a simply connected domain $\Omega \subset \mathbb{C}$. Then $d^c u$ is a closed 1-form. It essentially follows from Green's Theorem and the simple connectivity of $\Omega$ that, given $z_o \in \Omega$ and a curve $\gamma_{z_o,z} : [0,1] \to \Omega$ satisfying $\gamma_{z_o,z}(0) = z_o$ and $\gamma_{z_o,z}(1) = z$, the function

$$v(z) := 2 \int_{\gamma_{z_o,z}} d^c u$$

is well defined independent of the choice of $\gamma_{z_o,z}$. (If we change $z_o$, the function $v(z)$ changes by an additive constant.)

We claim that the function $f = u + \sqrt{-1}v$ is holomorphic. Indeed, since $v$ is independent of the curve $\gamma_{z_o,z}$, by choosing the $z$-end of $\gamma_{z_o,z}$ to be horizontal or vertical, one obtains from the Fundamental Theorem of Calculus that

$$v_x = -u_y \quad \text{and} \quad v_y = u_x.$$

These equations mean precisely that $\bar{\partial} f = 0$, and our claim is proved.

### 1.7.3. Mean values on circles.

PROPOSITION 1.7.2 (Mean Value Property). *If $u \in H(\Omega)$ and $\mathbb{D}(p,r) \subset \Omega$ then*

$$u(p) = \frac{1}{2\pi} \int_0^{2\pi} u(p + re^{\sqrt{-1}\theta}) d\theta.$$

**Proof.** Let

$$A_r(u) := \frac{1}{2\pi} \int_0^{2\pi} u(p + re^{\sqrt{-1}\theta}) d\theta.$$

By continuity, $\lim_{r \to 0} A_r(u) = u(p)$. By the divergence form of Green's Theorem,

$$\frac{dA_r(u)}{dr} = \frac{1}{2\pi r} \int_{|z-p|<r} \Delta u \, dx dy = 0.$$

The proof is complete. $\square$

COROLLARY 1.7.3. *Let $\psi$ be an integrable, compactly supported, radially symmetric function (i.e., $\psi(z) = \psi(|z|)$ for all $z \in \mathbb{C}$) such that $\int_{\mathbb{C}} \psi dA = 1$. Then*

$$u(x) = \int_{\mathbb{C}} \psi(z) u(z+x) dA(z)$$

*for all $u \in H(\Omega)$ and all $x \in \Omega$ such that $x + \text{Supp}(\psi) \subset \Omega$.*

### 1.7.4. The Maximum Principle.
The proof of Theorem 1.5.1 for holomorphic functions goes through for harmonic functions.

COROLLARY 1.7.4. *Let $\Omega \subset \mathbb{C}$ be open and connected. If $u \in H(\Omega)$ has a local maximum in $\Omega$ then $u$ is constant.*

REMARK. Since $u \in H(\Omega) \Rightarrow -u \in H(\Omega)$, harmonic functions also satisfy a minimum principle. $\diamond$

## 1.7. Harmonic functions

**1.7.5. The Poisson Formula.** If $f \in \mathcal{O}(\overline{\mathbb{D}})$ then by the Cauchy Formula

$$f(z) = \int_{|\zeta|=1} \frac{f(\zeta)}{\zeta - z} \frac{d\zeta}{2\pi\sqrt{-1}}, \qquad z \in \mathbb{D}.$$

On the other hand, if $w \in \mathbb{C} - \overline{\mathbb{D}}$, then

$$0 = \int_{|\zeta|=1} \frac{f(\zeta)}{\zeta - w} \frac{d\zeta}{2\pi\sqrt{-1}}.$$

If we now take $w = 1/\bar{z}$, then by using repeatedly that on the circle $|\zeta| = 1$, $\bar{\zeta} = 1/\zeta$ we obtain

$$\begin{aligned}
f(z) &= \int_{|\zeta|=1} \frac{f(\zeta)}{\zeta - z} \frac{d\zeta}{2\pi\sqrt{-1}} - \int_{|\zeta|=1} \frac{f(\zeta)}{\zeta - \bar{z}^{-1}} \frac{d\zeta}{2\pi\sqrt{-1}} \\
&= \int_{|\zeta|=1} \frac{\zeta f(\zeta)}{\zeta - z} \frac{d\zeta}{2\pi\sqrt{-1}\zeta} - \int_{|\zeta|=1} \frac{f(\zeta)}{1 - \bar{\zeta}\bar{z}^{-1}} \frac{d\zeta}{2\pi\sqrt{-1}\zeta} \\
&= \int_{|\zeta|=1} f(\zeta) \left( \frac{\zeta}{\zeta - z} + \frac{\bar{z}}{\bar{\zeta} - \bar{z}} \right) \frac{d\zeta}{2\pi\sqrt{-1}\zeta} \\
&= \int_{|\zeta|=1} f(\zeta) \frac{1 - |z|^2}{|\zeta - z|^2} \frac{d\zeta}{2\pi\sqrt{-1}\zeta}.
\end{aligned}$$

Now, on any circle centered at the origin, $d\zeta/\zeta$ is pure imaginary, since $0 = d\log|\zeta|^2 = \frac{d\zeta}{\zeta} + \frac{d\bar{\zeta}}{\bar{\zeta}}$. Thus, by taking real parts of our calculation above, we obtain the following result.

THEOREM 1.7.5 (Poisson Integral Formula). *If $u \in H(\overline{\mathbb{D}})$, then for all $z \in \mathbb{D}$ one has the integral representation*

$$u(z) = \int_{|\zeta|=1} u(\zeta) \frac{1 - |z|^2}{|\zeta - z|^2} \frac{d\zeta}{2\pi\sqrt{-1}\zeta}.$$

COROLLARY 1.7.6. *Let $\{u_n\} \subset H(\mathbb{D})$ be a sequence of harmonic functions that converges uniformly on compact sets of $\mathbb{D}$ to a function $u$. Then $u$ is harmonic.*

The proof is left to the reader.

REMARK (Dirichlet Problem for the unit disk). Let $f : \partial \mathbb{D} \to \mathbb{R}$ be a continuous function. The function $u$ defined by

(1.6) $$u(z) := \int_{|\zeta|=1} f(\zeta) \frac{1 - |z|^2}{|\zeta - z|^2} \frac{d\zeta}{2\pi\sqrt{-1}\zeta}, \quad |z| < 1$$

and $u(z) = f(z)$ for $|z| = 1$ is a continuous solution of the *Dirichlet Problem*

$$\begin{cases} \Delta u = 0 & \text{in } \mathbb{D} \\ u = f & \text{on } \partial \mathbb{D} \end{cases}$$

for the unit disk. By the Maximum Principle, this Dirichlet Problem has a unique solution, evidently given by (1.6). We shall return to the Dirichlet Problem in due course. ◇

**1.7.6. Regularity of harmonic functions.** If we interpret the Laplacian classically, we must require that harmonic functions be a priori $\mathscr{C}^2$. However, even if we interpret the definition in the sense of distributions, harmonic functions are still smooth—a fact sometimes known as Weyl's Lemma, which we now prove.

Let $\chi : [0, 1) \to [0, \infty)$ be a smooth function with compact support. Define $\psi(z) := \chi(|z|)/\int_\mathbb{C} \chi(|z|)$. Then $\psi$ is radially symmetric and has integral 1. Now let $u$ be a harmonic distribution, i.e., $u(\Delta\varphi) = 0$ for all smooth compactly supported functions $\varphi$. Define the function

$$\tilde{u}(x) := u * \psi(x) = u(\psi(x - \cdot)).$$

Then $u$ is smooth, and since $\Delta_x \psi(x - y) = \Delta_y \psi(x - y)$, we have

$$\Delta \tilde{u}(x) = u(\Delta_x \psi(x - \cdot)) = u(\Delta . \psi(x - \cdot)) = 0.$$

It follows that $\tilde{u}$ is harmonic. But then

$$\begin{aligned}\int \tilde{u}(x)\varphi(x)dA(x) &= u\left(\int \psi(x-\cdot)\varphi(x)dA(x)\right) \\ &= u\left(\int \varphi(z+y)\psi(z)dA(z)\right) \\ &= \int u(\varphi(z+\cdot))\psi(z)dA(z).\end{aligned}$$

But $\Delta_z u(\varphi(z+\cdot)) = u(\Delta_z \varphi(z+\cdot)) = u(\Delta.\varphi(z+\cdot)) = 0$, and so $u(\varphi(z+\cdot))$ is also harmonic. Since $\psi$ is radial, we have

$$\int u(\varphi(z+\cdot))\psi(z)dA(z) = u(\varphi).$$

It follows that $u = \tilde{u}$ is smooth and harmonic. □

## 1.8. Subharmonic functions

For the rest of this section, let $\Omega$ be an open subset of $\mathbb{C}$.

### 1.8.1. Subharmonic.

DEFINITION 1.8.1. A function $u : \Omega \to [-\infty, \infty]$ is *upper semi-continuous* if for every $s \in (-\infty, +\infty]$ the set $u^{-1}[-\infty, s)$ is open. ◇

Equivalently, $u$ is upper semi-continuous if

$$\limsup_{\zeta \to z} u(\zeta) \leq u(z).$$

## 1.8. Subharmonic functions

Note that every upper semi-continuous function $u$ is Lebesgue measurable. In fact, for any measurable subset $U \subset \mathbb{C}$

$$\int_U f d\mathrm{A} := \inf \left\{ \int_U f d\mathrm{A} \ ; \ f \in \mathscr{C}_0^\infty(U) \text{ and } f \geq u \right\}.$$

If $X$ is a Hausdorff space and $f : X \to \mathbb{R}$, the *upper regularization* of $f$ is the function

$$f^*(x) := \limsup_{y \to x} f(y).$$

It follows that $f^*$ is upper semi-continuous, $f^* \geq f$, and that if $g \geq f$ is upper semi-continuous then $g \geq f^*$. In particular, $f$ is upper semi-continuous if and only if $f^* = f$.

DEFINITION 1.8.2. A function $u : \Omega \to [-\infty, +\infty)$ is *subharmonic* (written $u \in SH(\Omega)$) if

(i) $u$ is upper semi-continuous, and
(ii) if $K \subset\subset \Omega$ and $h \in H(\text{interior}(K)) \cap \mathscr{C}^0(\overline{K})$ satisfy $u|_{\partial K} \leq h|_{\partial K}$, then $u \leq h$. ◇

### 1.8.2. Basic properties.

PROPOSITION 1.8.3. *Let $\Omega \subset \mathbb{C}$ be an open set.*

(1) *If $u \in SH(\Omega)$ and $c > 0$ is a constant, then $cu \in SH(\Omega)$.*
(2) *If $\{u_\alpha| \ \alpha \in A\} \subset SH(\Omega)$ and $u := \sup_\alpha u_\alpha$ is finite and upper semi-continuous, then $u \in SH(\Omega)$.*
(3) *If $u_1 \geq u_2 \geq ...$ is a sequence of subharmonic functions, then $u = \lim u_j$ is subharmonic.*

**Proof.** Statements (1) and (2) are trivial. To show statement (3), note first that since $\{u_j\}$ is a decreasing sequence of functions, $u^{-1}[-\infty, s) = \bigcup_j u_j^{-1}[-\infty, s)$ and so $u$ is upper semi-continuous. Let $K \subset\subset \Omega$ and let $h$ be a continuous function on $K$ that is harmonic on the interior of $K$ and majorizes $u$ on the boundary of $K$. Fix $\varepsilon > 0$ and let $A_j := \{z \in \text{bdry } K \mid u_j(z) \geq h(z) + \varepsilon\}$. Then $A_j$ is closed (hence compact) and $A_{j+1} \subset A_j$. Since $\bigcap_j A_j = \emptyset$, it follows that for $j_0$ large enough, $A_j = \emptyset$ for all $j \geq j_0$. Thus $u_j \leq h + \varepsilon$ in $K$ for all sufficiently large $j$, and hence $u \leq h$ in $K$. □

REMARK. In statement (2) of Proposition 1.8.3 one can drop the requirement that $u$ is upper semi-continuous, but then the conclusion is that $u^*$ is subharmonic. ◇

### 1.8.3. Subharmonic again.
There are several useful characterizations of subharmonic functions. We collect some of the most useful characterizations in the following theorem.

THEOREM 1.8.4. *Let $u : \Omega \to [-\infty, \infty)$ be upper semi-continuous. The following are equivalent.*

(1) $u \in SH(\Omega)$.

(2) *If $f$ is a holomorphic polynomial and $D \subset \Omega$ is a disk with $u \leq \operatorname{Re} f$ on $\partial D$, then $u \leq \operatorname{Re} f$ on $D$.*

(3) *If $\delta > 0$, $z \in \Omega$ is of distance more than $\delta$ from the boundary of $\Omega$, and $d\mu$ is a positive measure on $[0, \delta]$, then*

$$(\mu SMV) \qquad u(z) \leq \frac{1}{2\pi \int_0^\delta d\mu(r)} \int_0^{2\pi} \int_0^\delta u(z + re^{\sqrt{-1}\theta}) d\mu(r) d\theta.$$

(4) *For each $\delta > 0$ and $z \in \Omega$ of distance more than $\delta$ from the boundary of $\Omega$, there exists a positive measure $d\mu$ on $[0, \delta]$ that is not supported on $\{0\}$, such that $(\mu SMV)$ holds.*

**Proof.** Clearly statement (2) follows from statement (1), and statement (4) follows from statement (3). We will show that statement (2) $\Rightarrow$ statement (3) and statement (4) $\Rightarrow$ statement (1).

Assuming that statement (2) holds, let $z \in \Omega$ have distance at least $\delta$ from the boundary of $\Omega$, let $0 < r < \delta$, and let $\varphi$ be a continuous function on $\partial \mathbb{D}(z, r)$ majorizing $u$ there. By approximating $\varphi$ by a finite part of its Fourier series, we may assume $\varphi$ is a trigonometric polynomial, which we can extend into $\mathbb{D}(z, r)$ (by replacing $e^{\sqrt{-1}\theta}$ by $z$ and $e^{-\sqrt{-1}\theta}$ by $\bar{z}$) as a harmonic polynomial, hence the real part $h$ of a holomorphic polynomial. Thus $u \leq h$ on $\mathbb{D}(z, r)$, and we have,

$$u(z) \leq h(z) = \frac{1}{2\pi} \int_0^{2\pi} \varphi(z + re^{\sqrt{-1}\theta}) d\theta.$$

Since we can approximate $u$ from above, in $L^1_{\ell oc}$, by continuous functions, we see that

$$u(z) \leq \frac{1}{2\pi} \int_0^{2\pi} u(z + re^{\sqrt{-1}\theta}) d\theta,$$

and integration over $[0, \delta]$ with respect to $d\mu$ shows that statement (3) holds.

Assuming that statement (4) holds, let $K \subset\subset \Omega$ and suppose

$$h \in H(\operatorname{interior} K) \cap \mathscr{C}^0(\overline{K})$$

is such that $h \geq u$ on $\operatorname{bdry} K$. Let $v = u - h$ and set $M := \sup_K v$. Being an upper semi-continuous function, $v$ achieves its maximum on $K$. Assuming, for the sake of contradiction, that $M > 0$, we find that there exists a non-empty compact $F \subset K$ such that $v = M$ on $F$. Let $z_0$ be the point in $F$ which is of minimal distance $\delta$ to the boundary of $K$. Note that $\delta > 0$ because $v$ is non-positive on the

## 1.8. Subharmonic functions

boundary of $K$. Since $v$ is upper semi-continuous, $v < M$ on an open subset of $\mathbb{D}(z_0, \delta)$. Thus

$$M = v(z_0) \le \frac{1}{2\pi \int_0^\delta d\mu(r)} \int_0^{2\pi} \int_0^\delta v(z_0 + re^{\sqrt{-1}\theta}) d\mu(r) d\theta < M,$$

and this is the desired contradiction. □

COROLLARY 1.8.5.  (1) *The sum of two subharmonic functions is subharmonic.*

(2) *Subharmonicity is a local property:* $u \in SH(\Omega)$ *if and only if every point* $p \in \Omega$ *has a neighborhood* $U$ *such that* $u|_U$ *is subharmonic.*

(3) *If* $f \in \mathcal{O}(\Omega)$, *then* $\log|f|$ *is subharmonic (and, in fact, harmonic away from the zeros of* $f$).

(4) *If* $\varphi : \mathbb{R} \to \mathbb{R}$ *is a convex increasing function then with*

$$\varphi(-\infty) := \lim_{x \to -\infty} \varphi(x),$$

$\varphi \circ u$ *is subharmonic whenever* $u$ *is.*

(5) *If* $u_1$ *and* $u_2$ *are subharmonic functions then* $\log(e^{u_1} + e^{u_2})$ *is subharmonic.*

**Proof.** Statements (1) and (2) follow from statements (4) and (2) of Theorem 1.8.4 respectively, while statement (3) follows from statement (2) of Theorem 1.8.4 and the Maximum Principle. Statement (4) follows from Jensen's inequality. To see statement (5), let $D \subset \Omega$ be a closed disc, and let $f$ be a holomorphic polynomial such that $\log(e^{u_1} + e^{u_2}) \le \operatorname{Re} f$ on $\partial D$. Then $(e^{u_1} + e^{u_2})e^{-\operatorname{Re} f} \le 1$ on $\partial D$. Since $u_j - \operatorname{Re} f$ is subharmonic and the exponential function is convex, statement (4) implies that $e^{u_j - \operatorname{Re} f}$ is subharmonic for $j = 1, 2$. Hence by statement (1), $(e^{u_1} + e^{u_2})e^{-\operatorname{Re} f}$ is subharmonic, and therefore $e^{u_1} + e^{u_2} \le e^{\operatorname{Re} f}$ on $D$. Then statement (5) follows from statement (2). □

**1.8.4. Local integrability.** Another important corollary of Theorem 1.8.4 is the following result.

THEOREM 1.8.6. *If* $\Omega \subset \mathbb{C}$ *is connected and* $u \in SH(\Omega)$, *then either* $u \equiv -\infty$ *or* $u \in L^1_{loc}(\Omega)$.

**Proof.** Let

$$X := \{z \in \Omega \mid \text{there exists } r > 0 \text{ with } D(z, r) \subset\subset \Omega \text{ and } u \in L^1(D(z, r))\}.$$

Then $X$ is clearly open, and by the upper semi-continuity (hence boundedness from above) of $u$ and statement (3) of Theorem 1.8.4, each $z$ with $u(z) > -\infty$ belongs to $X$. We claim also that $X$ is closed. Indeed, if $p \in \overline{X}$ then there exist points $z$ arbitrarily close to $p$ such that $u(z) > -\infty$. One of these points is the center of a disc that contains $p$ and is contained in $\Omega$, and again by statement (3) of

Theorem 1.8.4, $u$ is integrable on a neighborhood of $p$. Thus, since $X$ is both open and closed and since $\Omega$ is connected, the theorem is proved. □

**1.8.5. Regularity.** Finally, we come to the most important characterization of subharmonicity.

THEOREM 1.8.7. *Let $u$ be a distribution. Then $u \in SH(\Omega)$ if and only if, in the sense of distributions,*
$$\Delta u \geq 0.$$

To prove this result, we need the following two lemmas.

LEMMA 1.8.8. *Let $u \in \mathscr{C}^\infty(\Omega)$. Then $u$ is subharmonic if and only if $\Delta u \geq 0$.*

**Proof.** By Theorem 1.8.4, it suffices to show that the sub-mean value property holds precisely when $\Delta u \geq 0$. Let us compute the Taylor series of $u$ to order 2 near a given point $p$. With $(x,y)$ as real coordinates in $\mathbb{C}$, we have

$$\begin{aligned} u(p + re^{\sqrt{-1}\theta}) &= u(p) + r\left(u_x(p)\cos\theta + u_y(p)\sin\theta\right) \\ &\quad + \frac{r^2}{2}\left(u_{xx}(p)\cos^2\theta + u_{yy}(p)\sin^2\theta + u_{xy}\sin 2\theta\right) + O(r^3), \end{aligned}$$

so that
$$\int_0^{2\pi} u(p + re^{\sqrt{-1}\theta})\frac{d\theta}{2\pi} - u(p) = \frac{r^2}{4}\Delta u + O(r^3).$$

Thus, if the mean value property holds, then $\Delta u \geq 0$.

Conversely, if $\Delta u \geq 0$, then with
$$f(r) = \int_0^{2\pi} u(p + re^{\sqrt{-1}\theta})\frac{d\theta}{2\pi}$$
we have, by the divergence form of Green's Theorem,
$$f'(r) = \frac{1}{2\pi r}\int_{D_r(0)} \Delta u \, dA \geq 0.$$

Since $u$ is continuous,
$$\lim_{r \to 0} f(r) = u(p).$$

Thus the sub-mean value property is established, and the proof is complete. □

LEMMA 1.8.9. *Let $u \in \mathscr{C}^\infty(\Omega)$ be subharmonic, and let $\psi \in \mathscr{C}_0^\infty(\mathbb{D})$ be a function such that $\psi \geq 0$ and $\psi(x) = \psi(e^{\sqrt{-1}\theta}x)$ for all real $\theta$. Write*
$$\psi^\varepsilon(z) = \varepsilon^{-2}\psi(\varepsilon^{-1}z) \quad \text{and} \quad u_\varepsilon = u * \psi^\varepsilon.$$

*Then $u_\varepsilon$ is subharmonic and decreases with $\varepsilon$.*

**Proof.** First, note that $\Delta(u*\psi^\varepsilon) = (\Delta u)*\psi^\varepsilon$ and the latter is non-negative because this is the case for both $\Delta u$ and $\psi$. Next,

$$u_\varepsilon(z) = \int_\mathbb{D} u(z-\varepsilon\zeta)\psi(\zeta)dA(\zeta) = \int_0^1 r\psi(r)dr \int_0^{2\pi} u(z-\varepsilon re^{\sqrt{-1}\theta})d\theta,$$

and since the mean value integral

$$\int_0^{2\pi} u(z-\varepsilon re^{i\theta})d\theta$$

is an increasing function of $\varepsilon$, the lemma is proved. $\square$

**Proof of Theorem 1.8.7.** Suppose first that $\Delta u \geq 0$ in the sense of distributions. Let $u_\varepsilon$ be as in Lemma 1.8.9. We claim that $u_\varepsilon$ is a smooth family of subharmonic functions decreasing with $\varepsilon$. Indeed, $u_\varepsilon$ is clearly smooth, and $\Delta u_\varepsilon(z) = u(\Delta\psi^\varepsilon(z-\cdot)) \geq 0$ because $\Delta u$ is positive in the sense of distributions and $\psi \geq 0$. To see that $u_\varepsilon$ is decreasing, we use a "double smoothing trick": let $\varphi \in \mathscr{C}_0^\infty(\mathbb{D})$ be non-negative. Then $u*\varphi^\delta$ is smooth and subharmonic (as we just argued), and hence $u_\varepsilon * \varphi^\delta = u * \varphi^\delta * \psi^\varepsilon$ is a subharmonic family decreasing with $\varepsilon$. Letting $\delta \to 0$ shows that $u_\varepsilon$ is decreasing. It now follows from statement (3) of Proposition 1.8.3 that $u$ is subharmonic.

Conversely, suppose $u$ is subharmonic. Define $u_\varepsilon = u * \psi^\varepsilon$ with $\psi$ as above. Then, using the notation

$$\fint_D f d\mu := \frac{1}{\mu(D)} \int_D f d\mu,$$

one has

$$\fint_{\mathbb{D}(z,r)} u_\varepsilon(\zeta)dA(\zeta) = \int_\mathbb{C} \left( \fint_{\mathbb{D}(z,r)} u(\zeta-\varepsilon\eta)dA(\zeta) \right) \psi(\eta)dA(\eta)$$
$$\geq \int_\mathbb{C} u(z-\varepsilon\eta)\psi(\eta)dA(\eta) = u_\varepsilon(z),$$

and therefore $u_\varepsilon$ is subharmonic by Theorem 1.8.4. It follows from Lemma 1.8.8 that $\Delta u_\varepsilon \geq 0$, and therefore if $h \geq 0$ has compact support then

$$\int u\Delta h = \lim \int u_\varepsilon \Delta h = \lim \int \Delta u_\varepsilon h \geq 0.$$

Thus $\Delta u \geq 0$ in the sense of distributions, as claimed. The proof is finished. $\square$

## 1.9. Exercises

1.1. Prove the formula (1.1) from Green's Theorem.

1.2. Prove that if $f \in \mathcal{O}(\mathbb{C})$ and $|f(z)|^2 \leq C(1+|z|^2)^N$ then $f$ is a polynomial in $z$, of degree at most $N$.

1.3. Prove that if $f \in \mathcal{O}(\mathbb{C})$ and $\int_\mathbb{C} |f|^2 dA < +\infty$ then $f \equiv 0$.

1.4. Prove that if $f$ is holomorphic on the punctured unit disk $\mathbb{D} - \{0\}$ and $\int_{\mathbb{D}-\{0\}} |f|^2 dA < +\infty$ then $f \in \mathcal{O}(\mathbb{D})$.

1.5. Let $\mathscr{F} \subset \mathcal{O}(\mathbb{C})$ be a family of entire functions such for each $R > 0$ there is a constant $C_R$ such that
$$\int_{\mathbb{D}(0,R)} |f|^2 dA \leq C_R \quad \text{for all } f \in \mathscr{F}.$$
Show that every sequence in $\mathscr{F}$ has a convergent subsequence.

1.6. Find a harmonic function in the punctured unit disk that is not the real part of a holomorphic function.

1.7. Let $u$ be a subharmonic function in the unit disk. The goal of this exercise is to show that the function $\psi : [0, \varepsilon) \to \mathbb{R} \cup [-\infty)$ defined by
$$\psi(r) := \frac{1}{2\pi} \int_0^{2\pi} u(e^r e^{\sqrt{-1}\theta}) d\theta$$
is convex and increasing.

(a) Show that $r \to \psi(r)$ is increasing.

(b) Show that if $I, J \subset \mathbb{R}$ are intervals and $v$ is a subharmonic function on $U = I + \sqrt{-1}J$ that depends only on $\operatorname{Im} z$ for all $z \in U$ then $v$ is convex.

(c) Show that the function
$$\Psi(\zeta) := \frac{1}{2\pi} \int_0^{2\pi} u\left(e^{\zeta + \sqrt{-1}\theta}\right) d\theta$$
is upper semi-continuous and has the mean value property.

(d) Show that $\psi$ is convex.

# Chapter 2

# Riemann Surfaces

## 2.1. Definition of a Riemann surface

### 2.1.1. Complex charts.

DEFINITION 2.1.1. Let $X$ be a topological space. A *complex chart* is a homeomorphism from an open set $U \subset X$ onto an open set $V \subset \mathbb{C}$:

$$\begin{array}{ccc} X & & \mathbb{C} \\ \cup & & \cup \\ U & \xrightarrow[\cong]{\varphi} & V \end{array}$$

$\diamond$

EXAMPLE 2.1.2. Take $U = X = \mathbb{R}^2$, $V = \mathbb{C}$, and $\varphi(x,y) = x + \sqrt{-1}y$. $\diamond$

EXAMPLE 2.1.3. Let $X = S^2$, realized as

$$X = \{(u,v,w) \in \mathbb{R}^3 \,;\, u^2 + v^2 + w^2 = 1\}.$$

Define the maps $\varphi_N : U_N \to \mathbb{C}$ and $\varphi_S : U_S \to \mathbb{C}$ by the formulae

$$\varphi_N(u,v,w) := \frac{u}{1-w} + \sqrt{-1}\frac{v}{1-w}$$

and

$$\varphi_S(u,v,w) := \frac{u}{1+w} - \sqrt{-1}\frac{v}{1+w}.$$

Here $U_N = S^2 - \{(0,0,1)\}$ and $U_S = S^2 - \{(0,0,-1)\}$.

Let $x + \sqrt{-1}y = \varphi_N(u,v,w)$. We have $u = x(1-w)$ and $v = y(1-w)$, and thus from $u^2 + v^2 + w^2 = 1$ we obtain $w^2 = 1 - (1-w)^2(x^2+y^2)$. This

quadratic equation in $w$ has the two solutions
$$w = 1 \quad \text{or} \quad w = \frac{x^2 + y^2 - 1}{x^2 + y^2 + 1}.$$
But on $U_N$ we have excluded the first solution, so we have the formula
$$\varphi_N^{-1}(x + \sqrt{-1}y) = \left(\frac{2x}{x^2+y^2+1}, \frac{2y}{x^2+y^2+1}, \frac{x^2+y^2-1}{x^2+y^2+1}\right).$$
We note that, with $z = x + \sqrt{-1}y$,
$$\varphi_S \circ \varphi_N^{-1}(z) = \frac{1}{z}$$
for all $z \in \varphi_N(U_N \cap U_S)$. ◇

**2.1.2. Riemann surface.**

DEFINITION 2.1.4. Let $X$ be a topological Hausdorff space.

(a) Two complex charts $\varphi : U \to \mathbb{C}$ and $\psi : V \to \mathbb{C}$ are said to be compatible if the map
$$\varphi \circ \psi^{-1} : \psi(U \cap V) \to \varphi(U \cap V)$$
is (bi)holomorphic. The map $\varphi \circ \psi^{-1}$ is called a transition map, or a change of coordinates.

(b) A complex atlas for $X$ is a collection of charts $\{\varphi_j : U_j \to \mathbb{C} \; ; \; j \in J\}$ any two of which are compatible and whose domains form an open cover of $X$:
$$X = \bigcup_{j \in J} U_j.$$

(c) Two complex atlases are equivalent if their union is also a complex atlas. An atlas is maximal if it contains every equivalent atlas. (Thus every atlas is contained in a unique maximal atlas.)

(d) A Riemann surface is a topological Hausdorff space together with a maximal complex atlas. Specifying a maximal complex atlas is often also referred to as specifying a complex structure.

(e) (Classical terminology) A non-compact Riemann surface is called an *open Riemann surface*. ◇

EXAMPLE 2.1.5. Let
$$\mathbb{C} := (\mathbb{R}^2, \{\varphi : (x,y) \mapsto x + \sqrt{-1}y\}).$$
The Riemann surface $\mathbb{C}$ is called the affine line or the complex plane. ◇

EXAMPLE 2.1.6. Let
$$\Delta := (\{(x,y) \in \mathbb{R}^2 \; ; \; x^2 + y^2 < 1\}, \{\varphi : (x,y) \mapsto x + \sqrt{-1}y\}).$$
The Riemann surface $\Delta$ is called the unit disc. ◇

EXAMPLE 2.1.7. Let
$$\mathbb{P}_1 := (S^2, \{\varphi_N, \varphi_S\}),$$
where $\varphi_N$ and $\varphi_S$ are the maps defined in Example 2.1.3. As we showed in that example, the pair $\{\varphi_N, \varphi_S\}$ is a complex atlas. The Riemann surface $\mathbb{P}_1$ thus obtained is called the projective line or the Riemann sphere. In complex analysis, other common notations are $\overline{\mathbb{C}}$ and $\mathbb{C} \cup \{\infty\}$. ◇

REMARK. We have $\Delta \subset \mathbb{C} \subset \mathbb{P}_1$. The second inclusion is not as clear, but follows from what was done in Example 2.1.3. In general, any open subset of a Riemann surface is a Riemann surface. ◇

## 2.2. Riemann surfaces as smooth 2-manifolds

In view of the Cauchy Formula, every holomorphic function on a domain in $\mathbb{C}$ is real analytic. Thus every Riemann surface is a real analytic manifold of dimension 2. But in fact, Riemann surfaces have a little more structure.

THEOREM 2.2.1. *Every Riemann surface is oriented.*

**Proof.** Let $f = f_{\alpha\beta} := \varphi_\alpha \circ \varphi_\beta^{-1}$ be a typical transition map. Write $f = u + \sqrt{-1}v$. Then the Jacobian determinant of $f$ is
$$\det(Df) = u_x v_y - u_y v_x.$$
On the other hand, since $f$ is holomorphic, we have $v_x = -u_y$ and $v_y = u_x$. Thus
$$\det(Df) = u_x^2 + u_y^2 = \left|\frac{\partial f}{\partial z}\right|^2 > 0.$$
This completes the proof. □

Compact oriented topological 2-manifolds are completely characterized; any such manifold is homeomorphic to a sphere with $g$ handles for some positive integer $g$, and this number is the only homeomorphism invariant. A compact oriented surface with one handle is called a torus.

DEFINITION 2.2.2. Let $X$ be a compact oriented topological 2-manifold. The number $g$ of handles of $X$ is called the genus of $X$. ◇

As it turns out, any oriented surface admits a complex structure.

THEOREM 2.2.3. *Let $k \geq 2$. Every oriented 2-manifold of class $\mathscr{C}^k$ has a complex structure.*

The key idea behind the proof of Theorem 2.2.3 is the following theorem.

THEOREM 2.2.4. *Let $g$ be a $\mathscr{C}^2$-smooth Riemannian metric in a neighborhood of $0$ in $\mathbb{R}^2$. There exists a positively oriented change of coordinates $z = u(x,y) + \sqrt{-1}v(x,y)$ fixing the origin such that*

$$g = a^{(z)} dz d\bar{z}.$$

*Such coordinates are called "isothermal coordinates".*

REMARK. The local notation $g = a^{(z)} dz d\bar{z}$ is classical and means the following. If $w_1 = b_1 \frac{\partial}{\partial u} + b_2 \frac{\partial}{\partial v}$ and $w_2 = c_1 \frac{\partial}{\partial u} + c_2 \frac{\partial}{\partial v}$ are tangent vectors to $X$ at some point, where $z = u + \sqrt{-1}v$, then $g(w_1, w_2) = a^{(z)}(b_1 c_1 + b_2 c_2)$. In other words, $dz d\bar{z} = (du + \sqrt{-1}dv)(du - \sqrt{-1}dv)$. Later we will see that the notation is related to the fact that the tangent bundle of a Riemann surface is also a *complex line bundle*. ◇

Theorem 2.2.4, which is a classical result in partial differential equations, is called the Korn-Lichtenstein Theorem. Though it is not yet clear here, the sought-after function $z$ can be obtained by solving a partial differential equation that is closely related to the Cauchy-Riemann equations—a system of partial differential equations that figures prominently in this book. We shall give a proof of Theorem 2.2.4 in Chapter 11, after we learn how to solve the Cauchy-Riemann equations using Hilbert space methods.

**Proof of Theorem 2.2.3 given Theorem 2.2.4.** Fix a Riemannian metric $g$ on $X$ and let $\{z_\alpha := \varphi_\alpha \, ; \, \alpha \in A\}$ be an oriented atlas of isothermal coordinates for $g$. If $z$ and $w$ are two such coordinates, then they are related by a smooth map

$$w = F(z).$$

Now

$$\begin{aligned} g &= a^{(w)} dw d\bar{w} \\ &= a^{(w)}(F(z))(F_z dz + F_{\bar{z}} d\bar{z})(\overline{F_z} d\bar{z} + \overline{F_{\bar{z}}} dz) \\ &= a^{(w)}(F(z))(|F_z|^2 + |F_{\bar{z}}|^2) dz d\bar{z} + F_z \overline{F_{\bar{z}}} dz^2 + F_{\bar{z}} \overline{F_z} d\bar{z}^2. \end{aligned}$$

On the other hand, $g = a^{(z)} dz d\bar{z}$. It follows that either $F_z = 0$ or $F_{\bar{z}} = 0$. But if the former holds, then $F$ is not positively oriented. Thus $F$ is a holomorphic function, which completes the proof. □

REMARK. Differential geometrically speaking, the idea behind the proof of Theorem 2.2.3 is the following. First, we put a Riemannian metric on an oriented surface $X$. This Riemannian metric gives us a notion of rotation through an angle of $\frac{\pi}{2}$ in each tangent space. Infinitesimally, this means we have a notion of multiplying by $\sqrt{-1}$ in the tangent space. Theorem 2.2.4 implies that this infinitesimal multiplication by $\sqrt{-1}$ is actually locally constant in the right coordinate system. Thus we obtain the structure of a Riemann surface. ◇

## 2.3. Examples of Riemann surfaces

**2.3.1. Complex manifolds.** Riemann surfaces are the simplest non-trivial examples of more general spaces called complex manifolds. (More precisely, Riemann surfaces are complex manifolds having complex dimension 1.) Since we are going to use some basic complex manifolds, we give their definition here, mostly for the sake of completeness; we shall not use any deep facts about complex manifolds in general.

DEFINITION 2.3.1. Let $B \subset \mathbb{C}^n$ be an open set.
   (1) A function $f : B \to \mathbb{C}$ is *holomorphic* if it is holomorphic in each variable separately.
   (2) A map $F = (F^1, ..., F^m) : B \to \mathbb{C}^m$ is *holomorphic* if each component $F^i : B \to \mathbb{C}$ is holomorphic.
   (3) A holomorphic map $F : B \to B' \subset \mathbb{C}^n$ is *biholomorphic* if it has a holomorphic inverse. ◇

REMARK. Regarding statement (1) of Definition 2.3.1, it is a deep result of Hartogs that a holomorphic function is automatically continuous as a function of several variables. This property has no real analog: the function $f(x, y) = (x^2 + y^2)^{-1} xy$ is real-analytic as a function of $x$ when $y$ is held fixed, and vice versa, but $f$ is not continuous.

Regarding statement (3) of Definition 2.3.1, as in the case of 1 complex variable, an injective holomorphic map between domains in $\mathbb{C}^n$ automatically has full rank, meaning that at each point the (complex) derivative matrix has rank $n$. Note however that this is not true if the dimension of the domain is not the same as the dimension of the range. Indeed, the map $f : \mathbb{C} \to \mathbb{C}^2$ sending $t$ to $(t^2, t^3)$ is injective and holomorphic, but its derivative has rank 0 at the origin. ◇

DEFINITION 2.3.2. Let $X$ be a topological Hausdorff space.
   (1) A *complex chart* of dimension $n$ on $X$ is a homeomorphism $\varphi : U \to V$ of open sets $U \subset X$ and $V \subset \mathbb{C}^n$.
   (2) Two complex charts $\varphi_1 : U_1 \to V_1$ and $\varphi_2 : U_2 \to V_2$ are said to be *compatible* if the map $\varphi_2 \circ \varphi_1^{-1}|_{\varphi_1(U_1 \cap U_2)} : \varphi_1(U_1 \cap U_2) \to \varphi_2(U_1 \cap U_2)$ is biholomorphic.
   (3) A complex atlas is a collection of compatible complex charts whose domains cover $X$. Two complex atlases are said to be equivalent if their union is also a complex atlas. A maximal complex atlas is an equivalence class of complex atlases. A maximal complex atlas is also called a *complex structure*.
   (4) A topological space together with a maximal complex atlas is called a *complex manifold*. ◇

REMARK. Thus a complex manifold is by definition equipped with charts taking values in complex Euclidean space, and the transition functions are complex differentiable. Of course, removing all occurrences of the word "complex" in the previous sentence also results in a meaningful statement; it gives the definition of a smooth manifold. ◇

DEFINITION 2.3.3. (1) Let $M$ be a complex manifold. A function $f : M \to \mathbb{C}$ is said to be holomorphic if for any chart $\varphi : U \to V \subset \mathbb{C}^n$, the function $f \circ \varphi^{-1} : V \to \mathbb{C}$ is holomorphic.

(2) Let $M_1, M_2$ be complex manifolds. A function $f : M_1 \to M_2$ is said to be holomorphic if for every pair of charts $\varphi_1 : U_1 \subset M_1 \to V_1 \subset \mathbb{C}^{n_1}$ and $\varphi_2 : U_2 \subset M_2 \to V_1 \subset \mathbb{C}^{n_2}$ the function $\varphi_2 \circ f \circ \varphi_1^{-1}$ is holomorphic.

(3) A holomorphic function $f : M_1 \to M_2$ between two complex manifolds is said to be an embedding if it is one-to-one and its derivative $Df$ has maximal rank at each point.

(4) Two complex manifolds $M_1$ and $M_2$ are said to be biholomorphic if there exists a holomorphic function $f : M_1 \to M_2$ that is both one-to-one and onto, and whose derivative $Df$ is invertible at each point. ◇

THEOREM 2.3.4. *A holomorphic function on a connected compact complex manifold is constant.*

**Proof.** Suppose $f : M \to \mathbb{C}$ is holomorphic. Then $|f|$ is continuous, and thus there is a point $p \in M$ at which $|f|$ takes on its maximum. Let $\varphi : U \to V \subset \mathbb{C}^n$ be a complex chart containing $p$ in its interior. Then $|f \circ \varphi^{-1}|$ attains its maximum in the interior of $V$. Applying the Maximum Principle to each variable separately, we see that $f$ is constant on $U$. Since $U$ is an open set and $M$ is connected, the Identity Theorem implies that $f$ is constant on $M$. □

COROLLARY 2.3.5. *If $M$ is a compact complex submanifold of $\mathbb{C}^n$, then $M$ is zero-dimensional.*

**Proof.** One applies Theorem 2.3.4 to the restriction of the coordinate functions $z_j$ to $M$. □

**2.3.2. Examples of quotient Riemann surfaces.** The examples of Riemann surfaces we considered until now were either trivially subsets of $\mathbb{C}$ or were constructed by "bare hands". But there are other methods of obtaining examples of Riemann surfaces. One approach is to obtain a Riemann surface as a quotient of a space by an equivalence relation.

EXAMPLE 2.3.6 (The projective line again). Previously we put a complex structure directly on $S^2$ using the stereographic projections $\varphi_N$ and $\varphi_S$. But there is another way to obtain $\mathbb{P}_1$.

## 2.3. Examples of Riemann surfaces

Consider the set $\mathbb{P}_1$ of lines through the origin in $\mathbb{C}^2$:

$$\mathbb{P}_1 := \{\text{1-dimensional subspaces of } \mathbb{C}^2\}.$$

There is a map

$$\pi : \mathbb{C}^2 - \{0\} \to \mathbb{P}_1$$

sending each non-zero vector to the unique 1-dimensional subspace containing it. We denote the image of $\pi$ using square brackets:

$$\pi(z_0, z_1) := [z_0, z_1].$$

The map $\pi$ is the projection associated to the equivalence relation

$$(z_0, z_1) \sim (w_0, w_1) \iff \exists \lambda \in \mathbb{C}^* \text{ such that } (z_0, z_1) = \lambda(w_0, w_1).$$

Specifying the point in projective space corresponding to the line passing through $(z_0, z_1)$ by the notation $[z_0, z_1]$ of points is abusively referred to as giving the "homogeneous coordinates" of the point.

Next we define the topology on $\mathbb{P}_1$ as follows: $U \subset \mathbb{P}_1$ is open if and only if $\pi^{-1}(U)$ is open. Note that any topology on $\mathbb{P}_1$ for which $\pi$ is continuous must contain this topology. Observe that since $\pi(S^3) = \mathbb{P}_1$, where $S^3 = \{|z| = 1\}$ is the unit sphere in $\mathbb{C}^2$, $\mathbb{P}_1$ is compact.

We now define charts for $\mathbb{P}_1$. Let $(z_0, z_1)$ be Euclidean coordinates in $\mathbb{C}^2$. We let

$$U_0 := \{[1, z_1] \, ; \, z_1 \in \mathbb{C}\} \quad \text{and} \quad U_1 := \{[z_0, 1] \, ; \, z_0 \in \mathbb{C}\},$$

i.e., $U_0 = \mathbb{P}_1 - \{[0, 1]\}$ and $U_1 = \mathbb{P}_1 - \{[1, 0]\}$. Define the maps $\varphi_0 : U_0 \to \mathbb{C}$ and $\varphi_1 : U_1 \to \mathbb{C}$ by

$$\varphi_0[z_0, z_1] := \frac{z_1}{z_0} \quad \text{and} \quad \varphi_1[z_0, z_1] := \frac{z_0}{z_1}.$$

Observe that

$$\varphi_0(U_0 \cap U_1) = \mathbb{C} - \{0\}$$

and that

$$\varphi_1 \circ \varphi_0^{-1}(\zeta) = \frac{1}{\zeta} \quad \text{and} \quad \varphi_0 \circ \varphi_1^{-1}(\eta) = \frac{1}{\eta}.$$

Thus we have defined a complex atlas, and therefore a complex structure, on $\mathbb{P}_1$. ◇

REMARK. We leave it as an exercise for the reader to show that $\mathbb{P}_1$, as defined here, is diffeomorphic to $S^2$, the unit sphere in $\mathbb{R}^3$. Later we shall see that the Riemann surface so obtained is the same as the one defined previously in Example 2.1.7. In fact, there is only one way to define a complex structure on $S^2$. ◇

REMARK (Blowup and tautological line bundle). We take this opportunity to describe an important construction. Let

$$\mathbb{U} := \{(z, \ell) \in \mathbb{C}^2 \times \mathbb{P}_1 \, ; \, z \in \ell\}.$$

Let $B\ell_0 : \mathbb{U} \to \mathbb{C}^2$ and $\pi : \mathbb{U} \to \mathbb{P}_1$ denote the restriction to $\mathbb{U}$ of the projections from $\mathbb{C}^2 \times \mathbb{P}_1$ to the first and second factors, respectively.

DEFINITION 2.3.7.  (1) $B\ell_0 : \mathbb{U} \to \mathbb{C}^2$ is called the blowup of $0$ in $\mathbb{C}^2$.

(2) $\pi : \mathbb{U} \to \mathbb{P}_1$ is called the tautological line bundle.  ◇

(Line bundles, of which $\pi : \mathbb{U} \to \mathbb{P}_1$ is an example, are defined in Chapter 4.)

The reader should ponder the following facts:

• Statement (1) corresponds to obtaining $\mathbb{U}$ from $\mathbb{C}^2$ by replacing the origin with a copy of $\mathbb{P}_1$. Here $\mathbb{P}_1$ should be interpreted as the set of possible slopes of the straight lines along which one can approach the origin. Pulling back a function from a neighborhood of $0$ in $\mathbb{C}^2$ to $\mathbb{U}$ is analogous to computing the derivative of that function.

• Statement (2) corresponds to defining $\mathbb{U}$ as the set of all points on the various lines through the origin in $\mathbb{C}^2$ that make up the projective line $\mathbb{P}_1$.  ◇

EXAMPLE 2.3.8 (Complex tori). Let $\omega_1, \omega_2 \in \mathbb{C}$ be $\mathbb{R}$-independent. Consider the lattice
$$L := \mathbb{Z}\omega_1 + \mathbb{Z}\omega_2 = \{n\omega_1 + m\omega_2 \; ; \; n, m \in \mathbb{Z}\}.$$
The set $L$ is a discrete subgroup of the additive group $\mathbb{C}$, so the quotient space
$$X = X_L := \mathbb{C}/L$$
is well-defined. We denote the projection by $\pi : \mathbb{C} \to X$. We therefore have a set (actually, an additive group) $X$, but in order to endow it with the structure of a Riemann surface, we must first give it the structure of a topological space.

We define a set $U \subset X$ to be open if its inverse image $\pi^{-1}(U) \subset \mathbb{C}$ is open. Then $\pi$ is continuous in this topology, and in fact any topology that makes $\pi$ continuous must be contained in this topology.

The map $\pi$ is an open mapping. Indeed, if $U \subset \mathbb{C}$ is open, then $\pi(U)$ is open, because
$$\pi^{-1}(\pi(U)) = \bigcup_{\omega \in L} U + \omega.$$
Observe that $X$ is compact. Indeed, the image of the compact set
$$P_z := \{z + t\omega_1 + s\omega_2 \; ; \; s, t \in [0,1]\}$$
under the continuous map $\pi$ covers $X$.

Next we will build charts on $X$. To this end, observe that since $L \subset \mathbb{C}$ is discrete, there is a positive number $\varepsilon$ such that the only point of $L$ that is within $2\varepsilon$ of the origin is the origin itself:
$$L \cap \{z \in \mathbb{C} \; ; \; |z| < 2\varepsilon\} = \{0\}.$$
By definition, for any such $\varepsilon$, any $\omega \in L - \{0\}$, and any $z \in \mathbb{C}$, we find that $(\mathbb{D}(z, \varepsilon) + \omega) \cap \mathbb{D}(z, \varepsilon) = \emptyset$. We define for every $z \in \mathbb{C}$ the map
$$\pi_z := \pi|_{\mathbb{D}(z,\varepsilon)} : \mathbb{D}(z, \varepsilon) \to \pi(\mathbb{D}(z, \varepsilon)).$$

## 2.3. Examples of Riemann surfaces

We leave it as an exercise for the reader to show that the maps $\pi_z$ are homeomorphisms and the collection
$$\mathcal{A} := \{\varphi_z := \pi_z^{-1} \, ; \, z \in \mathbb{C}\}$$
is an atlas. The interested reader may also wish to prove that $X_L$ is diffeomorphic to the torus $S^1 \times S^1$. ◇

REMARK. Of course, for every choice of vectors $\omega_1$ and $\omega_2$ we obtain a different construction of a torus $X_L$. While all of these constructions lead to a single smooth manifold, it is not clear whether all of these constructions lead to the same Riemann surface. In fact, they do not. ◇

REMARK. All Riemann surfaces can be constructed as quotients of either $\mathbb{P}_1$, $\mathbb{C}$, or $\Delta$. This deep and important fact, known as the Uniformization Theorem, will be proved in Chapter 10. ◇

**2.3.3. Implicitly defined Riemann surfaces.** In this paragraph, we discuss the possibility of defining a Riemann surface as a subset of $\mathbb{C}^2$ defined implicitly by a single equation $F(x, y) = 0$. We begin by describing the local version of this process.

2.3.3.1. GRAPHS OF HOLOMORPHIC FUNCTIONS. Let $D \subset \mathbb{C}$ be an open set and $f : D \to \mathbb{C}^n$ a holomorphic mapping. That is to say, $f = (f_1, ..., f_n)$ and each $f_j$ is holomorphic on $D$. We define the surface
$$S_f := \{(z, w) \in D \times \mathbb{C}^n \, ; \, w = f(z)\}.$$
We endow $S_f$ with the relative topology: $U \subset S_f$ is open if and only if there is an open subset $V \subset D \times \mathbb{C}^n$ such that $U = V \cap S_f$.

Denote by $p_1 : \mathbb{C} \times \mathbb{C}^n \to \mathbb{C}$ the projection to the first factor, i.e., $p_1(z, w) = z$. We define the following complex charts of $S_f$: if $V \subset D \times \mathbb{C}^n$ is open and $U = V \cap S_f$, then $\varphi : U \to \mathbb{C}$ is defined to be the restriction of $p_1$ to $U$:
$$\varphi := p_1|_U.$$
We define $\mathcal{A}$ to be the set of all such maps $\varphi$. It is not hard to show that $\mathcal{A}$ is a complex atlas for $S_f$ (and that $S_f$ is biholomorphic to $D$).

DEFINITION 2.3.9. Let $f : D \to \mathbb{C}^n$ be a holomorphic function. The Riemann surface $(S_f, \mathcal{A})$ is called the graph of $f$. ◇

2.3.3.2. THE IMPLICIT FUNCTION THEOREM. It is not hard to use the real Implicit Function Theorem to prove a holomorphic version. However, elementary complex analysis can be used to give a direct and short proof of the holomorphic Implicit Function Theorem.

THEOREM 2.3.10 (Implicit Function Theorem). *Let $U$ be an open subset of $\mathbb{C}^2$ (with coordinates $(x, y)$) containing the origin, and let $F : U \to \mathbb{C}$ be a continuous* function that is holomorphic in each variable separately. Suppose that $F(0,0) = 0$ and $F_y(0,0) \neq 0$. Then there exists a neighborhood $V$ of the origin in $\mathbb{C}^2$ such that for each $x_0 \in p_1(V)$ the equation $F(x_0, y) = 0$ has a unique solution $y_0 = \varphi(x_0)$ satisfying $(x_0, y_0) \in V$. Moreover, $x \mapsto \varphi(x)$ is holomorphic.

**Proof.** Since $F$ is continuous, there exists $\varepsilon > 0$ such that $F_y(x, y) \neq 0$ whenever $|x|, |y| \leq \varepsilon$. Let
$$n(x) := \frac{1}{2\pi\sqrt{-1}} \int_{|y|=\varepsilon} \frac{F_y(x,y)}{F(x,y)} dy.$$
By the Argument Principle $n(x)$ is an integer. Our hypotheses imply that $F(0, y) = yg(y)$ for some holomorphic function $g$ that does not vanish at the origin. Thus $n(0) = 1$ and by continuity there exists $\varepsilon' < \varepsilon$ such that $n(x) = 1$ for all $|x| \leq \varepsilon'$. Thus, by the Argument Principle, for each such $x$ there is a unique solution $y = f(x)$ of the equation $F(x, y) = 0$. Moreover, by the Residue Theorem we have the formula
$$f(x) = \frac{1}{2\pi i} \int_{|y|=\varepsilon} y \frac{F_y(x,y)}{F(x,y)} dy,$$
which shows that $f$ is holomorphic. $\square$

2.3.3.3. PLANE CURVES.

**DEFINITION 2.3.11.** An affine plane curve is the zero locus of a holomorphic function $f \in \mathcal{O}(\mathbb{C}^2)$. ◇

A plane curve is not necessarily a Riemann surface.

**DEFINITION 2.3.12.** (1) A holomorphic function $f = f(z, w) \in \mathcal{O}(\mathbb{C}^2)$ is non-singular at a point $p \in \mathbb{C}^2$ if either $\frac{\partial f}{\partial z}(p) \neq 0$ or $\frac{\partial f}{\partial w}(p) \neq 0$.
(2) The plane curve $X = \{f = 0\}$ is non-singular (or smooth) at $p \in X$ if $f$ is non-singular at $p$, and $X$ is non-singular (or smooth) if it is non-singular at each of its points. ◇

**THEOREM 2.3.13.** *A non-singular affine plane curve is a Riemann surface.*

**Proof.** Let $X = \{f = 0\}$ be an a non-singular affine plane curve. We must specify an atlas for $X$. To this end, let $p \in X$. Since $X$ is non-singular at $p$, one of the partial derivatives $\frac{\partial f}{\partial z}, \frac{\partial f}{\partial w}$ does not vanish at $p$. Without loss of generality, we may assume $\frac{\partial f}{\partial w}(p) \neq 0$. Let $p = (z_0, w_0)$. By the Implicit Function Theorem, there is a neighborhood $U$ of $z_0$ in $\mathbb{C}$ and a holomorphic function $g_p : U \to \mathbb{C}$ such that $X$ is the graph of $g_p$ over $U$. Thus the projection $\pi_p^1 : (z, w) \mapsto z$ onto the first variable gives a homeomorphism $(\pi_p^1)^{-1}(U) \to U$. We take this to be our complex chart.

We must now check the compatibility of these charts. Suppose $p_1$ and $p_2$ are two points of $X$ whose associated neighborhoods $U_1$ and $U_2$ contain a common point $p$. If the same partial derivative doesn't vanish at $p$, then the composition

$\pi_{p_1}^1 \circ (\pi_{p_2}^1)^{-1}$ is just the identity. On the other hand, if different partials do not vanish, say $\frac{\partial f}{\partial w}(p) \neq 0 \neq \frac{\partial f}{\partial z}(p)$, then in some neighborhood

$$\pi_{p_2}^2 \circ (\pi_{p_1}^1)^{-1}(z) = g_p(z),$$

which is holomorphic. The proof is complete. □

REMARK. As we shall prove in Chapter 12, every open Riemann surface can be embedded in $\mathbb{C}^3$. It is still unknown whether every open Riemann surface can be embedded in $\mathbb{C}^2$. ◇

2.3.3.4. EMBEDDED CURVES IN COMPLEX MANIFOLDS. We note that the construction of the previous paragraph is not limited to working in the affine plane. Let $M$ be a complex manifold of dimension $n$. Given a holomorphic function $f : M \to \mathbb{C}^{n-1}$, we can consider its zero set $X := \{p \in M \,;\, f(p) = 0\}$.

We leave it to the reader to state and prove a higher-dimensional version of the Implicit Function Theorem in $\mathbb{C}^n$ for $n > 2$ and then use it to prove that if, for any point $p \in X$, $df(p)$ has full rank $n - 1$ then $X$ is a Riemann surface.

One could have other complex manifolds constructed in this way, by considering maps $f : M \to \mathbb{C}^k$, $1 \leq k \leq \dim_\mathbb{C}(M)$. In general, when $f$ has full rank, the codimension of the resulting manifold is $k$.

REMARK. Of course, if $M$ is a compact complex manifold, then $M$ has no holomorphic functions on it other than constants. Thus the above construction cannot begin.

In fact, there are complex manifolds that have no closed complex submanifolds at all. But there are other manifolds, like the complex projective space, that have many complex submanifolds, none of which could be described as the common zero locus of some holomorphic functions. Nevertheless, in the case of projective space the implicit construction of plane and space curves has a natural analog. ◇

**2.3.4. Projective curves.**

2.3.4.1. THE PROJECTIVE PLANE. We define the projective plane to be the set

$$\mathbb{P}_2 := \{\text{1-dimensional subspaces of } \mathbb{C}^3\}.$$

If $(x, y, z) \in \mathbb{C}^3 - \{0\}$ then there is a unique line in $\mathbb{C}^3$ passing through $(x, y, z)$ and $(0, 0, 0)$. We denote this point of $\mathbb{P}_2$ by

$$[x, y, z].$$

We define the topology on $\mathbb{P}_2$ to be the coarsest topology such that the map $\pi : \mathbb{C}^3 - \{0\} \to \mathbb{P}_2$ sending $(x, y, z)$ to $[x, y, z]$ is continuous. That is to say, a set $U \subset \mathbb{P}_2$ is open if and only if $\pi^{-1}(U)$ is open in $\mathbb{C}^3$.

REMARK. When referring to points in $\mathbb{P}_2$, $x$, $y$, and $z$ are called homogeneous coordinates. Of course, homogeneous coordinates of a point are not unique. However, saying that a homogeneous coordinate (or, more generally, a homogeneous function) is zero or non-zero is meaningful. Later on, we will discuss an important interpretation of homogeneous functions as sections of certain line bundles on projective space. ◇

The projective plane is covered by the three rather large open sets
$$U_0 := \{[x,y,z] \; ; \; x \neq 0\}, \; U_1 := \{[x,y,z] \; ; \; y \neq 0\},$$
and
$$U_2 := \{[x,y,z] \; ; \; z \neq 0\}.$$
Each set $U_i$ is homeomorphic to $\mathbb{C}^2$ via the maps $\varphi_j : U_j \to \mathbb{C}^2$ given by
$$\begin{aligned} \varphi_0([x,y,z]) &= \left(\tfrac{y}{x}, \tfrac{z}{x}\right), \\ \varphi_1([x,y,z]) &= \left(\tfrac{x}{y}, \tfrac{z}{y}\right), \\ \varphi_2([x,y,z]) &= \left(\tfrac{x}{z}, \tfrac{y}{z}\right). \end{aligned}$$
It is easy to see that
$$\varphi_0 \circ \varphi_1^{-1}(\zeta, \eta) = \varphi_0([\zeta, 1, \eta]) = \left(\tfrac{1}{\zeta}, \tfrac{\eta}{\zeta}\right)$$
is well-defined and holomorphic on $\varphi_1(U_0 \cap U_1) = \{(\zeta, \eta) \; ; \; \zeta \neq 0\}$. Similar calculations for $\varphi_i \circ \varphi_j^{-1}$ show that $\mathbb{P}_2$ is a complex manifold of complex dimension 2.

Finally, we note that every line through the origin in $\mathbb{C}^3$ passes through the unit sphere
$$S^5 := \{|x|^2 + |y|^2 + |z|^2 = 1\} \subset \mathbb{C}^3$$
(in fact, it intersects $S^5$ in a circle) and thus $\mathbb{P}_2 = \pi(S^5)$ is compact.

2.3.4.2. HOMOGENEOUS POLYNOMIALS.

DEFINITION 2.3.14. A holomorphic function $F : \mathbb{C}^3 \to \mathbb{C}$ is said to be *homogeneous of degree $j$* if

(2.1) $$F(\lambda p) = \lambda^j F(p)$$

holds for any $p \in \mathbb{C}^3$ and any $\lambda \in \mathbb{C}$. ◇

LEMMA 2.3.15. *If $F$ is homogeneous of degree $j$, then*
$$jF(x,y,z) = x\frac{\partial F}{\partial x} + y\frac{\partial F}{\partial y} + z\frac{\partial F}{\partial z}.$$

**Proof.** Differentiate equation (2.1) with respect to $\lambda$ and then set $\lambda = 1$. □

REMARK. A homogeneous holomorphic function is automatically a polynomial. ◇

DEFINITION 2.3.16. A homogeneous polynomial $F(x,y,z)$ is said to be *non-singular* if the system of equations

(2.2) $$F = \frac{\partial F}{\partial x} = \frac{\partial F}{\partial y} = \frac{\partial F}{\partial z} = 0$$

## 2.3. Examples of Riemann surfaces

has no non-zero solutions. ◇

2.3.4.3. ZERO SETS OF HOMOGENEOUS POLYNOMIALS. As already mentioned, there are no non-constant holomorphic functions on $\mathbb{P}_2$ and thus one cannot imitate the construction of affine curves in projective space directly. Instead, one can define subsets of $\mathbb{P}_2$ using homogeneous polynomials in $\mathbb{C}^3$. Fix such a homogeneous polynomial $F$ and let $\tilde{X}$ denote its zero set in $\mathbb{C}^3$. Equation (2.1) shows that, with $\pi : \mathbb{C}^3 - \{0\} \to \mathbb{P}_2$ the natural projection, $\pi^{-1} \circ \pi(\tilde{X}) = \tilde{X}$. Thus no information is lost if we study the image $X = \pi(\tilde{X})$. Henceforth we shall say that $X$ *is cut out by* $F$.

Let
$$X_j := X \cap U_j \quad \text{and} \quad Y_j := \varphi_j(X_j), \quad j = 0, 1, 2.$$
Then each $Y_j \subset \mathbb{C}^2$ is an affine plane curve given by the polynomial $f_j(\zeta, \eta) = 0$, where
$$f_0(\zeta, \eta) = F(1, \zeta, \eta), \quad f_1(\zeta, \eta) = F(\zeta, 1, \eta), \quad \text{and} \quad f_2(\zeta, \eta) = F(\zeta, \eta, 1).$$

LEMMA 2.3.17. *If $F$ is non-singular and homogeneous then each $Y_j$ is a smooth affine plane curve.*

**Proof.** By symmetry it suffices to consider $Y_0$. Suppose $Y_0$ is not smooth. Then by the Implicit Function Theorem there exists a solution $(\zeta_0, \eta_0)$ to the system of equations
$$f_0(\zeta, \eta) = \frac{\partial f_0}{\partial \zeta} = \frac{\partial f_0}{\partial \eta} = 0.$$
But then $(1, \zeta_0, \eta_0)$ is a solution to the system of equations (2.2). Indeed,
$$\begin{aligned} F(1, \zeta_0, \eta_0) &= f_0(\zeta_0, \eta_0) = 0, \\ \frac{\partial F}{\partial y}(1, \zeta_0, \eta_0) &= \frac{\partial f_0}{\partial \zeta}(\zeta_0, \eta_0) = 0, \\ \frac{\partial F}{\partial z}(1, \zeta_0, \eta_0) &= \frac{\partial f_0}{\partial \eta}(\zeta_0, \eta_0) = 0, \\ \frac{\partial F}{\partial x}(1, \zeta_0, \eta_0) &= jF(1, \zeta_0, \eta_0) - \zeta_0 \frac{\partial F}{\partial y}(1, \zeta_0, \eta_0) - \eta_0 \frac{\partial F}{\partial z}(1, \zeta_0, \eta_0) = 0, \end{aligned}$$
where the last equality follows from Lemma 2.3.15. This completes the proof. □

Lemma 2.3.17 shows that if $X$ is the projectivization of the zero set of a non-singular homogeneous polynomial, then $X$ can be written as a union of three open sets, each of which is a Riemann surface. It is not hard to convince oneself that each $X_j$ is dense in $X$ and that $X$ is a smooth manifold. But we want to know more, namely, that $X$ is itself a Riemann surface.

We now endow $X$ with the unique maximal atlas $\mathcal{A}$ containing the following charts. If $p \in X$, then $p \in X_j$ for some $j$. It follows that $y_j := \varphi_j(p)$ is in $Y_j$, and since the latter is a smooth affine plane curve, there is a chart $\chi_j : U \to V$ containing $y_j$. We define a chart $\psi : W \to V$ at $p$ by

$$\psi := \chi_j \circ \varphi_j \quad \text{and} \quad W := \varphi_j^{-1} \circ \chi_j^{-1}(U).$$

In fact, we have the following theorem.

THEOREM 2.3.18. *The set $X$ with the maximal atlas $\mathcal{A}$ is a Riemann surface.*

The proof of Theorem 2.3.18 is a simple combination of the holomorphicity of the overlap maps defining an affine curve and the holomorphicity of the transition functions defining $\mathbb{P}_2$. We leave it to the reader as an exercise.

2.3.4.4. PROJECTIVE $n$-SPACE. We define $\mathbb{P}_n$ to be the set of all 1-dimensional subspaces of $\mathbb{C}^{n+1}$.

Let $(z_0, ..., z_n)$ be coordinates in $\mathbb{C}^{n+1}$. Again, to each point $z \in \mathbb{C}^{n+1} - \{0\}$ we can associate the unique line $[z] \in \mathbb{P}_n$ passing through $z$ and $0$. As in the case $n = 2$, the numbers $z_0, ..., z_n$ are called *homogeneous coordinates* of the point $[z_0, ..., z_n] \in \mathbb{P}_n$.

We can define a map $\pi : \mathbb{C}^{n+1} - \{0\} \to \mathbb{P}_n$ by $\pi(z) := [z]$. We then define the topology of $\mathbb{P}_n$ to be the coarsest topology for which $\pi$ is continuous. We observe again that, with $S^{2n+1} = \{z \in \mathbb{C}^{n+1} \; ; \; |z|^2 = 1\}$, $\pi(S^{2n+1}) = \mathbb{P}_n$. Thus $\mathbb{P}_n$ is compact.

We define open sets

$$U_j := \{[z] \in \mathbb{P}_n \; ; \; z_j \neq 0\}, \quad j = 0, ..., n,$$

and homeomorphisms $\varphi_j : U_j \to \mathbb{C}^n$ by

$$\varphi_j([z]) := \left( \frac{z_0}{z_j}, ..., \frac{z_{j-1}}{z_j}, \frac{z_{j+1}}{z_j}, ..., \frac{z_n}{z_j} \right), \quad j = 0, ..., n.$$

One sees that

$$\begin{aligned} & \varphi_j \circ \varphi_i^{-1}(\zeta_1, ..., \zeta_n) \\ = \; & \varphi_j[\zeta_1 : ... : \zeta_{i-1} : 1 : \zeta_{i+1} : ... : \zeta_n] \\ = \; & \begin{cases} \left( \frac{\zeta_1}{\zeta_j}, ..., \frac{\zeta_{i-1}}{\zeta_j}, \frac{1}{\zeta_j}, \frac{\zeta_{i+1}}{\zeta_j}, ..., \frac{\zeta_{j-1}}{\zeta_j}, \frac{\zeta_{j+1}}{\zeta_j}, ... \frac{z_n}{z_j} \right), & i < j \\ \left( \frac{\zeta_1}{\zeta_j}, ..., \frac{\zeta_{j-1}}{\zeta_j}, \frac{\zeta_{j+1}}{\zeta_j}, ..., \frac{\zeta_{i-1}}{\zeta_j}, \frac{1}{\zeta_j}, \frac{\zeta_{i+1}}{\zeta_j}, ... \frac{z_n}{z_j} \right), & i > j \end{cases} \end{aligned}$$

is well-defined and holomorphic on $\varphi_i(U_i \cap U_j) = \{\zeta \in \mathbb{C}^n \; ; \; \zeta_j \neq 0\}$.

2.3.4.5. COMPLETE INTERSECTIONS. To define curves in a higher-dimensional projective space $\mathbb{P}_n$, one can again consider zero sets of homogeneous polynomials. However, a single polynomial will only cut out a set whose dimension at a smooth point is $n - 1$. In order to cut out a curve, we must have $n - 1$ homogeneous polynomials.

## 2.3. Examples of Riemann surfaces

DEFINITION 2.3.19. Let $F_1, ..., F_{n-1}$ be homogeneous polynomials (possibly of different degrees) in $n+1$ variables $z_0, ..., z_n$, and let $X$ be the subset of $\mathbb{P}_n$ cut out by $F_1, ..., F_{n-1}$. We call $X$ a *complete intersection curve*. We further say that $X$ is smooth if for each $[x] \in X$ the $(n-1) \times (n+1)$ matrix of partial derivatives

$$\left( \frac{\partial F_i}{\partial z_j}(x) \right)$$

has maximal rank $n-1$. ◇

Using the higher-dimensional Implicit Function Theorem, one has the following proposition.

PROPOSITION 2.3.20. *A smooth complete intersection in $\mathbb{P}_n$ is a Riemann surface.*

LOCAL COMPLETE INTERSECTIONS

If $n \geq 3$, not all 1-dimensional complex submanifolds of $\mathbb{P}_n$ are smooth complete intersections. The simplest example is the so-called *twisted cubic* in $\mathbb{P}_3$, defined to be the image of the map

$$\iota_V : \mathbb{P}_1 \to \mathbb{P}_3; [z_0 : z_1] \mapsto [z_0^3 : z_0^2 z_1 : z_0 z_1^2 : z_1^3].$$

(Note that in the chart $U_0 \subset \mathbb{P}_1$ we have $\iota_V \circ \varphi_0^{-1}(t) = \iota_V([1,t]) = [1 : t : t^2 : t^3]$, hence the name.) Observe that $\iota_V(\mathbb{P}_1)$ is cut out by the three equations

$$z_0 z_3 = z_1 z_2, \quad z_0 z_2 = z_1^2, \quad z_1 z_3 = z_2^2.$$

It turns out that $\iota_V(\mathbb{P}_1)$ cannot be cut out by just two homogeneous polynomials.

Note that if $[z_0, z_1, z_2, z_3] \in \iota_V(\mathbb{P}_1)$ is a point at which $z_0 \neq 0$, then in a neighborhood of this point $\iota_V(\mathbb{P}_1)$ can be cut out by the first two equations, since near such a point the third equation follows from the first two.

DEFINITION 2.3.21. A *local complete intersection curve* is a subset $X \subset \mathbb{P}_n$ cut out by a family $\mathcal{F} = \{F_\alpha\}_{\alpha \in A}$ of homogeneous polynomials such that at each point $[x] \in X$ there is a neighborhood $U$ of $p$ in $\mathbb{P}_n$ and $n-1$ polynomials $F_{\alpha_1}, ..., F_{\alpha_{n-1}} \in \mathcal{F}$ so that $X \cap U = \{F_{\alpha_1} = ... = F_{\alpha_{n-1}} = 0\}$.

We say that $X$ is smooth if at each $[x] \in X$ there exist $F_{\alpha_1}, ..., F_{\alpha_{n-1}} \in \mathcal{F}$ such that the $(n-1) \times (n+1)$ matrix

$$\left( \frac{\partial F_{\alpha_i}}{\partial x_j}(x) \right)$$

has full rank $n-1$. ◇

As in the case of complete intersections, one can use a higher-dimensional version of the Implicit Function Theorem to prove the following proposition.

PROPOSITION 2.3.22. *Every smooth local complete intersection curve $X$ in $\mathbb{P}_n$ is a Riemann surface.*

It is a fact that every Riemann surface that can be holomorphically embedded in $\mathbb{P}_n$ (and must thus be compact) is a local complete intersection. As we will show in Chapter 12, every compact Riemann surface can be holomorphically embedded in $\mathbb{P}_n$ for some $n$. (In fact, $n = 3$ will do.)

## 2.4. Exercises

2.1. Show that $\mathbb{P}_1$, as it is defined in Example 2.3.6, is diffeomorphic to the unit sphere in $\mathbb{R}^3$.

2.2. Consider the singular curve $\Gamma := \{(z, w) \; ; \; z^2 = w^3\}$ in $\mathbb{C}^2$. Find (i) a singular curve $C$ that is a family of four straight lines meeting pairwise transversely in such a way that each point of $\mathbb{C}^2$ is contained in at most two of these lines and (ii) a holomorphic map $F : \mathbb{C}^2 \to \mathbb{C}^2$, such that $F(C) = \Gamma$ and for the generic point $p \in C$, $F^{-1}(p)$ consists of a single point.

Hint: Try to construct the map $F$ as a composition of transformations of the form $(x, y) \mapsto (x, xy)$ or $(xy, y)$.

2.3. Give the structure of a complex manifold to the set of all $k$-dimensional complex subspaces of the complex vector space $\mathbb{C}^n$. (The resulting manifold, called the Grassmannian, is denoted $G(k, n)$.)

2.4. Show that the set $\{[x, y, z] \in \mathbb{P}_2 \; ; \; zx^2 = y(y-z)(y-2z)\}$ is a Riemann surface.

2.5. On $S^2 = \{x \in \mathbb{R}^3 \; ; \; ||x|| = 1\}$ consider the map $A : x \mapsto -x$. Let $X$ be the quotient of $S^2$ by the action of the group $\{I, A\} \subset \text{Diffeo}(S^2)$. Show that $X$ is not orientable and thus does not admit the structure of a Riemann surface.

2.6. Show that $\mathbb{C}$ and the upper half-plane are not biholomorphic.

2.7. Show that every homogeneous holomorphic function on $\mathbb{C}^n$ is a polynomial.

2.8. Prove Theorem 2.3.18.

*Chapter 3*

# Functions on Riemann Surfaces

We have already defined holomorphic functions and maps on complex manifolds, thus on Riemann surfaces. In this chapter we dig a little deeper into the notion of functions and obtain more detailed local and global information.

## 3.1. Holomorphic and meromorphic functions

**3.1.1. Holomorphic functions.** Let $X$ be a Riemann surface, $p \in X$ a point, and $f$ a function defined in a neighborhood $W$ of $p$. Recall that we defined $f$ to be holomorphic at $p$ if there is a chart $\varphi : U \to V$ with $p \in U$ such that $f \circ \varphi^{-1}$ is holomorphic at $\varphi(p)$, and holomorphic on $W$ if it is holomorphic at each point of $W$. We leave the proof of the following elementary lemma as an exercise to the reader.

LEMMA 3.1.1. *With $X$, $p$, $W$, and $f$ as above:*

(a) *$f$ is holomorphic at $p$ if and only if for every chart $\varphi : U \to V$ with $p \in U$, $f \circ \varphi^{-1}$ is holomorphic at $\varphi(p)$.*

(b) *$f$ is holomorphic on $W$ if and only if for every atlas $\{\varphi_j : U_j \to \varphi_j\}_{j \in J}$ on $W$, we have that for any $p \in U_j$, $f \circ \varphi_j^{-1}$ is holomorphic at $\varphi_j(p)$.*

EXAMPLE 3.1.2. (a) Any complex chart is holomorphic on its domain.

(b) If $X = \mathbb{C}$, the notion of holomorphic function agrees with the classical notion.

(c) If $f$ and $g$ are holomorphic (at a point $p$ or on $W$) then so are $f \pm g$ and $fg$. Moreover, $f/g$ is holomorphic if $g$ is non-zero (at $p$ or at any point of $W$).

(d) Let $X = \mathbb{P}_1$, let $[z_0, w_0] \in \mathbb{P}_1$, and let $p, q \in \mathbb{C}[z, w]$ be two homogeneous polynomials of the same degree such that $q(z_0, w_0) \neq 0$. Then $f([z, w]) := p(z, w)/q(z, w)$ is holomorphic in a neighborhood of $[z_0, w_0]$. In fact, $f$ is holomorphic on the (rather large) open subset $A_q := \{[z, w] \,;\, q(z, w) \neq 0\}$.

(e) Consider a torus $X = \mathbb{C}/L$, and let $W \subset X$ be an open subset. Then $f$ is holomorphic at $p \in W$ if and only if there is a point $z \in \pi^{-1}(p)$ such that $f \circ \pi$ is holomorphic at $z$. In fact, $f$ is holomorphic on $W$ if and only if $f \circ \pi$ is holomorphic on $\pi^{-1}(W)$. (Here $\pi : \mathbb{C} \to X$ is the projection defining the quotient.)

We can, for instance, take $W = X$. In this case, a function $f$ on $X$ determines $\tilde{f}$ on $\mathbb{C}$ by pullback: $\tilde{f} = f \circ \pi$. To go in the other direction one needs a function on $\mathbb{C}$ that is invariant under the action of the lattice. More precisely, if $\tilde{f}(z + \omega) = \tilde{f}(z)$ for all $\omega \in L$, then there exists $f : X \to \mathbb{C}$ such that $\tilde{f} = f \circ \pi$. (Traditionally, one says that $\tilde{f}$ is doubly periodic.) Since we know that every holomorphic function on a compact complex manifold is constant, we see that every doubly periodic function on $\mathbb{C}$ is constant. (One can also prove the constancy of doubly periodic functions using Liouville's Theorem.)

(f) Let $X$ be an affine plane curve. The restriction of the coordinate functions to $X$ defines two holomorphic functions on $X$. In fact, the restriction of any holomorphic function on $\mathbb{C}^2$ to $X$ is holomorphic.

More generally, if $X$ is a curve in a complex manifold $M$, then the restriction to $X$ of every holomorphic function on $M$ is holomorphic.

It turns out that if $M = \mathbb{C}^n$ then the converse is also true: if $f$ is a holomorphic function on a space curve $X \subset \mathbb{C}^n$ then there exists a holomorphic function $F$ on all of $\mathbb{C}^n$ such that $F|_X = f$.

(g) Let $X \subset \mathbb{P}_n$ be a projective algebraic curve that is a local complete intersection. Suppose $F, G \in \mathbb{C}[x_0, ..., x_n]$ are homogeneous polynomials of the same degree. Then $f[x] = F(x)/G(x)$ is holomorphic on $W := \{[x] \in X \,;\, G(x) \neq 0\}$. ◇

DEFINITION 3.1.3. Let $X$ be a Riemann surface and let $W \subset X$ be an open set. We define

$$\mathcal{O}_X(W) := \{\text{holomorphic } f : W \to \mathbb{C}\}.$$

We might write $\mathcal{O}(W)$ when $X$ is clear from the context. ◇

**3.1.2. Meromorphic functions.** Recall that, in Chapter 1, we defined the notions of *pole* and *essential singularity* for a holomorphic function on a punctured neighborhood. These notions are invariant under local biholomorphic functions and thus can be extended to Riemann surfaces.

## 3.1. Holomorphic and meromorphic functions

DEFINITION 3.1.4. Let $f$ be holomorphic in a punctured neighborhood of a point $p \in X$. We say that $f$ has a removable singularity (resp. pole of order $n$, essential singularity, zero of order $n$) at $p$ if there exists a complex chart $\varphi : U \to V$ containing $p$ such that $f \circ \varphi^{-1}$ has a removable singularity (resp. pole of order $n$, essential singularity, zero of order $n$) at $\varphi(p)$. ◇

We leave it to the reader to check that these notions are well-defined.

As for holomorphic functions on domains in $\mathbb{C}$, we have the following properties of singularities.

THEOREM 3.1.5. *Let $f$ be holomorphic in a punctured neighborhood of a point $p \in X$.*

  (a) *If $f$ is bounded in a neighborhood of $p$ then $f$ has a removable singularity at $p$.*
  (b) *If $\lim_{z \to p} |f(z)| = +\infty$ then $f$ has a pole at $p$.*
  (c) *If $\lim_{z \to p} |f(z)|$ does not exist then $f$ has an essential singularity at $p$.*

DEFINITION 3.1.6. Let $X$ be a Riemann surface and let $W \subset X$ be an open subset. We say that a function $f$ on $W$ is meromorphic at $p \in W$ if $f$ is holomorphic in a punctured neighborhood of $p$ and has either a pole at $p$ or a removable singularity at $p$. We say that $f$ is meromorphic on $W$ if $f$ is meromorphic at every point of $W$. ◇

In other words, if we take a chart $\varphi : U \to V$ containing $p$ and expand $f \circ \varphi^{-1}$ in a Laurent series in a neighborhood of $\varphi(p)$, then there is a largest integer $n$ (depending on $p$ but not on $\varphi$) such that the Laurent series of $f \circ \varphi^{-1}(p)$ has no coefficients of order less than $n$.

We now have a well-defined notion of a meromorphic function on a Riemann surface and also a well-defined notion of the order of a meromorphic function at a point. We write

$$\mathrm{Ord}_p(f)$$

for the order of the meromorphic function $f$ at the point $p$.

EXAMPLE 3.1.7. Here are several examples of meromorphic functions.

  (a) Any holomorphic function is also a meromorphic function. Indeed, every point in the domain of definition is a removable singularity.
  (b) If $X = \mathbb{C}$ then the notion of meromorphic function agrees with the classical notion.
  (c) If $f$ and $g$ are meromorphic (at a point $p$ or on $W$) then so are $f \pm g$, $fg$, and $f/g$, the latter provided that $g \not\equiv 0$.

(d) Let $X = \mathbb{P}_1$. If $p, q \in \mathbb{C}[z, w]$ are homogeneous polynomials of the same degree such that $q \not\equiv 0$, then $f([z, w]) := p(z, w)/q(z, w)$ is a meromorphic function on $\mathbb{P}_1$.

(e) Consider a torus $X = \mathbb{C}/L$. We define a meromorphic function $\wp : \mathbb{C} \to \mathbb{C}$ as follows:
$$\wp(z) := \frac{1}{z^2} + \sum_{0 \neq \omega \in L} \left( \frac{1}{(z+\omega)^2} - \frac{1}{\omega^2} \right).$$
Ignoring issues of convergence, observe that $\wp(z + \omega) = \wp(z)$ for any $\omega \in L$, and thus $\wp$ determines a unique meromorphic function $f$ on the quotient $X$. Both $f$ and $\wp$ are called the Weierstrass $\wp$-function.

(f) Let $X$ be an affine plane curve. If $g$ and $h$ are holomorphic functions on $\mathbb{C}^2$ such that $X \not\subset \{h = 0\}$ then, after restricting $g$ and $h$ to $X$, the function $f = g/h$ is meromorphic on $X$. In fact, $\mathrm{Ord}_p(f) = \mathrm{Ord}_p(g) - \mathrm{Ord}_p(h)$.

More generally, let $X$ be a curve in a complex manifold $M$. Then the restriction to $X$ of the quotient of any two holomorphic functions on $M$ is a meromorphic function on $X$ provided the denominator does not vanish identically on $X$.

It turns out that if $M = \mathbb{C}^n$ then the converse is also true: if $f$ is a meromorphic function on $X$ then there exist holomorphic functions $F$ and $G$ on $\mathbb{C}^n$ such that $F/G = f$ on $X$.

(g) Let $X \subset \mathbb{P}_n$ be a projective algebraic curve that is a local complete intersection. Suppose $F, G \in \mathbb{C}[x_0, ..., x_n]$ are homogeneous polynomials of the same degree such that $G \not\equiv 0$ on $\pi^{-1}(X) \subset \mathbb{C}^{n+1}$. Then $f[x] = F(x)/G(x)$ is meromorphic on $X$. ◇

DEFINITION 3.1.8. Let $X$ be a Riemann surface and let $W \subset X$ be an open set. We define
$$\mathscr{M}_X(W) := \{\text{meromorphic } f : W \to \mathbb{C}\}.$$
We may write $\mathscr{M}(W)$ when $X$ is clear from the context. ◇

**3.1.3. The Argument Principle.** Let $\Gamma$ be a smooth embedded curve (i.e., an embedding of $[0, 1]$) in a Riemann surface $X$. Let $\varphi_j : U_j \to V_j$, $j = 1, ..., N$, be $N$ coordinate charts such that $\Gamma \subset \bigcup_{j=1}^N U_j$. For each $j$, let $W_j \subset\subset U_j$ such that $\Gamma \subset \bigcup_{j=1}^N W_j$, and let $\psi_j$ be a smooth function that is $\equiv 1$ on $W_j$ and whose support is in $U_j$. Put $\chi_j := \left( \sum_{k=1}^N \psi_k \right)^{-1} \psi_j$. We define

(3.1) $$\int_\Gamma \frac{df}{f} := \sum_{j=1}^N \int_{\varphi_j(\Gamma \cap U_j)} \chi_j(\varphi_j^{-1}(z)) \frac{(f \circ \varphi_j^{-1})'(z)}{f \circ \varphi_j^{-1}(z)} dz.$$

## 3.1. Holomorphic and meromorphic functions

It is a standard exercise in advanced calculus to show that the definition is independent of the choice of functions $\chi_j$. Moreover, if we choose different coordinate charts, say $\tilde{\varphi}_j : \tilde{U} \to \tilde{V}$, and let $\tilde{z}$ be the corresponding variable in $\tilde{V}$ then on $V \cap \tilde{V}$ we have

$$\frac{(f \circ \tilde{\varphi}^{-1})'(\tilde{z})}{f \circ \tilde{\varphi}^{-1}(\tilde{z})} d\tilde{z}$$
$$= \frac{(f \circ \varphi^{-1} \circ (\varphi \circ \tilde{\varphi}^{-1}))'(\tilde{\varphi} \circ \varphi^{-1}(z))}{f \circ \tilde{\varphi}^{-1}(\tilde{\varphi} \circ \varphi^{-1}(z))} d(\tilde{\varphi} \circ \varphi^{-1}(z))$$
$$= \frac{(f \circ \varphi^{-1})'(\varphi \circ \tilde{\varphi}^{-1} \circ \tilde{\varphi} \circ \varphi^{-1}(z))}{f \circ \varphi^{-1}(z)} d(\varphi \circ \tilde{\varphi}^{-1})(\tilde{\varphi} \circ \varphi^{-1}(z)) d(\tilde{\varphi} \circ \varphi^{-1}(z))$$
$$= \frac{(f \circ \varphi^{-1})'(z)}{f \circ \varphi^{-1}(z)} dz,$$

where the last two equalities follow from the chain rule. Thus the definition (3.1) is well posed, and we can state the following theorem, which is just the Argument Principle on Riemann surfaces.

THEOREM 3.1.9. *Let $X$ be a Riemann surface and $D \subset X$ an open subset whose closure is compact and whose boundary $\partial D$ is piecewise smooth. If $f$ is a meromorphic function on $X$ with no zeroes or poles on $\partial D$, then*

$$\frac{1}{2\pi\sqrt{-1}} \oint_{\partial D} \frac{df}{f} = \sum_{x \in D} \mathrm{Ord}_x(f).$$

**Proof.** The set

$$A := \{x \in D \; ; \; \mathrm{Ord}_x(f) \neq 0\}$$

is locally finite and thus finite. Let $x_1, ..., x_k$ be the points of $A$, and let $D_1, ..., D_k$ be relatively compact coordinate disks in $D$ with centers at $x_1, ..., x_k$ respectively. Consider the domain $E := D - \bigcup_{j=1}^{k} \overline{D_j}$. The boundary $\partial E$ of $E$ is piecewise smooth, and $f'/f$ is holomorphic in $E$. Thus

$$\int_{\partial D} \frac{df}{f} - \sum_{j=1}^{k} \int_{\partial D_j} \frac{df}{f} = \int_{\partial E} \frac{df}{f} = 0.$$

(The last equality is a consequence of Cauchy's Theorem; the simplest adaptation of Cauchy's Theorem to Riemann surfaces can be achieved using Stokes' Theorem, which will be recalled in Chapter 5.) On the other hand, by the argument principle

$$\int_{\partial D_j} \frac{df}{f} = 2\pi\sqrt{-1} \mathrm{Ord}_{x_j}(f).$$

This completes the proof. □

## 3.2. Global aspects of meromorphic functions

### 3.2.1. Meromorphic functions on compact Riemann surfaces.

COROLLARY 3.2.1. *Let $X$ be a compact Riemann surface. If $f \in \mathscr{M}(X)$, then*

$$\sum_{p \in X} \mathrm{Ord}_p(f) = 0.$$

**Proof.** Let $p \in X$ such that $f(p) \neq 0$ or $\infty$. Let $D$ be a small neighborhood of $p$ such that (i) $f$ has no zeros or poles in $D$ and (ii) $D$ is the image, under the inverse of some complex chart $\varphi$, of a disk in the plane with center $\varphi(p)$. Since $f$ has no zeros or poles in $D$, Theorem 3.1.9 shows that

$$\oint_{\partial D} \frac{df}{f} = 0 \quad \text{and} \quad \sum_{p \in X} \mathrm{Ord}_p(f) = \sum_{p \in X - D} \mathrm{Ord}_p(f) = \oint_{\partial(X-D)} \frac{df}{f}.$$

But

$$\oint_{\partial(X-D)} \frac{df}{f} = -\oint_{\partial D} \frac{df}{f}.$$

This completes the proof. $\square$

### 3.2.2. Meromorphic functions on $\mathbb{P}_1$.

Example 3.1.7 (d) above says that the quotient of any two homogeneous polynomials of the same degree in $\mathbb{C}^2$ descends to a meromorphic function on $\mathbb{P}_1$. In fact, we have the following theorem.

THEOREM 3.2.2. *Any meromorphic function on $\mathbb{P}_1$ is a quotient of two homogeneous polynomials of the same degree.*

**Proof.** Let $f$ be a meromorphic function on $\mathbb{P}_1$. It follows from the compactness of $\mathbb{P}_1$ that $f$ has only a finite number of zeroes and poles. Let $[z_j, w_j]$, $j = 1, ..., N$, be the points of $\mathbb{P}_1$ where the order $n_j := \mathrm{Ord}_{[z_j, w_j]}(f)$ of $f$ is non-zero. Consider the rational function

$$R(z, w) := \prod_{i=1}^{N} (zw_j - wz_j)^{n_j}.$$

Since $f \in \mathscr{M}(\mathbb{P}_1)$, Corollary 3.2.1 implies that $\sum_{j=1}^{N} n_j = 0$. Hence $R(tz, tw) = R(z, w)$ for all $t \in \mathbb{C}$ and $(z, w) \in \mathbb{C}^2$. Thus $R$ descends to a meromorphic function on $\mathbb{P}_1$ and by construction has the same zeroes and poles as $f$, counting multiplicity. In other words, $f/R$ is a meromorphic function on $\mathbb{P}_1$ with no zeroes or poles. Thus $f/R$ is holomorphic, and so constant. It follows that $f$ is a rational function, as desired. $\square$

### 3.2.3. Meromorphic functions on complex tori.
We saw a construction of a function on the torus in part (e) of Example 3.1.7. We shall now construct all functions on the torus. We specialize to a lattice of the form $L = \mathbb{Z} \oplus \tau\mathbb{Z}$ for some $\tau$ with positive imaginary part. Later we will show that every complex torus is biholomorphic to such a torus.

Fixing $\tau$ with $\operatorname{Im}(\tau) > 0$, we define

$$\theta(z) := \sum_{n=-\infty}^{\infty} e^{\sqrt{-1}\pi(n^2\tau + 2nz)}.$$

This series converges absolutely and uniformly on any compact subset of $\mathbb{C}$, and moreover $\theta(z+1) = \theta(z)$. On the other hand,

$$\theta(z + \tau) = e^{-\sqrt{-1}\pi(\tau + 2z)}\theta(z).$$

It follows that the zeroes of $\theta$ remain invariant after translation by any element of $L$, counting multiplicity. Moreover,

$$\theta'(z+1) = \theta'(z)$$

and

$$\theta'(z+\tau) = e^{-\sqrt{-1}\pi(\tau+2z)}\theta'(z) - 2\pi\sqrt{-1}e^{-\sqrt{-1}\pi(\tau+2z)}\theta(z).$$

PROPOSITION 3.2.3. *Let $\Pi_{1,\tau}$ be the closed parallelogram in $\mathbb{C}$ spanned by $1$ and $\tau$. Then $\theta(z)$ has exactly one zero, counting multiplicity, in $\Pi_{1,\tau}$.*

**Sketch of proof.** Observe first that for $x \in [0,1]$,

$$\theta(x) = \sum_{n \in \mathbb{Z}} e^{2\pi\sqrt{-1}nx} e^{\pi\sqrt{-1}n^2\tau} \quad \text{and} \quad \theta(x\tau) = e^{-\pi\sqrt{-1}x^2\tau} \sum_{n \in \mathbb{Z}} e^{(n+x)^2\sqrt{-1}\pi\tau}.$$

A careful analysis shows that these two sums do not vanish. Thus by periodicity, $\theta$ does not vanish on $\partial\Pi_{1,\tau}$.

Next, note that

$$\frac{\theta'(z+1)}{\theta(z+1)} = \frac{\theta'(z)}{\theta(z)} \quad \text{and} \quad \frac{\theta'(z+\tau)}{\theta(z+\tau)} = \frac{\theta'(z)}{\theta(z)} - 2\pi\sqrt{-1}.$$

The first claim is thus a direct application of the argument principle together with the fact that

$$\oint_{\partial\Pi_{1,\tau}} = \int_{[0,1]} + \int_{1+\tau\cdot[0,1]} - \int_{\tau+[0,1]} - \int_{\tau\cdot[0,1]}.$$

The proof is complete. $\square$

Using Jacobi's triple product formula, one has the following product representation for the function $\theta$:

$$\theta(z) = \prod_{m=1}^{\infty} (1-w^m)(1+q^{2m-1}w^2)(1+q^{2m-1}w^{-2}),$$

where $w = e^{\sqrt{-1}\pi z}$ and $q = e^{\sqrt{-1}\pi \tau}$. Looking at the third factor when $m = 1$, we find a zero of $\theta$ when $w^2 = -q$, i.e., when $2z = (1+\tau)$. Thus, modulo proof of the product formula, one has the following fact.

PROPOSITION 3.2.4. *The zero of $\theta$ in $\Pi_{1,\tau}$ is at $\frac{1+\tau}{2}$.*

Now let
$$\theta^{(x)}(z) := \theta(z - \tfrac{1}{2} - \tfrac{\tau}{2} - x).$$

Observe that $\theta^{(x)}$ is still periodic with period 1 but that
$$\theta^{(x)}(z+\tau) = -e^{2\pi\sqrt{-1}(z-x)}\theta^{(x)}(z) \quad and \quad \theta^{(x)}(x) = 0.$$

Thus, for any finite collection $x_1, ..., x_N, y_1, ..., y_M \in \mathbb{C}$, perhaps with repetitions, the ratio
$$R(z) := \frac{\prod_i \theta^{(x_i)}(z)}{\prod_j \theta^{(y_j)}(z)}$$

is meromorphic in $\mathbb{C}$, has period 1, and satisfies
$$R(z+\tau) = (-1)^{N-M} \exp\left(-2\pi\sqrt{-1}\left((N-M)z + \sum_j y_j - \sum_i x_i\right)\right) R(z).$$

In particular, we see that if
$$N = M \quad \text{and} \quad \sum_j y_j - \sum_i x_i \in \mathbb{Z},$$

then $R$ descends to a meromorphic function on $\mathbb{C}/L$. We thus have the following result.

THEOREM 3.2.5. *Any meromorphic function on a complex torus, whose zeros $[x_1], ..., [x_k]$ and poles $[y_1], ..., [y_k]$ (possibly with repetitions) satisfy $\sum(x_j - y_j) \in \mathbb{Z}$ is a quotient of two products of $\theta$ functions.*

The proof of Theorem 3.2.5 is left to the reader.

REMARK. We have already seen that the zeros $[x_1], ..., [x_M]$ and poles $[y_1], ..., [y_N]$ of a meromorphic function on a compact Riemann surface $X$, listed with repetition but multiplicity 1, must satisfy $M = N$. In Chapter 14 we will show that if $X = \mathbb{C}/L$, then
$$\sum_j x_j - y_j \in L.$$

Under this necessary condition, we can then modify $x_1$ by adding to it an integer multiple of $\tau$, so as to achieve that $\sum_j x_j - y_j \in \mathbb{Z}$. Thus, modulo the results of Chapter 14, every meromorphic function on $\mathbb{C}/L$ is a quotient of products of theta functions. ◇

## 3.3. Holomorphic maps between Riemann surfaces

In this section, we will study the basic properties of holomorphic maps between Riemann surfaces.

**3.3.1. Basic definitions and properties.** Let $X$ and $Y$ be Riemann surfaces. Recall that if $p \in X$ and $W$ is a neighborhood of $p$, we say that a map $F : W \to Y$ is holomorphic at $p$ if there are coordinate charts $\varphi_1 : U_1 \to V_1$ and $\varphi_2 : U_2 \to V_2$ such that $p \in U_1 \subset W$, $F(p) \in U_2$, and $\varphi_2 \circ F \circ \varphi_1^{-1} : V_1 \to V_2$ is holomorphic at $\varphi_1(p)$.

As in the definition of holomorphic functions, there are some immediate consequences stemming from the local biholomorphic invariance of Riemann surfaces. We state these in the following lemma, whose proof is omitted.

LEMMA 3.3.1. *Let $F : X \to Y$ be a mapping between Riemann surfaces.*

(1) *$F$ is holomorphic at $p$ if and only if for any pair of charts $\varphi_1 : U_1 \to V_1$ and $\varphi_2 : U_2 \to V_2$ with $p \in U_1$ and $F(p) \in U_2$, $\varphi_2 \circ F \circ \varphi_1^{-1}$ is holomorphic at $\varphi_1(p)$.*

(2) *$F$ is holomorphic on $W$ if and only if there is a collection of charts $\{\varphi_{1,j} : U_{1,j} \to V_{1,j}\}_{j \in J_1}$ of $X$ and $\{\varphi_{2,j} : U_{2,j} \to V_{2,j}\}_{j \in J_2}$ of $Y$ such that $W \subset \bigcup_{j \in J_1} U_{1,j}$, $F(W) \subset \bigcup_{j \in J_2} U_{2,j}$, and $\varphi_{2,i} \circ F \circ \varphi_{1,j}$ is holomorphic for every $i \in J_1$ and $j \in J_2$ such that it is well-defined.*

Of course, every holomorphic function is a holomorphic map into $\mathbb{C}$.

The proofs of the following facts are left to the reader.

(1) Every holomorphic mapping is $\mathcal{C}^\infty$-smooth.
(2) The composition of holomorphic maps is holomorphic.
(3) If $F : X \to Y$ is holomorphic and $W \subset Y$ is open, then
    (a) for any $f \in \mathcal{O}_Y(W)$, $F^*f := f \circ F \in \mathcal{O}_X(F^{-1}(W))$, and
    (b) for any $f \in \mathcal{M}_Y(W)$, $F^*f := f \circ F \in \mathcal{M}_X(F^{-1}(W))$.

DEFINITION 3.3.2. An isomorphism between two Riemann surfaces $X$ and $Y$ is a bijective holomorphic map $F : X \to Y$ whose inverse is also holomorphic. ◇

REMARK. We know from the Normal Form Theorem that a bijective holomorphic map between Riemann surfaces has a holomorphic inverse. Thus, in the definition of isomorphism, one can omit the requirement of holomorphicity of the inverse without gain of generality. ◇

EXAMPLE 3.3.3. We leave it as an exercise to check that the map
$$[z, w] \mapsto \left( \frac{2\mathrm{Re}\,(z\bar{w})}{|z|^2 + |w|^2}, \frac{2\mathrm{Im}\,(z\bar{w})}{|z|^2 + |w|^2}, \frac{|z|^2 - |w|^2}{|z|^2 + |w|^2} \right)$$
is an isomorphism from $\mathbb{P}_1$ to $S^2$ with the atlas $\{\varphi_N, \varphi_S\}$, defined in Chapter 1. ◇

As an immediate consequence of the Open Mapping Theorem, we have the following corollary.

PROPOSITION 3.3.4. *A non-constant holomorphic mapping between Riemann surfaces maps open sets to open sets.*

A corollary of the Identity Theorem is the following result.

PROPOSITION 3.3.5. *Let $X$ and $Y$ be two Riemann surfaces such that $X$ is connected, and let $F, G : X \to Y$ be holomorphic maps. If $F = G$ on some open subset of $X$, then $F \equiv G$.*

The next proposition is a topological fact.

PROPOSITION 3.3.6. *If $X$ and $Y$ are two connected Riemann surfaces with $X$ compact, and $F : X \to Y$ is a non-constant holomorphic map, then $Y$ is compact and $F$ is surjective.*

**Proof.** Since $X$ is connected and $F$ is non-constant, $F$ is an open mapping. As $X$ is compact, $F(X)$ is both open and closed. Since $Y$ is connected, $F(X) = Y$. $\square$

PROPOSITION 3.3.7. *Let $F : X \to Y$ be a non-constant holomorphic map from a connected Riemann surface $X$. Then for each $y \in Y$, $F^{-1}(y)$ is a discrete set. In particular, if $X$ is compact then $F^{-1}(y)$ is finite for each $y$.*

**Proof.** Let $y$ be a point in $Y$ and choose a chart $\varphi$ such that $\varphi(y) = 0$. Then $\varphi \circ F$ is a non-constant holomorphic function, and since $X$ is connected, the zero set of $\varphi \circ F$ is discrete. But the zero set of this function is precisely $F^{-1}(y)$. $\square$

**3.3.2. Meromorphic functions as maps to the Riemann sphere.** In this section we discuss the correspondence between meromorphic functions and maps to $\mathbb{P}_1$.

THEOREM 3.3.8. *Let $f \in \mathscr{M}(X)$. Then there exists a unique holomorphic map $F_f : X \to \mathbb{P}_1$ such that for every $p \in X$ there are holomorphic functions $g_p$ and $h_p$ near $p$, such that*

$$F_f = [g_p, h_p] \quad \text{and} \quad f = \frac{h_p}{g_p}.$$

**Proof.** Let $f \in \mathscr{M}(X)$. If $\text{Ord}_p(f) \geq 0$ then we take $g_p = 1$ and $h_p = f$. If $\text{Ord}_p(f) \leq 0$, we take $g_p = 1/f$ and $h_p = 1$. Observe that if $\text{Ord}_p(f) = 0$, then $[f, 1] = [1, 1/f]$. Since this set of points is open, the identity theorem implies that $F$ is well-defined. To see uniqueness, suppose $F_f = [g'_p, h'_p]$. Then, $h'_p/g'_p = h_p/g_p$. If $f \equiv 0$, then $g_p \equiv g'_p \equiv 0$ and we are done. Otherwise, we see that $g'_p/h'_p = g_p/h_p$. It follows that at the points where $h_p \neq 0 \neq h'_p$, $[g'_p, h'_p] = [g_p, h_p]$. Since the set of such points is open, uniqueness follows from the Identity Theorem. $\square$

## 3.3. Holomorphic maps between Riemann surfaces

THEOREM 3.3.9. *Let $F : X \to \mathbb{P}_1$ be a holomorphic map of Riemann surfaces. Then there exists a unique $f \in \mathcal{M}(X)$ such that $F = F_f$.*

**Proof.** Suppose $p \in X$ and $F(p) \in U_0$. Then there is a neighborhood $U$ of $p$ such that $F(q) \in U_0$ for all $q \in U$. It follows that $\varphi_0 \circ F$ is holomorphic on $U$, and in fact $F = [1, \varphi_0 \circ F]$. By a similar argument when $F(p) \in U_1$, we see that $F = [\varphi_1 \circ F, 1]$ near $p$. Since $\frac{1}{\varphi_0 \circ F} = \varphi_1 \circ F$ on their common domain of definition, we set

$$f(p) := \begin{cases} \varphi_1 \circ F(p), & F(p) \in U_1 \\ \frac{1}{\varphi_0 \circ F(p)}, & F(p) \in U_0 \end{cases}.$$

It is clear from the definition of $\mathbb{P}_1$ that $f$ is a well-defined meromorphic function, and $F = F_f$. □

REMARK. In view of the results of this section, there is no need to distinguish between meromorphic functions and holomorphic functions to $\mathbb{P}_1$. In the text, we will often abuse notation by using phrases like "... a meromorphic function $f : X \to \mathbb{P}_1$ ...". ◇

### 3.3.3. Multiplicity.

THEOREM 3.3.10 (Normal Form Theorem). *Let $F : X \to Y$ be a holomorphic map between Riemann surfaces, and let $x \in X$. Then there exist two coordinate charts $\varphi_1 : U_1 \to V_1$ and $\varphi_2 : U_2 \to V_2$ at $x$ and $F(x)$, respectively, and a unique integer $m = m_x$ such that $\varphi_1(x) = \varphi_2(F(x)) = 0$ and*

$$\varphi_2 \circ F \circ \varphi_1^{-1}(z) = z^m.$$

**Proof.** To see uniqueness, suppose $\varphi_2 \circ F \circ \varphi_1^{-1}(z) = z^m$ and $\tilde{\varphi}_2 \circ F \circ \tilde{\varphi}_1^{-1}(z) = z^{\tilde{m}}$. Then, with $h_1 = \varphi_1 \circ \tilde{\varphi}_1^{-1}$ and $h_2 = \varphi_2 \circ \tilde{\varphi}_2^{-1}$, we see that

$$h_2((h_1^{-1}(z))^{\tilde{m}}) = z^m.$$

Since $h_1$ and $h_2$ are biholomorphic, a power series expansion shows that $\tilde{m} = m$.

For existence, choose any pair of coordinate charts. After translations, we can assume that $\tilde{\varphi}_1(x) = \varphi_2(F(x)) = 0$. It follows that for some integer $m$, $\varphi_2 \circ F \circ \tilde{\varphi}_1^{-1}(\zeta) = \zeta^m e^{h(\zeta)}$. Let $\psi(\zeta) := \zeta e^{m^{-1}h(\zeta)}$. Then $\psi'(0) \neq 0$, and thus $\psi$ has an inverse in some neighborhood of the origin. We let $\varphi_1 := \psi \circ \tilde{\varphi}_1$. □

DEFINITION 3.3.11. Let $X$ be a Riemann surface, let $x \in X$, and let $F : X \to Y$ be a holomorphic map to another Riemann surface $Y$.

(1) We call the integer $m =: \text{Mult}_x(F)$ the multiplicity of $F$ at $x$.

(2) If $p \in X$ and $\text{Mult}_p(F) \geq 2$, we say that $F$ is ramified at $p$ and that $p$ is a ramification point for $F$.

(3) If $p \in X$ is a ramification point of $F$, we call $F(p)$ a branch point of $F$. ◇

LEMMA 3.3.12. (1) *Let $X$ be a smooth affine plane curve cut out by a non-singular polynomial $f(x, y)$, and let $\pi : X \to \mathbb{C}$ be the restriction to $X$ of the map $(x, y) \mapsto x$. Then $\pi$ is ramified at $p \in X$ if and only if*

$$\left.\frac{\partial f(x,y)}{\partial y}\right|_{(x,y)=p} = 0.$$

(2) *Let $X$ be a smooth projective plane curve cut out by a non-singular homogeneous polynomial $F(x, y, z)$, and let $G : X \to \mathbb{P}_1$ be the restriction to $X$ of the map $[x, y, z] \mapsto [x, z]$. Then $G$ is a holomorphic map, and $G$ is ramified at $[x_0, y_0, z_0]$ if and only if*

$$\left.\frac{\partial F(x,y,z)}{\partial y}\right|_{(x,y,z)=(x_0,y_0,z_0)} = 0.$$

**Proof.** If $\left.\frac{\partial f}{\partial y}\right|_p \neq 0$ then, by the Implicit Function Theorem, $\pi$ is a chart. Thus, since a chart is a local isomorphism, and thus not ramified, $p$ is not a ramification point of $\pi$.

Conversely, suppose $\left.\frac{\partial f}{\partial y}\right|_p = 0$. Since $f$ is non-singular, we must have $\left.\frac{\partial f}{\partial x}\right|_p \neq 0$. It follows from the Implicit Function Theorem that $X$ is cut out, near $p$, by the equation $x - g(y) = 0$ for some function $g$. Now, along $X$ we have $f(g(y), y) \equiv 0$. Differentiation with respect to $y$ at $p = (x_0, y_0)$ gives us that $f_x(p)g'(y_0) + f_y(p) = 0$. Since $f_y(p) = 0 \neq f_x(p)$, we must have $g'(y_0) = 0$. But $g$ is precisely the local expression for $\pi$, and thus (1) holds.

The proof of part (2) is similar and is left as an exercise. □

Finally, we leave the reader to check the following assertions.

PROPOSITION 3.3.13. *Let $f \in \mathscr{M}(X)$ and $F = F_f : X \to \mathbb{P}_1$ its associated holomorphic map.*

(1) *If $p$ is not a pole of $f$ then $\mathrm{Mult}_p(F) = \mathrm{Ord}_p(f - f(p))$.*

(2) *If $p$ is a pole of $f$, then $\mathrm{Mult}_p(F) = -\mathrm{Ord}_p(f)$.*

**3.3.4. Degree of a holomorphic map.** We begin with our first main theorem of this section.

THEOREM 3.3.14. *Let $F : X \to Y$ be a holomorphic mapping between compact connected Riemann surfaces. Then the function $\mathrm{Deg}(F) : Y \to \mathbb{Z}$ defined by*

$$\mathrm{Deg}(F)(y) = \sum_{x \in F^{-1}(y)} \mathrm{Mult}_x(F)$$

*is constant.*

## 3.3. Holomorphic maps between Riemann surfaces

**Proof.** Let $Y_n := \{y \in Y \ ; \ Deg(F)(y) \geq n\}$. We claim that $Y_n$ is open in $Y$. Indeed, by compactness, the number of preimages of $y$ is finite. Then, using the normal form of a map, we see that if $\text{Mult}_x(F) = m$ for $x \in F^{-1}(y)$, then all points $y' \neq y$ sufficiently near $y$ have $m$ preimages near $x$. Since Mult is a positive integer, we see that $\text{Deg}(F)(y') \geq \text{Deg}(F)(y)$ and $Y_n$ is indeed open.

To show that $Y_n$ is closed, let $(y_k)$ be a sequence in $Y_n$, converging to a point $y \in Y$. Since there are only finitely many ramification points in $Y$, we may assume that for any $y_k$, none of the preimages $f^{-1}(y_k)$ contain any ramification points. Thus $f^{-1}(y_k)$ contains $n$ distinct points $x_{k,1}, ..., x_{k,n}$. Since $X$ is compact, we may extract from $(x_{k,n})_{k \geq 1}$ a convergent subsequence and thus obtain $n$ points $x_1, ..., x_n \in X$. Now, the points $x_1, ..., x_n$ may not be distinct. But if, say, $x = x_{i_1} = x_{i_2} = ... = x_{i_j}$, then we claim that the multiplicity of $x$ is at least $j$. Indeed, by the Normal Form Theorem, the restriction of $F$ to a neighborhood of $x$ is an $m$-to-one map for some integer $m$. Moreover, if $k$ is sufficiently large, this map takes the $j$ distinct points $x_{k,i_1}, ..., x_{k,i_j}$ to the single point $y_k$. Thus $m \geq j$. It follows from this argument that the sum of the multiplicities at the points of the set (of $\leq n$) points $\{x_1, ..., x_n\}$ is $\geq n$, and thus $y \in Y_n$, i.e., $Y_n$ is closed.

We conclude that, for each $n \geq 1$, either $Y_n = Y$ or $Y_n = \emptyset$.

Now, choose any $y \in Y$. If $n = \text{Deg}(F)(y)$, then $y \in Y_n$ but $y \notin Y_{n+1}$. It follows that $Y_n = Y$ and $Y_{n+k} \subset Y_{n+1} = \emptyset$ for all $k \geq 1$. Thus $\text{Deg}(F) \equiv n$ on $Y$, as desired. □

REMARK. Recall that a map is proper if the inverse image of any compact set is compact. Theorem 3.3.14 holds even if $X$ and $Y$ are non-compact, so long as the map $F : X \to Y$ is proper. The proof is left as an exercise to the reader. ◇

DEFINITION 3.3.15. The constant value of $\text{Deg}(F)$ is called the *degree of F*. ◇

COROLLARY 3.3.16. *A holomorphic map between compact connected Riemann surfaces is an isomorphism if and only if it has degree* 1.

COROLLARY 3.3.17. *If $f \in \mathscr{M}(X)$ has exactly one pole and this pole is simple then $X$ is isomorphic to $\mathbb{P}_1$.*

**Proof.** By hypothesis, $\text{Deg}(F) = \#(F^{-1}(\infty)) = 1$, and thus by the previous corollary, $F$ is an isomorphism. □

REMARK. We can give another proof that the orders of a meromorphic function on a compact Riemann surface sum to 0. Indeed, according to the theorem on degree,

$$\text{Deg}(f) = \sum_{f(x)=0} \text{Mult}_x(f) = \sum_{f(y)=\infty} \text{Mult}_y(f).$$

But then,

$$\sum_{p \in X} \mathrm{Ord}_p(f) = \sum_{f(x)=0} \mathrm{Ord}_x(f) + \sum_{f(y)=\infty} \mathrm{Ord}_y(f)$$
$$= \sum_{f(x)=0} \mathrm{Mult}_x(f) - \sum_{f(y)=\infty} \mathrm{Mult}_y(f) = 0,$$

where the last equality follows from Proposition 3.3.13. ◇

### 3.3.5. The Riemann-Hurwitz Formula.

DEFINITION 3.3.18. Let $S$ be a compact 2-manifold with smooth (possibly empty) boundary.

(1) A 0-simplex, or *vertex*, is a point. A 1-simplex, or *edge*, is a set homeomorphic to a closed interval. A 2-simplex, or *face*, is a set homeomorphic to the triangle $\{(x,y) \in [0,1] \times [0,1] \, ; \, x+y \leq 1\}$.

(2) A triangulation of $S$ is a decomposition of $S$ into faces, edges, and vertices, such that the intersection of any two faces is a union of edges and the intersection of any two edges is a union of vertices.

(3) Let $S$ have a triangulation with total number of faces equal to $F$, total number of edges equal to $E$, and total number of vertices equal to $V$. The number

$$\chi(S) := F - E + V$$

is called the Euler characteristic of $S$. ◇

A fundamental theorem about the Euler characteristic is the following result, stated here without proof.

THEOREM 3.3.19. *Every compact 2-manifold has a triangulation, and the Euler characteristic $\chi(S)$ is independent of the triangulation. In particular, if $S$ is an orientable 2-manifold without boundary and of genus $g$, then*

$$\chi(S) = 2 - 2g.$$

REMARK. If one believes the first two statements, it is easy to prove the third. Indeed, observe that the Euler characteristic of a sphere is 2, that of a cylinder is 0 and that of a disk is 1. To increase the genus of a given surface by 1, remove two disks and attach a cylinder to the boundary of the resulting surface.

For the reader versed in topology, the existence of a triangulation of a compact Riemann surface can be established by using the representation of the surface as a quotient of a $4g$-gon. The independence of the Euler characteristic of a triangulation is proved by showing that two triangulations can be refined, by adding vertices and edges, to form a common triangulation. The process of refinement is then checked to see that it leaves the Euler characteristic unchanged. ◇

## 3.3. Holomorphic maps between Riemann surfaces

If $X$ is a compact Riemann surface, we denote by $g(X)$ the genus of $X$. We are now ready to discuss the second main theorem of this section.

THEOREM 3.3.20 (RIEMANN-HURWITZ FORMULA). *Let $f : X \to Y$ be a non-constant holomorphic map of compact connected Riemann surfaces. Then*

$$2g(X) - 2 = \mathrm{Deg}(f)(2g(Y) - 2) + \sum_{x \in X}(\mathrm{Mult}_x(f) - 1).$$

**Proof.** Let $d = \mathrm{Deg}(f)$. Take a triangulation of $Y$ such that every branch point is a vertex. (There may, of course, be other vertices.) Suppose this triangulation has $F$ faces, $E$ edges, $V_u$ unbranched vertices, and $V_b$ branched vertices.

Since the preimage of every unbranched point has $d$ points, we obtain a triangulation of $X$ with $dF$ faces, $dE$ edges, and $W$ vertices. To express $W$ in terms of $V$ and $f$, we observe that if $x \in X$ is a ramification point for $f$, then $\mathrm{Mult}_x(f)$-many point are collapsed into one point, so that we have

$$W = dV - \sum_{y \in V_b} \sum_{x \in f^{-1}(y)} \mathrm{Mult}_x(f) - 1 = dV - \sum_{x \in X} \mathrm{Mult}_x(f) - 1.$$

The last equality follows because $\mathrm{Mult}_x(f) = 1$ for all unramified points $x$. The theorem follows from Theorem 3.3.19 and the definition of the Euler characteristic. □

REMARK. Observe that the number

$$R(f) := \sum_{x \in X}(\mathrm{Mult}_x(f) - 1),$$

called the ramification degree of $f$, is even. ◇

COROLLARY 3.3.21. *Let $F : X \to Y$ be a non-constant holomorphic map of compact connected Riemann surfaces.*

(1) $g(Y) \leq g(X)$.
(2) *If $F$ is unramified and $g(X) = 1$ then $g(Y) = 1$.*
(3) *If $F$ is unramified, $g(X) > 1$ and $\mathrm{Deg}(F) > 1$ then $g(X) > g(Y) > 1$ and $\mathrm{Deg}(F) | (g(X) - 1)$.*
(4) *If $g(X) = g(Y) \geq 1$ then either $\mathrm{Deg}(F) = 1$ or $g(X) = 1$. In either case, $F$ is unramified.*

**Proof.** Use either the formula

$$g(Y) = \frac{g(X)}{\mathrm{Deg}(F)} + \frac{\mathrm{Deg}(F) - 1}{\mathrm{Deg}(F)} - \frac{\sum_{x \in X}(\mathrm{Mult}_x(F) - 1)}{2\mathrm{Deg}(F)}$$

or the Riemann-Hurwitz Formula, from which the former is derived. □

**3.3.6. An example: Maps between complex tori.** Let $L$ be a lattice in $\mathbb{C}$ and let $X = \mathbb{C}/L$ be the associated complex torus. For each $a \in \mathbb{C}$, the translation map $\tau_a : \mathbb{C} \to \mathbb{C}$ defined by $\tau_a(z) = z + a$ descends to a holomorphic self-map of the torus $X$. This map is an automorphism, with $\tau_a^{-1} = \tau_{-a}$.

Now let $L'$ be another lattice and $X'$ the associated torus. Suppose $F : X \to X'$ is a holomorphic map. We claim there exists a map $\tilde{F} : \mathbb{C} \to \mathbb{C}$ such that the diagram

$$\begin{array}{ccc} \mathbb{C} & \xrightarrow{\tilde{F}} & \mathbb{C} \\ \pi \downarrow & & \downarrow \pi \\ X & \xrightarrow{F} & X' \end{array}$$

commutes, where the vertical arrows are the quotient maps. Indeed, because $\mathbb{C}$ is simply connected, the map $\pi \circ F : \mathbb{C} \to X$ lifts to the universal cover of $X$. (The basic ideas of covering spaces will be reviewed in Chapter 10.) Moreover, the map $\tilde{F}$ is holomorphic because $\pi$ and $F$ are holomorphic and $\pi$ is a covering map, hence a local isomorphism.

After composing $F$ and $\tilde{F}$ with translations, we may assume that $F([0]) = [0]$ and that $\tilde{F}(0) = 0$. In this case, there exists a holomorphic map $\omega' : \mathbb{C} \times L \to \mathbb{C}$ whose values lie in $L'$, such that $\tilde{F}(z + \omega) = \tilde{F}(z) + \omega'(z, \omega)$. By the continuity of $\omega'$ and the discreteness of $L'$, it follows that $\omega'$ does not depend on $z$. Thus, for each $\omega \in L$, the function $\tilde{F}(z + \omega) - \tilde{F}(z)$ is constant. In particular, $\tilde{F}'(z)$ is doubly periodic and thus constant. It follows that $\tilde{F}(z) = \alpha z$ for some $\alpha \in \mathbb{C}$. Thus $\omega'(z, \omega) = \alpha \omega$, and since we must have $\alpha \omega \in L'$ for all $\omega \in L$, $\alpha$ is somewhat restricted. We summarize everything in the following proposition.

PROPOSITION 3.3.22. *Let $X = \mathbb{C}/L$ and $X' = \mathbb{C}/L'$ be two complex tori, and let $F : X \to X'$ be a holomorphic map. Then there exists an affine linear holomorphic function $\tilde{F} : \mathbb{C} \to \mathbb{C}$ sending $z$ to $\alpha z + a$ for some $\alpha, a \in \mathbb{C}$. The number $a$ is determined modulo $L'$, and the number $\alpha$ must satisfy $\alpha L \subset L'$. Moreover, $F$ is an isomorphism if and only if $\alpha L = L'$.*

REMARK. It is now obvious that every complex torus is biholomorphic to a torus whose lattice is generated by the complex numbers 1 and $\tau$, where $\operatorname{Im} \tau > 0$.  ◇

REMARK. Let $F : X \to X'$ be the holomorphic map of tori induced by the linear map $\tilde{F}(z) = \alpha z$, and let $d := [L' : \alpha L]$ be the index of $\alpha L$ in $L'$. Then $\deg(F) = d$.  ◇

Let us now determine the automorphisms of a complex torus $X = \mathbb{C}/L$. To this end, let $F : X \to X$ be an automorphism. After composing with translation maps, we may assume that $F([0]) = [0]$. Let $\tilde{F}(z) = \alpha z$ be the associated lift. Since $F$ is an automorphism, $\alpha L = L$. We claim that any such $\alpha$ has unit modulus. Indeed, if $\omega \in L - \{0\}$ is the closest non-zero lattice element to the origin, then $\alpha \omega$ must also be the closest element of $L - \{0\}$ to the origin. It follows that $|\alpha| = 1$.

Clearly we can have $\alpha = \pm 1$. But there may be other possibilities. Indeed, suppose $\alpha \in \mathbb{C} - \mathbb{R}$. Let $\omega \in L - \{0\}$ be a number of minimal length. As we saw, $\alpha\omega$ must also be of minimal length, and thus if $\alpha \neq \pm 1$, $\omega$ and $\alpha\omega$ must generate $L$. It follows that $\alpha^2\omega = m\alpha\omega + n\omega$ for some integers $m$ and $n$. But then $\alpha$ must satisfy a quadratic equation

$$\alpha^2 - m\alpha - n = 0$$

for some integers $m$ and $n$. Solving this equation gives

$$\alpha = \frac{m \pm \sqrt{m^2 + 4n}}{2}.$$

Since $\alpha$ is a (non-real) root of unity, we must have $m^2 + 4n < 0$, and thus

$$1 = |\alpha|^2 = \frac{m^2}{4} - \left(n + \frac{m^2}{4}\right) = -n.$$

We find that $m^2 < 4$, so the only possibilities for $m$ are $0$ or $\pm 1$. If $m = 0$ we obtain $\alpha = \pm\sqrt{-1}$ and if $m = \pm 1$ we obtain $\alpha = \pm\frac{1}{2} \pm \frac{\sqrt{-3}}{2}$. (This is six possibilities in total.) In the cases $\alpha = \pm\sqrt{-1}$ we see that $L$ must be a square lattice, and in the other cases a hexagonal lattice. Thus we have the following proposition.

PROPOSITION 3.3.23. *Let $X = \mathbb{C}/L$ be a complex torus, and let $F \in \mathrm{Aut}(X)$ be an automorphism fixing the origin. Then $F$ is induced by the linear map $z \mapsto \alpha z$ where*

(1) *if $L$ is a square lattice then $\alpha$ is a $4^{th}$ root of unity,*

(2) *if $L$ is a hexagonal lattice then $\alpha$ is a $6^{th}$ root of unity, and*

(3) *otherwise $\alpha = \pm 1$.*

We can now classify all complex tori. As we have already remarked, every torus $X$ is isomorphic to a torus $X_\tau$, with lattice generated by $1$ and $\tau$, for some $\mathrm{Im}\,\tau > 0$. Now suppose $X_\tau \cong X_{\tau'}$. Then we have $a + b\tau = \tau'$ and $c + d\tau = 1$ for some integers $a, b, c, d$. Thus $\tau' = (a + b\tau)/(c + d\tau)$. Observe that, conversely, $\tau'$ determines $\tau$ from such a relation, and therefore the matrix

$$\begin{pmatrix} a & b \\ c & d \end{pmatrix}$$

lies in the group $SL(2,\mathbb{Z})$. We thus obtain the following proposition.

PROPOSITION 3.3.24. *Two complex tori $X_\tau$ and $X_{\tau'}$ are isomorphic if and only if there is a matrix*

$$\begin{pmatrix} a & b \\ c & d \end{pmatrix} \in SL(2,\mathbb{Z})$$

*such that $\tau' = (a + b\tau)/(c + d\tau)$.*

REMARK. It is a fact that $\{\tau \in \mathbb{C} \,;\, \operatorname{Im} \tau > 0\}/SL(2,\mathbb{Z})$, with its quotient complex structure, is a Riemann surface isomorphic to the unit disk. ◇

## 3.4. An example: Hyperelliptic surfaces

**3.4.1. Gluing surfaces.** Let $X$ and $Y$ be two Riemann surfaces, $U \subset X$ and $V \subset Y$ two open subsets, and $\varphi : U \to V$ a holomorphic isomorphism. We define an equivalence relation on the disjoint union $X \coprod Y$ of $X$ and $Y$ by identifying $u \in U$ with $\varphi(u) \in V$. The resulting space is denoted $X \coprod Y / \varphi$.

PROPOSITION 3.4.1. *There is a unique complex structure on $X \coprod Y / \varphi$ such that the natural inclusions $X \hookrightarrow X \coprod Y / \varphi$ and $Y \hookrightarrow X \coprod Y / \varphi$ are holomorphic. Moreover, if $X \coprod Y / \varphi$ is a Hausdorff space, then it is a Riemann surface.*

**Proof.** Simply take the union of the atlases for $X$ and $Y$. These atlases are compatible on $U$ and $V$ because $\varphi$ is an isomorphism. □

EXAMPLE 3.4.2. The space $X \coprod Y / \varphi$ need not be Hausdorff. Indeed, the map obtained by gluing together $\mathbb{C}$ and $\mathbb{C}$ along $\mathbb{C} - \{0\}$ and $\mathbb{C} - \{0\}$ via the map $\varphi(z) = z$ is not Hausdorff, as the origins of the two copies of $\mathbb{C}$ cannot be separated by open sets in the quotient.

On the other hand, if we take $\varphi(z) = 1/z$ then $\mathbb{C} \coprod \mathbb{C}/\varphi \cong \mathbb{P}_1$. ◇

REMARK. While we only defined complex structures for Hausdorff spaces, the definition makes sense more generally. In any case, we will only use them in situations where the underlying topological space is Hausdorff. ◇

**3.4.2. Hyperelliptic surfaces.** Let $h$ be a polynomial in one complex variable, of degree $2g + 1 + \varepsilon$, where $\varepsilon = 1$ or $0$, and define the polynomial $\hbar(x) = x^{2g+2} h(1/x)$. If $g = 0$, we assume $\varepsilon = 1$, so that $\deg h \geq 2$. Suppose all the roots of $h$ are simple. Then the roots of $\hbar$ are also simple. We define smooth affine plane curves

$$X_h := \{(x,y) \,;\, y^2 = h(x)\} \quad \text{and} \quad X_\hbar := \{(x,y) \,;\, y^2 = \hbar(x)\}$$

and also consider the open subsets

$$U_h = \{(x,y) \in X_h \,;\, x \neq 0\} \quad \text{and} \quad U_\hbar = \{(x,y) \in X_\hbar \,;\, x \neq 0\}.$$

Observe that $X_h$ and $X_\hbar$ are smooth plane curves, since all the roots of $h$ and $\hbar$ are simple. We define the isomorphisms $\varphi : U_h \to U_\hbar$ and $\varphi : U_\hbar \to U_h$ by

$$\varphi(x,y) = \left(\frac{1}{x}, \frac{y}{x^{g+1}}\right).$$

(We denote both by the same letter since they are given by the same formula. Note that $\varphi \circ \varphi = \operatorname{id}$.) Let

$$Z := X_h \coprod X_\hbar / \varphi.$$

## 3.4. An example: Hyperelliptic surfaces

LEMMA 3.4.3. *The set $Z$ is a compact Riemann surface of genus $g$, and the meromorphic function*
$$X_h \ni (x, y) \to x$$
*extends to a holomorphic map $\pi : Z \to \mathbb{P}_1$ whose degree is 2. Every root of $h$ is a branch point of $\pi$, and there is at most one other branch point, namely $\infty$, which is a branch point if and only if $\varepsilon = 0$.*

**Proof.** To see compactness, observe that $Z$ is the union of (the image in $Z$ of) the compact sets
$$\{(x, y) \in X_h \,;\, |x| \leq 1\} \quad \text{and} \quad \{(z, w) \in X_\hbar \,;\, |z| \leq 1\}.$$

We verify that $Z$ is Hausdorff as follows. Any two points of $X_h$ that lie in $U_h$ can be separated by open sets in $Z$, since they can be separated by open sets in $U_h$. The image in $Z$ of a point $p$ in $\{x = 0\} \subset X_h$ can be separated from any other point $q$ of $Z$: if $q \in X_h$ there is again nothing to prove. On the other hand, suppose $q \in X_\hbar$. Take a neighborhood of $p$ in $X_h$ so small that its image under $\varphi$ has very large first coordinates. Then the image of this neighborhood will miss $q$. Similar arguments applied to $X_\hbar$ show that $Z$ is Hausdorff, and thus $Z$ is a compact Riemann surface.

Observe that the complement in $Z$ of $X_h$ is a finite set. (In fact, it has either one or two points.) We claim that $\pi_h : (x, y) \mapsto x$ extends to a meromorphic function on $Z$. Indeed, on $X_\hbar$ we can define the meromorphic function $\pi_\hbar : (x, y) \mapsto 1/x$, and this map agrees with the former on $\varphi(U_h)$:
$$\pi_\hbar(\varphi(x, y)) = 1/(1/x) = x = \pi_h(x, y).$$
Alternatively, we can look at the associated map $\pi : Z \to \mathbb{P}_1$. In fact, let
$$\pi(x, y) := \begin{cases} [1, x], & (x, y) \in X_h \\ [1/x, 1], & (x, y) \in X_\hbar \end{cases}.$$
Clearly this map is consistent with the topological quotient and corresponds to the meromorphic function $(x, y) \mapsto x$ on $X_h$.

Finally we turn to the genus. First, it is clear that $\pi$ has degree 2, since at any non-root $x$ of $h$, the preimage of any point $x$ is the pair of points $(x, \pm\sqrt{h(x)})$. Moreover, ramification points occur either at the roots of $h$ or, if $\varepsilon = 0$, at $\infty$ (which is a root of $\hbar$ if and only if the degree of $h$ is odd), so there are precisely $g + 1$ ramification points. The multiplicity of the ramification points is precisely 2. Thus the ramification number is $R(f) = 2g + 2$, and by the Riemann-Hurwitz Formula we have
$$2g(Z) - 2 = 2(2g(\mathbb{P}_1) - 2) + 2g + 2 = 2g - 2,$$
and thus $g(Z) = g$. □

DEFINITION 3.4.4. A Riemann surface $X$ admitting a double cover $\pi : X \to \mathbb{P}_1$ is called *hyperelliptic*. The map $\pi$ is called the *hyperelliptic projection*. In such a surface, the involution $\sigma$ permuting the fibers of $\pi$ (and so fixing the ramification points) is called the *hyperelliptic involution*. ◇

REMARK. In fact every hyperelliptic surface is of the form $Z$ for some polynomial $h$ with distinct roots. ◇

**3.4.3. Meromorphic functions on hyperelliptic surfaces.** We now describe all of the meromorphic functions on a hyperelliptic Riemann surface $(Z, \sigma, \pi)$. The essential point is the following lemma.

LEMMA 3.4.5. *Let $f : Z \to \mathbb{P}_1$ be a $\sigma$-invariant meromorphic function: $\sigma^* f = f$. Then there is a meromorphic function $g : \mathbb{P}_1 \to \mathbb{P}_1$ such that $f = g \circ \pi$.*

**Proof.** Simply define $g(p) = f(q)$ where $q \in \pi^{-1}(p)$. It is clear that $g$ is well-defined. We now show that $g$ is a holomorphic map. Since $Z$ has charts in either $X_h$ or $X_{\bar{h}}$ and there is a symmetry between these two charts, we will work only near the points of $X_h$.

The map $\pi$ is given on $X_h$ by $\pi(x, y) = x$, so we are trying to show that $f(x, y) = g(x)$ is holomorphic in $x$. (We know only that it is well-defined at this point.) Because of the structure of singularities of holomorphic functions, it suffices to work at the points of $X_h$ where the derivative of $h$ does not vanish.

First, observe that $Z$ is cut out locally near the point $p \in X_h$ at which $h'(p) \neq 0$ by a holomorphic function of the form $(x - p)e^{\psi(x)} + c = y^2$. In particular, $x$ is a holomorphic function of $y$ near $y = 0$. But then $g(x(y)) = f(x(y), y) = f(x(-y), -y) = g(x(-y))$. Thus the function $F = g \circ x$ is holomorphic in a neighborhood of $y = 0$ and satisfies $F(-y) = F(y)$. It follows (cf. Exercise 3.11) that $F(y) = h(y^2)$ for some holomorphic function $h$, and therefore

$$g(x) = h((x - p)e^{\psi(x)})$$

is holomorphic in $x$. The proof is finished. □

We can immediately understand the restriction to $X_h$ of meromorphic functions that are $\sigma$-anti-symmetric. Indeed, if $f : Z \to \mathbb{P}_1$ is a meromorphic function with $\sigma^* f = -f$, the function defined on $X_h$ by

$$(x, y) \mapsto \frac{f(x, y)}{y}$$

is $\sigma$-symmetric and, being a product of meromorphic functions, meromorphic. It follows that on $X_h$ every $\sigma$-anti-symmetric meromorphic function may be written as a product $f(x, y) = yg(x)$, where $g$ is a meromorphic function of $x$.

## 3.5. Harmonic and subharmonic functions

It is now easy to describe all meromorphic functions on a hyperelliptic surface. Indeed, let $f : Z \to \mathbb{P}_1$ be any meromorphic function. Then the functions
$$f^+ := \frac{1}{2}(f + \sigma^* f) \quad \text{and} \quad f^- := \frac{1}{2}(f - \sigma^* f)$$
are also meromorphic, and $f = f^+ + f^-$. Moreover, $f^+$ is symmetric and $f^-$ is anti-symmetric. We thus have the following proposition.

PROPOSITION 3.4.6. *Let $X \subset Z$ be the affine part of a hyperelliptic surface, i.e., the surface cut out by the equation $y^2 = h(x)$ where $h$ is a polynomial with only simple roots. Then every $f \in \mathcal{M}(Z)$ may be written, on $X$, as*
$$f(x, y) = g_1(x) + g_2(x) y$$
*for some rational functions $g_1$ and $g_2$.*

REMARK. The function $y$ extends to a meromorphic function on $Z$ (cf. Exercise 3.12). ◇

### 3.5. Harmonic and subharmonic functions

We introduce harmonic and subharmonic functions on Riemann surfaces. In this section we will focus on local aspects of these functions, but later on we will look at more global information.

#### 3.5.1. Definition of harmonic functions.
Suppose given a smooth function $f$ and a chart $\varphi : V \to U$ with domain $V \subset \mathbb{C}$. Let $z = x + \sqrt{-1} y$ be the coordinate on $V$. We can calculate
$$\Delta(f \circ \varphi) := \left( \frac{\partial^2}{\partial x^2} + \frac{\partial^2}{\partial y^2} \right) f \circ \varphi = 4 \frac{\partial^2 (f \circ \varphi)}{\partial z \partial \bar{z}}.$$
Observe that if $f$ and $g$ are complex-valued functions, then
$$\begin{aligned}
\Delta(f \circ g) &= 4 \frac{\partial}{\partial z} \left( \frac{\partial f(g)}{\partial \zeta} \frac{\partial g}{\partial \bar{z}} + \frac{\partial f(g)}{\partial \bar{\zeta}} \overline{\left( \frac{\partial g}{\partial z} \right)} \right) \\
&= 4 \left( \frac{\partial^2 f(g)}{\partial \zeta^2} \frac{\partial g}{\partial z} \frac{\partial g}{\partial \bar{z}} + \frac{\partial^2 f(g)}{\partial \bar{\zeta}^2} \overline{\left( \frac{\partial g}{\partial z} \frac{\partial g}{\partial \bar{z}} \right)} \right) + \left| \frac{\partial g}{\partial z} \right|^2 (\Delta f)(g).
\end{aligned}$$
In particular, if $g$ is holomorphic, we have
$$(3.2) \qquad \Delta(f \circ g) = \left| \frac{\partial g}{\partial z} \right|^2 (\Delta f)(g).$$
Thus we see that the vanishing of the "local function" $\Delta f$ is a property that is invariant under holomorphic changes of coordinates.

REMARK. In fact, the transformation rule for the local functions $\Delta(f \circ \varphi)$ shows that $\Delta f$ is not a function at all, but rather the coefficient of a differential 2-form, and even a $(1, 1)$-form—a concept we will discuss in Chapter 5. ◇

The invariance of the kernel of the operator $\Delta$ leads to the following definition.

DEFINITION 3.5.1. Let $X$ be a Riemann surface. We say that a function $f$ defined in a neighborhood of $p$ is said to be harmonic at $p$ if there is a complex chart $\varphi : U \to V$ such that $p \in U$ and
$$\Delta(f \circ \varphi^{-1}) \equiv 0 \quad \text{on } V.$$
If $W \subset X$, we say that $f$ is harmonic on $W$ if it is harmonic at each point of $W$. $\diamond$

**3.5.2. Harnack's Principle.** We conclude with a discussion of Harnack's Principle. First, we note that since harmonic functions are by definition locally harmonic, the following analogue of Corollary 1.7.6 holds, and with the same proof.

PROPOSITION 3.5.2. *If a sequence of harmonic functions $u_j$ on a Riemann surface $X$ converges locally uniformly to a function $u$, then $u$ is harmonic on $X$.*

We turn now to Harnack's Inequality. For its proof, it is convenient to note that, by replacing $z$ with $z/r$, we can translate the Poisson Formula to the disk of radius $r$:
$$P_r f(z) := \frac{1}{2\pi} \int_{|\zeta|=r} \frac{|\zeta|^2 - |z|^2}{|\zeta - z|^2} f(\zeta) \frac{d\zeta}{\sqrt{-1}\zeta}.$$
We can also translate the origin to another point $a$ by replacing $z$ and $\zeta$ with $z - a$ and $\zeta - a$, respectively.

Next, one has the following trivial estimate of the Poisson kernel:
$$|z| < r \quad \Rightarrow \quad \frac{r - |z|}{r + |z|} \leq \frac{r^2 - |z|^2}{|re^{\sqrt{-1}\theta} - z|^2} \leq \frac{r + |z|}{r - |z|}.$$
We can then easily prove the following result.

LEMMA 3.5.3 (Harnack's Inequality). *Let $X$ be a Riemann surface, let $D \subset X$ be an open subset, and let $K \subset\subset D$. Then there exists a constant $c = c(K, D)$ such that for any positive harmonic function $u$ on $D$ and all $z_1, z_2 \in K$,*
$$\frac{1}{c} \leq \frac{u(z_1)}{u(z_2)} \leq c.$$

**Proof.** Fix a small number $r$ with the property that for any $z \in K$, $\{\zeta \in X \,;\, |\zeta| < r\} \subset D$. (We fix here finitely many coordinate charts covering $K$.) In the disk of radius $r$ centered at $z_1$, for any $z_2$ with $|z_2 - z_1| \leq r/2$ and any positive harmonic function $u$, we have
$$\begin{aligned}
\tfrac{1}{3} u(z_1) &\leq \frac{r - |z_2 - z_1|}{r + |z_2 - z_1|} u(z_1) \\
&\leq u(z_2) = \frac{1}{2\pi} \int_{|\zeta - z_1| = r} \frac{|\zeta - z_1|^2 - |z_2 - z_1|^2}{|\zeta - z_2|^2} u(\zeta) \frac{d\zeta}{\sqrt{-1}\zeta} \\
&\leq \frac{r + |z_2 - z_1|}{r - |z_2 - z_1|} u(z_1) \leq 3 u(z_1).
\end{aligned}$$

By compactness of $K$, there is a positive integer $N$ such that for any two points $z_1, z_2 \in K$ there are $n \leq N$ distinct points $x_1 = z_1, x_2, ..., x_{n-1}, x_n = z_2$ such that $|x_i - x_{i+1}| \leq r/2$. Take $c = 3N$. The proof is complete. $\square$

As a corollary of Harnack's Inequality, we have the following result.

THEOREM 3.5.4 (Harnack Principle). *Let $u_1 \leq u_2 \leq ...$ be an increasing sequence of harmonic functions on an open subset $D$ of a compact Riemann surface $X$. Then $u_j \to u$ uniformly on compact sets, where $u$ is either harmonic on $X$ or else $\equiv +\infty$ on $X$.*

**Proof.** By considering $v_j := u_j - u_1$, we may assume that $u_j \geq 0$ for all $j$. Let $u := \lim u_j$. Since the sequence $u_j$ is increasing, $u$ is well-defined pointwise, provided we allow the value $+\infty$. Moreover, by Harnack's Inequality, the convergence is locally uniform.

Consider the decomposition
$$X := \{x \; ; \; u(x) = +\infty\} \cup \{x \; ; \; u(x) < +\infty\}.$$

By Harnack's Inequality, both sets are open, and thus both are closed. Since $X$ is connected, one of these two sets is empty. The result now follows from Proposition 3.5.2. $\square$

### 3.5.3. Subharmonic functions.

DEFINITION 3.5.5. *We say that an $L^1$-loc function $u$ is subharmonic if $\Delta u$, computed in the sense of distributions, defines a positive measure on the coordinate chart in question.* $\diamond$

The formula (3.2) shows that the definition of subharmonic makes sense. Moreover, by the local nature of the definition of subharmonic, the results about subharmonic functions proved in Chapter 1 carry over to their obvious global analogs.

## 3.6. Exercises

3.1. Consider the metric $g = (1 - |z|^2)^{-2}|dz|^2$ on the unit disk $\mathbb{D}$. Show that if $F : \mathbb{D} \to \mathbb{D}$ is holomorphic and an isometry of $g$ at one point $z_o \in \mathbb{D}$, i.e., $(1 - |F(z_o)|^2)^{-2}|F'(z_o)|^2 = (1 - |z_o|^2)^{-2}$, then $F$ is invertible and is furthermore an isometry of $g$ at all points of $\mathbb{D}$.

3.2. Find all subharmonic functions $u$ on $\mathbb{C}$ such that $u(z) = u(z+1) = u(z + \sqrt{-1})$ for all $z \in \mathbb{C}$.

3.3. Prove Theorem 3.2.5.

3.4. Write the Weierstrass $\wp$-function as a quotient of theta functions.

3.5. Show that the map
$$[z,w] \mapsto \left( \frac{2\text{Re}\,(z\bar{w})}{|z|^2 + |w|^2}, \frac{2\text{Im}\,(z\bar{w})}{|z|^2 + |w|^2}, \frac{|z|^2 - |w|^2}{|z|^2 + |w|^2} \right)$$
is an isomorphism from $\mathbb{P}_1$ to $S^2$ with the atlas $\{\varphi_N, \varphi_S\}$, defined in Chapter 1.

3.6. Show that $\mathbb{P}_1 = (\mathbb{C}^2 - \{0\})/\sim$, where $z \sim w$ if $z$ and $w$ lie on the same line through $0$ in $\mathbb{C}^2$, is isomorphic to the extended complex plane $\mathbb{C} \cup \{\infty\}$.

3.7. Prove part (2) of Lemma 3.3.12.

3.8. In the notation of Paragraph 3.4.2, show that the involution $\sigma$ for the hyperelliptic surface $Z$ is given on $X_h$ or $X_{\bar{h}}$ by $\sigma(x, y) = (x, -y)$.

3.9. Show that the closure $\overline{X_h} \subset \mathbb{P}_2$ of the plane curve $y^2 = h(x)$ in the projective plane is singular for any polynomial $h$.

3.10. Show that a hyperelliptic projection is necessarily ramified.

3.11. Show that if $F$ is a holomorphic function on the unit disk satisfying $F(-z) = F(z)$ then there is a holomorphic function $h$ such that $F(z) = h(z^2)$.

3.12. Again in the notation of Paragraph 3.4.2, show that the function $y$, defined on the subset $X_h \subset \mathbb{C}^2$ of $Z$ by restricting the coordinate $y$ to $X_h$, extends to a well-defined meromorphic function on $Z$.

3.13. Show there are no non-constant subharmonic functions on a compact Riemann surface.

# Chapter 4

# Complex Line Bundles

## 4.1. Complex line bundles

### 4.1.1. Basic definitions.

DEFINITION 4.1.1. Let $M$ be a manifold, assumed connected for simplicity. A complex line bundle on $M$ is a manifold $L$ together with a map $\pi : L \to M$ having the following properties.

(1) (Local triviality) For every $p \in M$ there is a neighborhood $U$ of $p$ and a map $f_U : \pi^{-1}(U) \to \mathbb{C}$ such that the map
$$F_U : \pi^{-1}(U) \ni v \mapsto (\pi(v), f_U(v)) \in U \times \mathbb{C}$$
is a diffeomorphism.

(2) (Global linear structure) For each pair of such neighborhoods $U_\alpha$ and $U_\beta$ there is a map
$$g_{\alpha\beta} : U_\alpha \cap U_\beta \to \mathbb{C}^*$$
such that $F_{U_\alpha} \circ F_{U_\beta}^{-1}(x, \lambda) = (x, g_{\alpha\beta}(x)\lambda)$. ◇

DEFINITION 4.1.2. (1) The map $F_U$ is also called a local trivialization of the line bundle.

(2) The maps $g_{\alpha\beta}$ are called transition functions.

(3) The sets $L_x := \pi^{-1}(x)$, $x \in M$, are called the fibers of the line bundle. ◇

EXAMPLE 4.1.3. An important example of a line bundle is the trivial bundle
$$\pi : M \times \mathbb{C} \to M; \quad \pi(x, v) = x.$$
By definition, every line bundle is locally the same as the trivial bundle. ◇

REMARK. We point out that

(4.1)   $g_{\alpha\alpha} = 1$ on $U_\alpha$   and   $g_{\alpha\beta} \cdot g_{\beta\gamma} \cdot g_{\gamma\alpha} = 1$ on $U_\alpha \cap U_\beta \cap U_\gamma$.

An open cover $\{U_\alpha\}$ together with functions $\{g_{\alpha\beta}\}$ satisfying (4.1) is called a 1-*cocycle*. ◇

DEFINITION 4.1.4. A section of a complex line bundle $\pi : L \to M$ is a map $s : M \to L$ such that
$$\pi \circ s = \mathrm{id}_M.$$
The set of all sections of $L$ is denoted $\Gamma(M, L)$. ◇

Observe that a section is automatically injective and therefore provides a copy of $M$ inside $L$. Conversely, given an injective map $M \hookrightarrow L$ such that $M \cap L_x$ is a single point for each $x \in M$, we have a section.

EXAMPLE 4.1.5. Consider the trivial line bundle $M \times \mathbb{C} \to M$. If $f : M \to \mathbb{C}$ is a function then the graph $\{(x, f(x)) \,;\, x \in M\}$ of $f$ defines a section. ◇

REMARK. Example 4.1.5 provides a lot of intuition for the notion of section. We shall see that, to some extent, sections are introduced because the collection of functions on a given manifold is deficient. ◇

Every line bundle has at least one section.

DEFINITION 4.1.6. Let $\pi : L \to M$ be a complex line bundle.

(1) If $x \in \pi(U)$ then the section $o_L$, assigning to $x$ the point $o_L(x) := F_U^{-1}(x, 0)$, is called the *zero section* of $L$. The point $o_L(x)$ is called *the origin in the fiber $L_x$*.

(2) There is an action of $\mathbb{C}^*$ defined on $L$ as follows. If $\pi(v) \in U$, then
$$\lambda \cdot v := F_U^{-1}(\pi(v), \lambda f_U(v)).$$

We leave it as an exercise to the reader to show that the $\mathbb{C}^*$-action and the zero section are well-defined. ◇

DEFINITION 4.1.7. Let $L$ and $L'$ be two complex line bundles over a manifold $M$.

(1) A morphism of line bundles, or line bundle map, from $L$ to $L'$ is a map $F : L \to L'$ such that
   (a) the diagram
$$\begin{array}{ccc} L & \xrightarrow{F} & L' \\ \downarrow & & \downarrow \\ M & \xrightarrow{\mathrm{id}} & M \end{array}$$
   commutes, and
   (b) $F_x := F|_{L_x} : L_x \to L'_x$ is a complex linear map of complex vector spaces.

## 4.1. Complex line bundles

(2) We say that $L$ and $L'$ are isomorphic if there are line bundle morphisms $F : L \to L'$ and $G : L' \to L$ such that $F \circ G = \mathrm{id}_{L'}$ and $G \circ F = \mathrm{id}_L$. ◇

**4.1.2. Description by transition functions.** A line bundle is completely determined by the following data: a set of neighborhoods $U_\alpha$, $\alpha \in A$, that cover $M$, and a set of transition functions $g_{\alpha\beta} : U_\alpha \cap U_\beta \to \mathbb{C}^*$ satisfying the cocycle conditions (4.1). Indeed, given a manifold $M$ and such data $g_{\alpha\beta}$, we define

$$L(\{g_{\alpha\beta}\}) := \left( \coprod_\alpha U_\alpha \times \mathbb{C} \right) / \sim$$

where as before $\coprod$ means disjoint union, and

$$(x, v) \sim (x, g_{\alpha\beta}(x)v) \quad \text{whenever } x \in U_\alpha \cap U_\beta.$$

The projection map $\pi(\{g_{\alpha\beta}\})$ is then defined by $\pi(\{g_{\alpha\beta}\})[x, v] = x$.

REMARK. The reader is invited to show that the relation $\sim$ defining the line bundle $L(\{g_{\alpha\beta}\})$ is an equivalence relation and that if $\pi : L \to M$ is a line bundle and $\{g_{\alpha\beta}\}$ is a complete set of transition functions, then $\pi(\{g_{\alpha\beta}\}) : L(\{g_{\alpha\beta}\}) \to M$ and $\pi : L \to M$ are isomorphic. ◇

**4.1.3. Description by local sections.** Let $L \to M$ be a complex line bundle and let $U \subset M$ be an open set. The set $\pi^{-1}(U) \to U$ is itself a line bundle, called the restriction of $L$ to $U$ and denoted $L|_U$. We abuse notation and denote the space of sections of this line bundle $\Gamma(U, L)$.

DEFINITION 4.1.8. *A frame for $L$ over $U$ is a section $\xi \in \Gamma(U, L)$ of $L$ over $U$ having no zeros.* ◇

If $L$ has a frame over $U$ then the map

$$v \mapsto (\pi(v), v/\xi(\pi(v)))$$

gives an isomorphism $L|_U \to U \times \mathbb{C}$. (Note that since $\xi(\pi(v)) \neq 0$, the vector $v$ is a multiple of $\xi(\pi(v))$, and we denote the factor $v/\xi(\pi(v))$.) It follows that $L|_U$ is trivial. Conversely, suppose $U$ is an open set in $M$ such that $L|_U$ is trivial. Let $\xi : U \to L$ be the section associated to the section $x \to (x, 1)$ of $U \times \mathbb{C}$ via the map $F_U$. Then $\xi$ is a frame for $L$ over $U$. The discussion can be summarized as follows.

PROPOSITION 4.1.9. *A line bundle is trivial if and only if it has a nowhere vanishing section.*

The following proposition is obvious.

PROPOSITION 4.1.10. *Let $\xi_U$ be a frame for $L$ over $U$. For each $s \in \Gamma(U, L)$ there exists a function $s_U : U \to \mathbb{C}$ such that*

$$s = s_U \xi_U.$$

Let $U_\alpha$ and $U_\beta$ be two open sets in $M$ such that $L|_{U_\alpha}$ and $L|_{U_\beta}$ are both trivial. Let $\xi_\alpha$ and $\xi_\beta$ be frames for $L|_{U_\alpha}$ and $L|_{U_\beta}$, respectively. Then these frames are related by

$$\xi_\alpha = g_{\beta\alpha}\xi_\beta$$

for some function $g_{\alpha\beta} : U_\alpha \cap U_\beta \to \mathbb{C}^*$. It is easy to see that $g_{\alpha\beta}$ satisfy the cocycle condition (4.1). If $x \in U_\alpha \cap U_\beta$, then for each $v \in L_x$ one has

$$v = v_\alpha \xi_\alpha = v_\beta \xi_\beta.$$

It follows that

$$g_{\alpha\beta}(x) v_\beta = v_\alpha.$$

From Proposition 4.1.10 and the definition of the transition functions associated to a frame it follows that a section is determined locally by a function, and the fact that the section is globally defined means that these local functions satisfy linear relations with a non-linear parameter. We can recap this discussion in one convenient proposition.

PROPOSITION 4.1.11. *Let $L \to M$ be a complex line bundle and $U_\alpha$, $\alpha \in A$, a collection of locally trivializing charts with frames $\xi_\alpha$ and transition functions $g_{\alpha\beta}$. Suppose there is a family of functions $\{s_\alpha\}_{\alpha \in A}$ such that $g_{\alpha\beta} s_\beta = s_\alpha$ for all $\alpha, \beta \in A$. Then there is a section $s \in \Gamma(M, L)$ such that*

$$s|_{U_\alpha} = s_\alpha \xi_\alpha.$$

**Proof.** Clearly $s_\alpha \xi_\alpha = s_\beta \xi_\beta$ on $U_\alpha \cap U_\beta$, so we set $s := s_\alpha \xi_\alpha$ on $U_\alpha$. □

We have introduced three realizations of the notion of line bundle:

- global realization,
- realization by transition functions, and
- realization by the spaces of local sections $\Gamma(U, L)$, $U \subset M$ open.

It is often important to know all three realizations intimately.

**4.1.4. Remark: Vector bundles.** In this book we will not use vector bundles in a serious way. However, in certain definitions it is natural to have available the notion of vector bundle—for example, when defining differential forms. (In the next chapter, the reader might notice that our expedient, *ad-hoc* approach to complex differential forms feels, at times, somewhat unnatural.)

The definition of complex vector bundle of rank $r$ mimics that of complex line bundle. The fibers are vector spaces and the transition functions take values in $GL(r, \mathbb{C})$. One important difference between vector bundles and line bundles comes from the non-commutativity of the group $GL(r, \mathbb{C})$ for $r \geq 2$. This difference makes the study of certain aspects of vector bundles rather more involved.

## 4.2. Holomorphic line bundles

Some of the results in this book, such as Hörmander's Theorem in Chapter 11, have analogs for sections of holomorphic vector bundles, and these analogs are no harder to prove. By contrast, classification of vector bundles analogous to the Abel-Jacobi Theory presented in Chapter 14 is significantly more complicated.

### 4.2. Holomorphic line bundles

#### 4.2.1. Definition of holomorphic line bundle.

DEFINITION 4.2.1 (Holomorphic line bundle). Let $L$ and $X$ be complex manifolds. A complex line bundle $\pi : L \to X$ is said to be a holomorphic line bundle if $\pi$ is holomorphic and one can choose the functions $f_{U_i} : \pi^{-1}(U_i) \to \mathbb{C}$ holomorphic for some atlas $\{U_i\}$. ◇

#### 4.2.2. Picard group.
If $L \to M$ and $L' \to M$ are complex line bundles, then we can define new complex line bundles $L^* \to M$ and $L \otimes L' \to M$ as follows. If $g^L_{\alpha\beta}$ and $g^{L'}_{\alpha\beta}$ are the transition functions for $L$ and $L'$, then we take transition functions

$$(g^{L^*})_{\alpha\beta} := g^L_{\beta\alpha} = \frac{1}{g^L_{\alpha\beta}} \quad \text{for} \quad L^*$$

and

$$g^{L \otimes L'}_{\alpha\beta} := g^L_{\alpha\beta} g^{L'}_{\alpha\beta} \quad \text{for} \quad L \otimes L'.$$

We claim that $L \otimes L^*$ is always the trivial bundle. The latter can be seen in several ways. For example, the transition functions are the identity. Or one could observe that $x \mapsto \mathrm{id}_{L_x}$ is a nowhere-zero section of $\mathrm{Hom}(L, L) \cong L \otimes L^*$.

It follows that the set of complex line bundles on a manifold is an Abelian group, with $\otimes$ denoting the group addition and $L \mapsto L^*$ the group negation. Clearly the holomorphic line bundles form a subgroup.

DEFINITION 4.2.2. The group of holomorphic line bundles on a complex manifold $X$ is called the Picard group $\mathrm{Pic}(X)$ of $X$. ◇

#### 4.2.3. Holomorphic and meromorphic sections.

DEFINITION 4.2.3. A section of a holomorphic line bundle $L \to X$ is called holomorphic if it is holomorphic as a map $X \to L$. The set of holomorphic sections of $L$ is denoted $\Gamma_{\mathcal{O}}(X, L)$. ◇

DEFINITION 4.2.4. A meromorphic section $s$ of a holomorphic line bundle $L \to X$ is a locally finite set $P := \{x_1, x_2, ...\} \subset X$ and a holomorphic section $s : X - P \to L$ such that if $E$ is a frame in a neighborhood of some $x_i \in P$ and $s = fE$, then $f$ does not have an essential singularity at $x_i$. The set of meromorphic sections of $L$ is denoted $\Gamma_{\mathcal{M}}(X, L)$. ◇

The notion of order of a meromorphic function can be extended to meromorphic sections as follows.

DEFINITION 4.2.5. Let $s \in \Gamma_{\mathcal{M}}(X,L)$, $x \in X$, let $U$ be a neighborhood of $X$, and let $\xi \in \Gamma_{\mathcal{O}}(U,L)$ such that $\xi(x) \neq 0$. Then
$$\mathrm{Ord}_x(s) := \mathrm{Ord}_x\left(\frac{s|_U}{\xi}\right).$$
(We leave it to the reader to check that $\mathrm{Ord}_x(s)$ does not depend on $U$ or $\xi$.) ◇

The following proposition is trivial.

PROPOSITION 4.2.6. *Let $s \in \Gamma_{\mathcal{M}}(X,L) - \{0\}$. Then for any $\tilde{s} \in \Gamma_{\mathcal{M}}(X,L)$ there exists a unique function $f \in \mathcal{M}(X)$ such that $\tilde{s} = f \cdot s$.*

It follows that the space of global holomorphic sections can be identified with a subspace of the space of all meromorphic functions. This subspace is defined as follows. If there is no non-identically zero holomorphic section, there is nothing to do. On the other hand, if $s$ is a global holomorphic section, then for any meromorphic function $f$, the section $fs$ is holomorphic if and only if for all $x \in X$,
$$\mathrm{Ord}_x(f) \geq -\mathrm{Ord}_x(s).$$

REMARK. On a compact complex manifold the ratio of two global holomorphic sections of a holomorphic line bundle is either constant or else has non-trivial poles. Indeed, if there are no poles then the ratio is a holomorphic function on a compact Riemann surface. ◇

## 4.3. Two canonically defined holomorphic line bundles

### 4.3.1. The canonical bundle of a Riemann surface.
Let $X$ be a Riemann surface. There are several ways to define the canonical bundle of $X$. We shall begin with a function-theoretic definition of the canonical bundle and then produce local sections and transition functions.

DEFINITION 4.3.1 (Germs of holomorphic functions). If $f_i : U_i \to \mathbb{C}$, $i = 1, 2$, are holomorphic functions on domains $U_i$ containing $x$, we say that $(f_1, U_1)$ and $(f_2, U_2)$ define the same germ if there is an open subset $U \subset U_1 \cap U_2$ such that $f_1|_U = f_2|_U$. The equivalence class of all $(f, U)$ that define the same germ is denoted $\underline{f}_x$ and is called a germ of a holomorphic function. The set of germs of holomorphic functions is denoted $\mathcal{O}_{X,x}$. ◇

The set of germs of holomorphic functions has a vector space structure: if $(f, U)$ and $(g, V)$ are two representatives of germs at $x$, then for any $W \subset U \cap V$, $(f+g, W)$ is an element of $\mathcal{O}_{X,x}$, and thus we define
$$\underline{f}_x + \underline{g}_x := \underline{f+g}_x.$$
Scalar multiplication is even easier, since the domain of the constant function can be taken to be all of $X$. We define
$$c \cdot \underline{f}_x := \underline{c \cdot f}_x.$$

## 4.3. Two canonically defined holomorphic line bundles

REMARK. More general than scalar multiplication, one can define products of germs in the obvious way: replace $+$ by $\cdot$ in the definition of addition of germs. In this way $\mathcal{O}_{X,x}$ can be given the structure of a ring. Since this ring naturally contains $\mathbb{C}$, we can also see it as an algebra over $\mathbb{C}$.

As a complex vector space, $\mathcal{O}_{X,x}$ is infinite-dimensional. Thus it may sometimes be better to treat $\mathcal{O}_{X,x}$ as an algebra. The collection of these algebras over all $x \in X$ is best treated as a sheaf, but we have decided to avoid sheaves in this book. ⋄

Since the vector space $\mathcal{O}_{X,x}$ is too large for some purposes, we cut it down to something more manageable by defining an equivalence relation on germs.

DEFINITION 4.3.2 ((1,0)-forms). Two germs $\underline{f}_x = [(f, U)]$ and $\underline{g}_x = [(g, V)]$ are said to be co-tangent, written $\underline{f}_x \sim_1 \underline{g}_x$, if there is a coordinate chart $\varphi : W \to \mathbb{C}$ such that $x \in W \subset U \cap V$ and

$$(f \circ \varphi^{-1})'(\varphi(x)) = (g \circ \varphi^{-1})'(\varphi(x)).$$

The chain rule implies that the equivalence relation is well-defined. The set of equivalence classes of mutually tangent germs is denoted $K_{X,x}$. The elements of $K_{X,x}$ are called $(1,0)$-forms. The equivalence class of germs tangent to $\underline{f}_x$ is denoted $df(x)$. ⋄

EXAMPLE 4.3.3. If $z : U \to \mathbb{C}$ is a complex chart, then $dz(x)$ is a 1-form. ⋄

The relation $\sim_1$ is clearly linear, in the sense that if $\underline{f}_x \sim_1 \underline{\tilde{f}}_x$ and $\underline{g}_x \sim_1 \underline{\tilde{g}}_x$ then $\underline{f}_x + a\underline{g}_x \sim_1 \underline{\tilde{f}}_x + a\underline{\tilde{g}}_x$. We can therefore define the linear map $d : \mathcal{O}_{X,x} \to K_{X,x}$ by

$$a \cdot df(x) := d(af)(x) \quad \text{and} \quad df(x) + dg(x) := d(f+g)(x).$$

PROPOSITION 4.3.4. *The vector space $K_{X,x}$ is 1-dimensional.*

**Proof.** This is an easy consequence of the chain rule: in fact, if we fix a coordinate chart $z : U \to \mathbb{C}$ at $x$, then any holomorphic function $f$ whose domain lies in $U$ is of the form $\tilde{f} \circ z$ for some holomorphic function $\tilde{f}$ of one complex variable. Thus

$$df(x) = \tilde{f}'(z(x))dz(x)$$

is a multiple of the $(1,0)$-form $dz(x)$. □

Let now

$$K_X := \bigcup_{x \in X} K_{X,x}.$$

We define the map $\pi : K_X \to X$ by $\pi^{-1}(x) = K_{X,x}$. We give $K_X$ the structure of a line bundle as follows. Let $z : U \to \mathbb{C}$ be a complex chart for $X$. Since $z$ is a homeomorphism, the map

$$dz : U \ni x \mapsto dz(x) \in K_X|_U$$

is injective. This map is by definition our non-vanishing section over $U$ and gives the description of $K_X$ by local sections: the map $U \times \mathbb{C} \ni (x, c) \mapsto c \cdot dz(x)$ is a complex chart for $K_X$.

Let $(z_\alpha, U_\alpha)$ and $(z_\beta, U_\beta)$ be two charts on $X$. If we define $h_{\alpha\beta}(x) := (z_\alpha \circ z_\beta^{-1})'(z_\beta(x))$ then by the chain rule we have

$$dz_\alpha(x) = d\left((z_\alpha \circ z_\beta^{-1}) \circ z_\beta\right)(x) = h_{\alpha\beta}(x) dz_\beta(x).$$

The cocycle condition (4.1) for transition functions of a line bundle is also satisfied.

REMARK. Since the transition functions for $K_X$ are holomorphic, $K_X$ is a holomorphic line bundle. ◇

DEFINITION 4.3.5. Let $X$ be a Riemann surface. The line bundle $K_X$ with transition functions $h_{\alpha\beta}$ defined above is called the canonical bundle of $X$. ◇

REMARK. Another name for the line bundle $K_X$ is the holomorphic cotangent bundle. If this name is used, one usually denotes $K_X$ by the symbol $T_X^{1,0*}$ or $T_X^{*1,0}$. Both $K_X$ and $T_X^{1,0*}$ generalize to higher dimensions. However, while the canonical bundle generalizes to a line bundle on higher-dimensional complex manifolds, the holomorphic cotangent bundle generalizes to a vector bundle. ◇

### 4.3.2. The holomorphic tangent bundle of a Riemann surface.

Let $x$ be a point in a Riemann surface $X$. As in the case of the cotangent bundle, one could consider germs of holomorphic functions $f : D \to X$ where $D \subset \mathbb{C}$ and $x \in f(D)$. Instead, for the sake of expedience, we go directly to tangency equivalence classes.

DEFINITION 4.3.6. We write $f_1 : D_1 \to X \sim_1 f_2 : D_2 \to X$ if there exists a complex chart $z_\alpha : U_\alpha \to \mathbb{C}$ such that $x \in U_\alpha$, $f_1(t_{o,1}) = f_2(t_{o,2}) = x$ and $(z_\alpha \circ f_1)'(t_{o,1}) = (z_\alpha \circ f_2)'(t_{o,2})$. An equivalence class is denoted $[f]_x$, and the collection of equivalence classes is denoted

$$T_{X,x}^{1,0}.$$

The set $T_{X,x}^{1,0}$ is called the holomorphic tangent space, or the $(1,0)$-tangent space, to $X$ at $x$, and its elements are called $(1,0)$-vectors. ◇

EXAMPLE 4.3.7. The equivalence class $[z^{-1}]_x$ of the inverse of a complex chart is also denoted $\frac{\partial}{\partial z}\big|_x$. ◇

We give $T_{X,x}^{1,0}$ the structure of a vector space, defining scalar multiplication by

$$a \cdot [f]_x := [f_a]_x, \quad \text{where } f_a(t) := f(a(t - t_o) + t_o),$$

and addition by

$$[f]_x + [g]_x := [h]_x,$$

where, if $f(t_o) = g(s_o) = x$,

$$h(t) := z_\alpha^{-1}(z_\alpha \circ f(t + t_o) + z_\alpha \circ g(t + s_o) - z_\alpha(x)).$$

## 4.3. Two canonically defined holomorphic line bundles

Observe that $h(0) = x$ and that $h$ is well-defined on a small neighborhood of $0$.

**PROPOSITION 4.3.8.** *Let $f : X \to Y$ be a holomorphic map between two Riemann surfaces. For each $x \in U$ the map*

$$f'(x) : T^{1,0}_{X,x} \ni [h]_x \mapsto [f \circ h]_{f(x)} \in T^{1,0}_{Y,f(x)},$$

*also denoted $Df(x)$, is linear.*

We leave the proof to the exercises.

Having defined the 1-dimensional complex vector spaces $T^{1,0}_{X,x}$, we glue them together into a line bundle. Letting

$$T^{1,0}_X := \bigcup_{x \in X} T^{1,0}_{X,x}$$

and defining $\pi : T^{1,0}_X \to X$ by $\pi^{-1}(x) = T^{1,0}_{X,x}$, we give $T^{1,0}_X$ a complex structure as follows. Let $x \in X$ and let $z : U \to \mathbb{C}$ be a complex chart for $X$ containing $x$. The map

$$\frac{\partial}{\partial z} : U \to T^{1,0}_X|_U$$

defined by $\frac{\partial}{\partial z}(x) := \frac{\partial}{\partial z}\big|_x$ is a nowhere-zero section, which we take to be a frame over $U$. Equivalently, we have a complex chart $U \times \mathbb{C} \to T^{1,0}_X|_U$ given by

$$(x, v) \mapsto v \frac{\partial}{\partial z}(x).$$

To compute the transition functions for $T^{1,0}_X \to X$, take an atlas $\{(z_\alpha : U_\alpha \to \mathbb{C})\}$ on $X$. Then the functions $z_\alpha \circ z_\beta^{-1} : z_\beta(U_\alpha \cap U_\beta) \to \mathbb{C}$ have nowhere-vanishing derivative. We define

$$g_{\alpha\beta}(x) : U_\alpha \cap U_\beta \to \mathbb{C}^*$$

by $g_{\alpha\beta}(x) := (z_\alpha^{-1} \circ z_\beta)'(x)$. Now

$$\frac{\partial}{\partial z_\alpha}(x) = [z_\alpha^{-1}]_x = [(z_\alpha^{-1} \circ z_\beta) \circ z_\beta^{-1}]_x = g_{\alpha\beta}(x) \frac{\partial}{\partial z_\beta}(x),$$

where the second to last equality follows from Proposition 4.3.8

**DEFINITION 4.3.9.** The line bundle $T^{1,0}_X$ with the transition functions $g_{\alpha\beta}$ defined above is called the *holomorphic tangent bundle* of $X$. ◇

**REMARK.** Since the transition functions of $T^{1,0}_X$ are holomorphic, $T^{1,0}_X$ is a holomorphic line bundle. ◇

**4.3.3. Duality of $K_X$ and $T_X^{1,0}$.** In this short section, we make explicit the observation that

(4.2) $$K_X^* = T_X^{1,0}.$$

There are several ways to see this fact. One such way is to observe that the transition functions are reciprocals. (To see this relation between the transitions functions, one uses the chain rule.)

A second way to prove (4.2) is to note that $K_X \otimes T_X^{1,0} \to X$ has a global section with no zeros. To construct this section, we may use the local sections $dz_\alpha$ and $\frac{\partial}{\partial z_\alpha}$. A calculation using the chain rule and the bilinearity of the tensor product shows that the nowhere-zero section $s$ given by

$$s := dz_\alpha \otimes \frac{\partial}{\partial z_\alpha} \quad \text{on} \quad U_\alpha$$

is globally defined.

A third way to construct the desired isomorphism stems from the possibility of composing local holomorphic maps from $\mathbb{C}$ into $X$ with holomorphic functions on $X$, provided their domains are compatible. To be more precise, let $\omega = df(x) \in K_{X,x}$ and $v := [g]_x \in T_{X,x}^{1,0}$, where $g(t_o) = x$. Define

$$\langle \omega, v \rangle := (f \circ g)'(t_o).$$

It is easy to show that the definition is independent of the representatives of the equivalence classes and that this pairing is bilinear and non-degenerate.

The pairing defines an endomorphism $T_X^{1,0} \to K_X^*$ by

$$T_{X,x}^{1,0} \ni v \mapsto \langle \cdot, v \rangle \in K_{X,x}^*,$$

and this endomorphism is non-degenerate by the non-degeneracy of the bilinear form.

The three methods of establishing the aforementioned duality (4.2) are closely related. The relation between the first two methods is obvious. The third method of establishing (4.2) is related to the second through the observation that the fiber of the line bundle $K_X \otimes T_X^{1,0} \to X$ over $x$ can be thought of as the collection of all endomorphisms from $T_{X,x}^{1,0}$ to $K_{X,x}^*$. Thus the non-zero section $s$ gives rise to a non-degenerate endomorphism.

## 4.4. Holomorphic vector fields on a Riemann surface

### 4.4.1. Definition.

DEFINITION 4.4.1. A global holomorphic section of $T_X^{1,0}$ is called a holomorphic vector field. ◇

## 4.4. Holomorphic vector fields on a Riemann surface

EXAMPLE 4.4.2. Let $X = \mathbb{C}$ with the global coordinate function $z$. Then $\frac{\partial}{\partial z}$ is a holomorphic vector field. Since $\frac{\partial}{\partial z}$ has no zeros, any holomorphic vector field on $\mathbb{C}$ is of the form $f\frac{\partial}{\partial z}$ for some entire holomorphic function. ◇

EXAMPLE 4.4.3. Let $X = \mathbb{P}_1$ with local coordinates $z_0$ on $U_0$ and $z_1$ on $U_1$. As we saw, $z_1 = \frac{1}{z_0}$ on $U_1 \cap U_0$. Consider the holomorphic vector fields

$$\xi_0 := \frac{\partial}{\partial z_0} \quad \text{and} \quad \eta_0 = z_0 \frac{\partial}{\partial z_0}$$

on $U_0$. Then on $U_0 \cap U_1$, the chain rule tells us that

$$\xi_0 = \frac{\partial z_1}{\partial z_0}\frac{\partial}{\partial z_1} = -(z_1)^2 \frac{\partial}{\partial z_1}$$

while

$$\eta_0 = \frac{\partial z_1}{\partial z_0}\frac{1}{z_1}\frac{\partial}{\partial z_1} = -z_1 \frac{\partial}{\partial z_1}.$$

If we define

$$\xi_1 := -(z_1)^2 \frac{\partial}{\partial z_1} \quad \text{and} \quad \eta_1 = -z_1 \frac{\partial}{\partial z_1} \quad \text{on } U_1$$

then $\xi_0 = \xi_1$ and $\eta_0 = \eta_1$ on $U_0 \cap U_1$. By setting $\xi = \xi_i$ and $\eta = \eta_i$ on $U_i$, we obtain two independent holomorphic vector fields on $\mathbb{P}_1$. (Independence here means linear independence in the vector space of sections of $T^{1,0}_{\mathbb{P}_1}$.)

Later we will see that any holomorphic vector field on $\mathbb{P}_1$ is a linear combination of $\xi$ and $\eta$. ◇

EXAMPLE 4.4.4. Let $X$ be the complex torus $\mathbb{C}/L$. Then on $\mathbb{C}$ we have $[z]_x = [z+a]_x$ for any constant $a$, which means that the vector field $\frac{\partial}{\partial z}$ on $\mathbb{C}$ descends to a well-defined vector field on $X$, often still denoted $\frac{\partial}{\partial z}$. Observe that since the vector field $\frac{\partial}{\partial z}$ has no zeros, $T^{1,0}_X$ is trivial. ◇

REMARK. In fact, we will see later that complex tori are the only compact Riemann surfaces with trivial tangent bundle. On the other hand, we will also see that every line bundle on an open Riemann surface is trivial. ◇

REMARK. We will see later on that among compact Riemann surfaces, only $\mathbb{P}_1$ and tori have non-zero holomorphic vector fields. ◇

**4.4.2. Flow.** In this section we will assume the existence and uniqueness theorem for solutions of complex differential equations. The form of the theorem we will use is the following.

THEOREM 4.4.5 (**Existence, uniqueness and regularity for ODE**). *Let $U \subset \mathbb{C}$ be an open set, let $U' \subset\subset U$ be an open subset, and let $v$ be a holomorphic function* on $U$. There exists a neighborhood $W \subset \mathbb{C}$ of $0$ and a holomorphic function $\gamma : U' \times W \to U$ such that, for all $(x,t) \in U' \times W$,

$$\gamma(x,0) = x \quad \text{and} \quad \frac{\partial \gamma(x,t)}{\partial t} = v\left(\gamma(x,t)\right).$$

Moreover, if $\tilde\gamma : U' \times \tilde W$ is another such function, then $\gamma$ and $\tilde\gamma$ agree on $U' \times (W \cap \tilde W)$.

REMARK. The real version of Theorem 4.4.5, which is often stated for Lipschitz ODE, is proved in most sufficiently advanced books on ODE. A proof of Theorem 4.4.5 can be obtained either by adapting the real proof to the holomorphic category, or by applying the real theorem and a regularity theorem to then show that the solutions are holomorphic. ◇

REMARK. In the notation of Theorem 4.4.5, if $s, t \in W$ and $\gamma(x,s) \in U'$, then by uniqueness $s + t \in W$ and

(4.3) $$\gamma(\gamma(x,s), t) = \gamma(x, s+t).$$

Indeed, one only has to differentiate both sides to see that they satisfy the same ODE and initial condition.

One often writes $\gamma_x(t) = \gamma^t(x) = \gamma(x,t)$ when one wants to fix $x$ or $t$, respectively. Then the differential equation becomes

$$\gamma'_x(t) = v(\gamma_x(t)),$$

while (4.3) reads $\gamma^t \circ \gamma^s = \gamma^{s+t}$. ◇

Let $X$ be a Riemann surface and let $\xi \in H^0(X, T_X^{1,0})$ be a holomorphic vector field. If $z : U \to \mathbb{C}$ is a complex chart on $X$, we define the holomorphic function $v = v^{(z)} : z(U) \to \mathbb{C}$ by

$$\xi = v \frac{\partial}{\partial z}.$$

In view of the transformation rules for the sections $\frac{\partial}{\partial z}$, we see that the functions $v_\alpha := v^{(z_\alpha)}$ transform as

$$v_\alpha(z_\alpha(x)) = (z_\alpha \circ z_\beta^{-1})'(x) v_\beta(z_\beta(x)).$$

If $x \in U_\alpha$, we define the family of curves $\gamma_{x,\alpha} : W_\alpha \to \mathbb{C}$ with holomorphic parameter $x$ to be the family of curves satisfying the initial value problem

$$\gamma'_{x,\alpha}(t) = v_\alpha(\gamma_{x,\alpha}(t)), \quad \gamma_{x,\alpha}(0) = z_\alpha(x).$$

We calculate, using the chain rule, that

$$\frac{d}{dt}\left(z_\alpha \circ z_\beta^{-1} \circ \gamma_{x,\beta}\right) = v_\alpha(z_\alpha \circ z_\beta^{-1} \circ \gamma_{x,\beta})$$

and also observe that

$$z_\alpha \circ z_\beta^{-1} \circ \gamma_{x,\beta}(0) = z_\alpha \circ z_\beta^{-1}(z_\beta(x)) = z_\alpha(x).$$

## 4.4. Holomorphic vector fields on a Riemann surface

It follows from the Existence and Uniqueness Theorem for ODE (particularly uniqueness) that

(4.4) $$\gamma_{x,\alpha} = z_\alpha \circ z_\beta^{-1} \circ \gamma_{x,\beta}.$$

We define a subset $\mathscr{D}_\xi \subset X \times \mathbb{C}$ as follows. A pair $(x,t)$ is in $\mathscr{D}_\xi$ if there exist pairs
$$(x,0) = (x_0, t_0),\ (x_1, t_1),\ ...,\ (x_N, t_N) = (x^t, t)$$
with
$$x_j \in U_{\alpha_j} \cap U_{\alpha_{j+1}} \quad \text{and} \quad z_{\alpha_j}(x_{j+1}) = \gamma_{x_j, \alpha_j}(t_j), \qquad j = 1, 2, ..., N.$$
We define the map $F_\xi : \mathscr{D}_\xi \to X$ by
$$F_\xi(x,t) = x_N = x^t.$$
Using (4.4), one can verify that $F_\xi$ is well-defined.

DEFINITION 4.4.6. The space $\mathscr{D}_\xi$ is called the fundamental domain of $\xi$, and the map $F_\xi : \mathscr{D}_\xi \to X$ is called the flow of $\xi$. One also writes
$$F_\xi^t(x) := F_\xi(x,t).$$
The map $F_\xi^t$ is called the time-$t$ map of $\xi$. ◇

REMARK. If $(x,s) \in \mathscr{D}_\xi$ and $(F_\xi^s(x), t) \in \mathscr{D}_\xi$ then $(x, s+t) \in \mathscr{D}_\xi$. ◇

DEFINITION 4.4.7. A vector field $\xi$ is said to be complete if $\mathscr{D}_\xi = X \times \mathbb{C}$. In this case, the previous remark implies that $F_\xi : X \times \mathbb{C} \to X$ is a holomorphic $\mathbb{C}$-action and that $\{F_\xi^t\ ;\ t \in \mathbb{C}\} \subset \mathrm{Aut}(X)$ is a one-parameter subgroup. ◇

Of course, the set $X \times \{0\}$ is always in $\mathscr{D}_\xi$, and moreover $F_\xi^0(x) = x$ for all $x \in X$.

LEMMA 4.4.8. *If there exists $\epsilon > 0$ such that $X \times \{t \in \mathbb{C}\ ;\ |t| \leq \varepsilon\} \subset \mathscr{D}_\xi$, then $\xi$ is complete.*

**Proof.** If $t \in \mathbb{C}$, then let $N$ be so large that $\delta_N := |t|/N < \varepsilon$. Then put $x_j = F_\xi(x_{j-1}, t\delta_N/|t|)$ for $j = 1, 2, ..., N$, where $x_0 = x$. Then inductively applying the above remark about $F_\xi$, we see that $(x_j, jt\delta_N/t) \in \mathscr{D}_\xi$ for all $j = 1, ..., N$. □

THEOREM 4.4.9. *If $X$ is a compact Riemann surface then each $\xi \in \Gamma(X, T_X^{1,0})$ is complete.*

**Proof.** By the Existence and Uniqueness Theorem for ODE, to each $x \in X$ there is associated a number $\varepsilon_x > 0$ and a neighborhood $U_x$ such that $U_x \times \{|t| < \varepsilon_x\} \subset \mathcal{D}_V$. By compactness, there exist $x_1, ..., x_k$ such that $\bigcup_{i=1}^k U_{x_i} = X$. Let $\varepsilon := \min\{\varepsilon_{x_1}, ..., \varepsilon_{x_k}\}$, and apply Lemma 4.4.8. □

REMARK. Lemma 4.4.8 and Theorem 4.4.9 hold for general smooth manifolds and vector fields. ◇

EXAMPLE 4.4.10. Let $L \subset \mathbb{C}$ be a lattice and let $X = \mathbb{C}/L$ be the complex torus associated to $L$. The flow of the vector field $c\frac{\partial}{\partial z}$ is $F^t_{c\frac{\partial}{\partial z}}([z]) = [z + ct]$. ◇

EXAMPLE 4.4.11. Consider the vector fields $\xi$ and $\eta$ on $\mathbb{P}_1$ defined in Example 4.4.3. If we use coordinates $z_0$ on $U_0$ and $z_1$ on $U_1$, then for $x \in U_0$,

$$z_0 \circ F^t_\xi(x) = z_0(x) + t \quad \text{and} \quad z_0 \circ F^t_\eta(x) = z_0(x)e^t,$$

while on $U_1$,

$$z_1 \circ F^t_\xi(x) = \frac{z_1(x)}{1 + z_1(x)t} \quad \text{and} \quad z_1 \circ F^t_\eta(x) = z_1(x)e^{-t}.$$

If we look only at real times, the closures of the integral curves of the vector field $\xi$ are horocycles touching at the fixed point $\infty$ while the integral curves of $\eta$ are lines of latitude of the sphere whose poles are $0$ and $\infty$.

Perhaps it is also interesting to note that while the flow of $\xi$ restricts to a complete flow on $U_0$, it does not restrict to a complete flow on $U_1$. ◇

## 4.5. Divisors and line bundles

In the study of compact Riemann surfaces, one of the most important concepts is the notion of divisor. The concept of divisor is also useful in the study of open Riemann surfaces, but it is not as important as in the compact case.

The geometry of divisors is nicely captured using line bundles. In this section we discuss the relationship between divisors and line bundles.

### 4.5.1. Divisors.

DEFINITION 4.5.1. Let $X$ be a Riemann surface.

(1) A divisor $D$ on a Riemann surface $X$ is a locally finite[1] subset $\{p_1, p_2, ...\}$ of distinct points of $X$, together with a collection of integers $m_1, m_2, ...$ with $m_i$ associated to $p_i$. The notation is

$$D = \sum_j m_j p_j.$$

(2) The divisor all of whose points are assigned the integer $0$ is written $0$.

(3) The set of points $\{p_1, p_2, ...\}$ is called the support of $D$.

(4) When the support of $D$ is finite, the number

$$\deg(D) := \sum_j m_j \in \mathbb{Z}$$

is called the degree of $D$. (The notion of degree is most useful on compact Riemann surfaces.)

---

[1]It is useful to note that locally finite is not the same as isolated. For example, the set $E = \{1/n \, ; \, n \geq 1\}$ in the complex plane is not locally finite, but $E \cap \mathbb{D}(\frac{1}{n}, \varepsilon) = \{\frac{1}{n}\}$ for any $\varepsilon < \frac{1}{n(n+1)}$.

## 4.5. Divisors and line bundles

(5) The collection of divisors on $X$ is denoted $\mathrm{Div}(X)$.

(6) Equivalently, a divisor is a map $D : X \to \mathbb{Z}$ whose support is a locally finite subset of $X$. We sometimes write
$$D = \sum_{p \in X} D(p)p$$
or $m_j = D(p_j)$.  ◇

REMARK. Perhaps it is useful to keep in mind that divisors are meant to capture the notion of zeros and poles of a meromorphic function locally. Globally a divisor is in general not the set of zeros and poles of a meromorphic function. For example, on a compact Riemann surface we will see that the degree of the divisor of a meromorphic function must be 0, but this necessary condition is still not sufficient. In Chapter 14 we will prove a theorem of Abel that identifies geometrically all those divisors that are the collections of zeros and poles of some meromorphic function on a compact Riemann surface. On the other hand, for open Riemann surfaces there is no obstruction: in Chapter 8 we will prove a theorem, due to Weierstrass, stating that, on an open Riemann surface, every divisor is the divisor of some meromorphic function.  ◇

The addition rule
$$D \pm D' := \sum_{p \in X} (D(p) \pm D'(p)) \cdot p$$
endows the set $\mathrm{Div}(X)$ with the structure of an Abelian group whose additive identity is 0. (If we view $D$ as a function, this group law is consistent with the usual addition of functions.)

If the Riemann surface $X$ is compact, then any locally finite subset must be finite, and thus in this case $\mathrm{Div}(X)$ is just the free Abelian group generated by the points of $X$.

EXAMPLE 4.5.2.  (1) Let $f$ be a meromorphic function on $X$. Then we have a divisor
$$\mathrm{Ord}(f) := \sum \mathrm{Ord}_p(f)p.$$
We may sometimes write
$$\mathrm{Ord}(f) = (f)_0 - (f)_\infty,$$
where
$$(f)_0 = \sum_{f(p)=0} \mathrm{Ord}_p(f) \cdot p$$
is the divisor of zeros of $f$ and
$$(f)_\infty = \sum_{f(p)=\infty} -\mathrm{Ord}_p(f) \cdot p$$

is the divisor of poles of $f$. Note that $\deg(\mathrm{Ord}(f)) = 0$.

(2) Let $L \to X$ be a holomorphic line bundle and $s$ a meromorphic section of $L$ over $X$. Then one has the order divisor

$$\mathrm{Ord}(s) := \sum \mathrm{Ord}_p(s) \cdot p = (s)_0 - (s)_\infty.$$

(3) On a Riemann surface the residue of a meromorphic function makes no sense, since to compute the residue of a function we need to make a choice of a local coordinate.

On the other hand, the residue of a meromorphic 1-form at a point $x \in X$ does make sense. The residue of a meromorphic 1-form $\omega$ at a point $x \in X$ is defined as follows: if $z$ is a local coordinate with $z(x) = 0$ and $\omega = f(z)dz$ locally, then

$$\mathrm{Res}_x(\omega) = \frac{1}{2\pi\sqrt{-1}} \int_{|z|=\varepsilon} f(z)dz.$$

Since $\omega$ is holomorphic away from 0, the integral is independent of the path of integration. If, after a change of coordinates $w = w(z)$, $\omega = g(w)dw$, then by definition $g(w)\frac{dw}{dz} = f(z)$. Thus $\mathrm{Res}_x(\omega)$ is independent of the choice of coordinates. (Later we will define integration of 1-forms in general using exactly the same principle.) We thus obtain the residue divisor

$$\mathrm{Res}(\omega) := \sum_{p \in X} \mathrm{Res}_p(\omega) \cdot p.$$

Note that since for a locally defined meromorphic function $g$ the derivative $g'$ has no residue,

$$\deg(\mathrm{Res}(df)) = 0$$

for a meromorphic function $f \in \mathscr{M}(X)$.

(4) Let $F : X \to Y$ be a holomorphic map. Then we have two divisors associated to $F$:

(a) The ramification divisor of $F$

$$\mathrm{Ram}(F) := \sum_{x \in X} (\mathrm{Mult}_x(F) - 1) \cdot x$$

in $\mathrm{Div}(X)$, whose support is called the ramification locus and whose degree is the previously introduced ramification number $R(F)$ of $F$.

(b) The branch divisor of $F$

$$\mathrm{Branch}(F) := \sum_{y \in Y} \left( \sum_{x \in f^{-1}(y)} (\mathrm{Mult}_x(F) - 1) \right) \cdot y$$

## 4.5. Divisors and line bundles

in $\mathrm{Div}(Y)$, whose support is called the branch locus of $F$, or the set of critical values of $F$. By the theorem on normal forms, the support of $\mathrm{Branch}(F)$ is locally finite. ◇

DEFINITION 4.5.3. Let $X$ be a Riemann surface and let $D, \Delta \in \mathrm{Div}(X)$.

(1) $D$ is said to be *effective* if $D(p) \geq 0$ for all $p \in X$. In this case, we write $D \geq 0$. If $D - \Delta$ is effective, we may write $D \geq \Delta$. (In particular, any divisor can be written as a difference of two effective divisors.)

(2) $D$ is said to be *linearly trivial* if $D = \mathrm{Ord}(f)$ for some meromorphic function $f$ on $X$. In this case, we write $D \equiv 0$. Classically, linearly trivial divisors are called *principal divisors*.

(3) $D$ and $\Delta$ are said to be *linearly equivalent* if $D - \Delta$ is linearly trivial. In this case we write $D \equiv \Delta$.

(4) $D$ is said to be *numerically trivial* if $\deg(D) = 0$. We write $D \equiv_{\mathrm{num}} 0$.

(5) $D$ and $\Delta$ are said to be *numerically equivalent* if $D - \Delta$ is numerically trivial. In this case we write $D \equiv_{\mathrm{num}} \Delta$. ◇

As we have shown, every linearly trivial divisor is numerically trivial. Indeed, this assertion is by definition equivalent to the statement that the number of zeros of a meromorphic function, counted with multiplicity, is equal to the number of poles of that function, also counted with multiplicity.

On the Riemann sphere $\mathbb{P}_1$, the converse assertion is true: every numerically trivial divisor is linearly trivial. This fact is seen as follows: by choosing projective coordinates wisely, we may guarantee that our divisor does not contain the point at infinity. That is to say, it lies entirely in $U_0$. Consider the polynomials

$$p(z_0) = \prod_{D(x)>0} (z_0 - z_0(x))^{D(x)} \quad \text{and} \quad q(z_0) = \prod_{D(x)<0} (z_0 - z_0(x))^{D(x)}.$$

By the hypothesis of numerical triviality,

$$\deg p = \deg q.$$

It follows that the rational function $f = p/q$ has no poles at $\infty$. Hence by Riemann's Removable Singularities Theorem, $f$ extends to a meromorphic function on $\mathbb{P}_1$ and $D = \mathrm{Ord}(f)$.

EXAMPLE 4.5.4. The difference between numerical and linear triviality of divisors on a compact Riemann surface occurs as soon as the genus is positive. In fact, if $X$ is a compact Riemann surface of positive genus, the numerically trivial divisor $p - q$ is never linearly trivial. Indeed, if a meromorphic function $f \in \mathscr{M}(X)$ has exactly one pole counting multiplicity, then $\mathrm{Deg}(f) = 1$, and thus $f : X \to \mathbb{P}_1$ is an isomorphism by Corollary 3.3.16. ◇

The result identifying exactly which numerically trivial divisors on a compact Riemann surface are linearly trivial is known as Abel's Theorem. It will be stated and proved in Chapter 14.

**4.5.2. The line bundle of a divisor.** Let $D \in \text{Div}(X)$ be a divisor, and fix an atlas $\{(z_\alpha, U_\alpha)\}$ for $X$ such that $U_\alpha \subset\subset X$ for all $\alpha$. For each $\alpha$, fix a function $f_\alpha \in \mathscr{M}_X(U_\alpha)$ such that
$$\text{Ord}(f_\alpha) = D|_{U_\alpha} := \sum_{p \in U_\alpha} D(p) \cdot p.$$
(For example, one could take $f_\alpha = \prod_{p \in U_\alpha}(z-p)^{D(p)}$.) Then we obtain a collection of functions
$$g_{\alpha\beta} := \frac{f_\alpha}{f_\beta} \in \mathcal{O}(U_\alpha \cap U_\beta)$$
with no zeros. Evidently $\{g_{\alpha\beta}\}$ satisfies the cocyle condition (4.1) and hence is the set of transition functions of a line bundle which we denote $L_D$. Moreover,
$$g_{\alpha\beta} f_\beta = f_\alpha,$$
and so the collection of functions $\{f_\alpha\}$ defines a meromorphic section $s_D \in \Gamma_{\mathscr{M}}(X, L_D)$. Furthermore, $\text{Ord}(s_D) = D$.

DEFINITION 4.5.5. The line bundle $L_D$ is called the line bundle associated to the divisor $D$, and the section $s_D$ is called the canonical meromorphic section of $D$. ◊

REMARK. When $X$ is compact, $s_D$ is uniquely determined up to a multiplicative constant. Indeed, if we have another such section $\tilde{s}_D$, then by the Riemann Removable Singularities Theorem the function $\tilde{s}_D/s_D$ is holomorphic on all of $X$, and thus constant. The same argument shows that in general $s_D$ is only determined up to a global, nowhere-zero holomorphic function. ◊

**4.5.3. The divisor of a line bundle with meromorphic section.** We have already seen that given a meromorphic section $s$ of a holomorphic line bundle $L$, we obtain a divisor
$$\text{Ord}(s).$$
Let $\{g_{\alpha\beta}\}$ be the transition functions of $L$. The section $s$ then provides us with local functions $\{f_\alpha\}$ such that $f_\alpha = g_{\alpha\beta} f_\beta$. Moreover, it is clear that the zeros and poles of $f_\alpha$ define the divisor $\text{Ord}(s)|_{U_\alpha}$. Thus the transition functions of $L$ are
$$g_{\alpha\beta} = \frac{f_\alpha}{f_\beta}.$$
In other words, we have the following result.

PROPOSITION 4.5.6. *If $s \in \Gamma_{sm}(X, L)$ is a meromorphic section, then $L_{\text{Ord}(s)} = L$ and $s_{\text{Ord}(s)}$ is a nowhere-zero multiple of $s$. In particular, if $X$ is compact then $s_{\text{Ord}(s)}$ is a constant multiple of $s$.*

## 4.5.4. Summary of the divisor-line bundle correspondence.
We leave the reader to show that
$$L_{D+D'} = L_D \otimes L_{D'} \quad \text{and} \quad L_{-D} = L_D^*.$$
Thus the map $\mathfrak{L} : D \mapsto L_D$ gives a homomorphism
$$\mathfrak{L} : \text{Div}(X) \to \text{Pic}(X).$$

Now, if $D \equiv 0$, then by definition $D$ is the divisor of a meromorphic function $f$, and thus the line bundle $L_D$ is trivial. Conversely, if $L_D$ is trivial then $D = \text{Ord}(f)$ for some meromorphic function $f$. Indeed, the triviality of $L_D$ means that, perhaps after shrinking our atlas, there are nowhere-zero holomorphic functions $h_\alpha$ such that the transition functions of $L_D$ are $g_{\alpha\beta} = h_\alpha/h_\beta$. Thus $f_\alpha/h_\alpha = f_\beta/h_\beta$ on $U_\alpha \cap U_\beta$, and therefore the function $f$ whose value on $U_\alpha$ is $f_\alpha/h_\alpha$ is a globally-defined meromorphic function, and $D = \text{Ord}(f)$. It follows that
$$\text{Kernel}(\mathfrak{L}) = \{D \in \text{Div}(X) \ ; \ D \equiv 0\}.$$
It is natural to ask whether $\mathfrak{L}$ is surjective. In view of Proposition 4.5.6, the surjectivity of $\mathfrak{L}$ is equivalent to having an affirmative answer to the following.

QUESTION 4.5.7. *Does every line bundle on a Riemann surface admit a meromorphic section that is not identically zero?*

The answer to this question, though not easy, is nevertheless affirmative.

THEOREM 4.5.8. *Every holomorphic line bundle on a Riemann surface has a non-identically zero meromorphic section.*

The typical proof of Theorem 4.5.8 uses either Green's Functions or vanishing theorems for sheaf cohomology. In this book we take a non-typical approach, which we postpone until Chapter 11, when we are able to solve the $\bar{\partial}$ equation with $L^2$ estimates.

Assuming Theorem 4.5.8, we can conclude this paragraph with the following statement.

THEOREM 4.5.9. *The set of holomorphic line bundles on a Riemann surface is in 1-1 correspondence with the set of divisors modulo linear equivalence.*

## 4.6. Line bundles over $\mathbb{P}_n$

**4.6.1. The tautological bundle.** Recall that in item (2) of Definition 2.3.7 we presented the tautological line bundle as the map $\pi : \mathbb{U} \to \mathbb{P}_1$ defined so that the preimage of a point $\ell \in \mathbb{P}_1$ is the set of points in $\mathbb{C}^2$ making up the line $\ell$. We can extend this idea from the Riemann sphere to any projective space.

Recall that $\mathbb{P}_n$ is the space of lines through $0$ in $\mathbb{C}^{n+1}$. Consider the set
$$\mathbb{U} := \{(z, \ell) \in \mathbb{C}^{n+1} \times \mathbb{P}_n \ ; \ z \in \ell\}.$$

Let $\pi : \mathbb{U} \to \mathbb{P}_n$ be the restriction to $\mathbb{U}$ of the projection $\mathrm{P} : \mathbb{C}^{n+1} \times \mathbb{P}_n \to \mathbb{P}_n$ to the second factor. We leave it to the reader to check that $\mathbb{U}$ is a manifold. (Hint: It is cut out by the $\binom{n+1}{2}$ functions
$$F_{i,j}(z, [\ell]) := z_i \ell_j - z_j \ell_i, \quad 0 \leq i < j \leq n,$$
of which only $n$ functions are independent. It follows that the dimension of $\mathbb{U}$ is $n + (n+1) - n = n+1$.) We can give local trivializations of $\mathbb{U} \to \mathbb{P}_n$ as follows. Choose a neighborhood $U_j := \{[\ell] \; ; \; \ell_j \neq 0\} \cong \mathbb{C}^n$ and consider the affine hyperplane $H_j := \{\ell_j = 1\} \subset \mathbb{C}^{n+1}$. The set of lines through the origin that meet $H_j$ is precisely $U_j$, and thus the local trivialization $\mathbb{U}|_{U_j} \cong \mathbb{C} \times U_j$ can be represented by the "point-slope" formula
$$\mathbb{U} \ni (z, [w_1, ..., w_{j-1}, 1, w_j, ..., w_n]) \mapsto (w, z_j) \in U_j \times \mathbb{C}.$$
The transition functions for $\mathbb{U}$ can now be easily worked out, but since we shall not need them, we leave the precise formula to the reader, who can also readily verify that these transition functions are holomorphic.

**4.6.2. The hyperplane bundle and its global sections.** We turn our attention now to the dual bundle $\mathbb{U}^* \to \mathbb{P}_n$. The fibers $\mathbb{U}^*_\ell$ consist of the set of linear functionals on the line $\ell$ through the origin in $\mathbb{C}^{n+1}$. Since every such linear functional is the restriction to $\ell$ of some linear functional on $\mathbb{C}^{n+1}$, we see that we can identify the space $(\mathbb{C}^{n+1})^*$ with a subspace of the space of global holomorphic sections of $\mathbb{U}^*$. Observe that the zero set of any such section is a hyperplane in $\mathbb{P}_n$. Because of this fact, $\mathbb{U}^*$ is often denoted $\mathbb{H}$ and called the hyperplane line bundle.

In fact, every section of $\mathbb{P}_n$ is associated to a linear functional on $\mathbb{C}^{n+1}$. To see this, we argue as follows. Let $s \in \Gamma_\mathcal{O}(\mathbb{P}_n, \mathbb{H})$. For any zero vector $v \in \mathbb{U} = \mathbb{H}^*$, we have $\langle s, v \rangle = 0$. On the other hand, if $v \in \mathbb{U}$ is non-zero, we can identify $v$ with exactly one vector in $\mathbb{C}^{n+1}$, which we also denote $v$. For any complex number $\lambda$, we have by the linearity of the pairing of points of $\mathbb{U}_{[v]}$ and points of $\mathbb{H}_{[v]}$ that
$$\langle s, \lambda v \rangle := \langle s([v]), \lambda v \rangle = \lambda \langle s([v]), v \rangle = \lambda \langle s, v \rangle.$$
Thus we can define the linear function
$$\ell_s : v \mapsto \langle s, v \rangle.$$
Observe that $s \mapsto \ell_s$ is injective and is clearly the inverse of the map associating points of $(\mathbb{C}^{n+1})^*$ to sections of $\mathbb{H}$.

The result of this discussion can be summarized in the following way.

PROPOSITION 4.6.1. $(\mathbb{C}^{n+1})^* \cong \Gamma_\mathcal{O}(\mathbb{P}_n, \mathbb{H})$.

**Alternate proof in the case $n = 1$.** Since the zero set of a hyperplane section is a hyperplane and thus exactly one point, any other non-trivial holomorphic section $s$ must have exactly one zero. By choosing a linear functional $F$ whose zero is exactly the point $\{s = 0\}$, we see that the meromorphic function $s/F$ has no poles (or zeros). It follows that this quotient is constant, and thus the section $s = \lambda F$ comes from a linear functional. □

REMARK. We have identified a subset of the set of global sections of $\mathbb{H} \to \mathbb{P}_n$ with the homogeneous polynomials of degree 1 on $\mathbb{C}^{n+1}$. It follows, by taking tensor products, that the homogeneous polynomials of degree $m$ form a subspace of the global sections of the line bundle $\mathbb{H}^{\otimes m} \to \mathbb{P}_n$. Thus we have found a way to identify homogeneous polynomials, which are not functions on $\mathbb{P}_n$, with a natural generalization of functions, namely sections of a holomorphic line bundle. ◊

A similar argument, which we leave to the reader in the form of an exercise at the end of the chapter, shows that in fact

$$\text{Sym}^m(\mathbb{C}^2)^* \cong \Gamma_{\mathcal{O}}(\mathbb{P}_1, \mathbb{H}^{\otimes m}).$$

The same result holds in higher dimensions:

$$\text{Sym}^m(\mathbb{C}^{n+1})^* \cong \Gamma_{\mathcal{O}}(\mathbb{P}_n, \mathbb{H}^{\otimes m}).$$

**4.6.3. Line bundles over $\mathbb{P}_1$.** Now let $L \to \mathbb{P}_1$ be a holomorphic line bundle. As we will see in Chapter 11, $L$ has a meromorphic section $\sigma$. Let $D = \sum_{i=1}^{m^+} p_i - \sum_{j=1}^{m^-} q_j$ be the divisor of $\sigma$, where $p_1, ..., p_k, q_1, ..., q_m \in \mathbb{P}_1$ are possibly non-distinct points. Then each $p_i$ (resp. $q_j$) is the zero set of some section $s_i$ (resp. $t_j$) of $\mathbb{H} \to \mathbb{P}_1$. It follows that

$$L = \mathbb{H}^{\otimes(m^+ - m^-)}$$

(and also that $\sigma$ is a multiple of $\frac{\prod s_i}{\prod t_j}$). Thus, modulo the existence of a meromorphic section, we have proved the following fact.

PROPOSITION 4.6.2. *Every line bundle on $\mathbb{P}_1$ is an integer tensor power of $\mathbb{H}$.*

REMARK. In fact Proposition 4.6.2 holds on projective spaces of all dimensions. ◊

## 4.7. Holomorphic sections and projective maps

Let $X$ be a Riemann surface and let $H \to X$ be a holomorphic line bundle. Assume that $\Gamma_{\mathcal{O}}(X, H)$, the space of global holomorphic sections of $H \to X$, is non-trivial. As we will show at the end of the present chapter, if $X$ is compact then $\Gamma_{\mathcal{O}}(X, H)$ is a finite-dimensional vector space.

**4.7.1. Definition of projective map and morphism.** Let us fix a non-zero subspace $W \subset \Gamma_{\mathcal{O}}(X, H)$ of finite dimension. Such a subspace is classically called a linear system. If $W = \Gamma_{\mathcal{O}}(X, H)$, one says $W$ is a complete linear system.

It may happen that at some point of $X$, all the members of $W$ vanish. Such a point is called a basepoint of $W$. The collection of basepoints of $W$ is denoted

$$\text{Bs}(W).$$

If $W = \Gamma_{\mathcal{O}}(X, H)$, it is common to write
$$\mathbf{Bs}(\Gamma_{\mathcal{O}}(X,H)) = \mathbf{Bs}(|H|).$$

For each point of $X - \mathbf{Bs}(W)$, we can define the subspace $\phi_W(x) \subset W$ by

(4.5)  $$\phi_W(x) := \{s \in W \; ; \; s(x) = 0\}.$$

Fixing an element $\xi \in H_x^* - \{0\}$, the map

(4.6) $$W \ni s \mapsto \langle s(x), \xi \rangle \in \mathbb{C}$$

is a non-trivial linear functional whose kernel is precisely $\phi_W(x)$. Thus $\phi_W(x)$ is a hyperplane in the projectivization $\mathbb{P}(W)$ of $W$.

The set of hyperplanes in a projective space $\mathbb{P}(W)$ is itself a projective space $\mathbb{P}(W)^\vee$, called the dual projective space, and is easily seen to be the projectivization $\mathbb{P}(W^*)$ of the dual vector space $W^*$. Thus for each $x \in X - \mathbf{Bs}(W)$ we obtain an element $\phi_W(x) \in \mathbb{P}(W^*)$. Again if $W = \Gamma_{\mathcal{O}}(X, H)$, it is common to write
$$\phi_{\Gamma_{\mathcal{O}}(X,H)} = \phi_{|H|}.$$

(Though the definition makes perfect sense if $\Gamma_{\mathcal{O}}(X,H)$ is infinite-dimensional, it is most often considered when $\Gamma_{\mathcal{O}}(X,H)$ is finite-dimensional, for example if $X$ is compact.) We usually omit reference to the base locus of $W$ and instead write
$$\phi_W : X \dashrightarrow \mathbb{P}(W^*).$$

Such a map is called a projective map. If $\mathbf{Bs}(W) = \emptyset$, then we call $\phi_W$ a projective morphism, and write
$$\phi_W : X \to \mathbb{P}(W^*).$$

**4.7.2. Description of $\phi_W$ in terms of a basis of $W$.** Suppose we fix a basis $s_0, ..., s_n \in W$. Since $W^{**} \cong W$, each element of $W$ is identified with a linear functional on $W^*$. It follows that $s_0, ..., s_n$ may serve as homogeneous coordinates on $\mathbb{P}(W^*)$. Thus a choice of basis for $W$ identifies the $n$-dimensional projective space $\mathbb{P}(W^*)$ with $\mathbb{P}_n$.

We'd like to think of $s \mapsto s(x)$ as a linear functional, but this is not the case since $s(x)$ is not a number. However, if we choose a vector $\xi \in H_x - \{0\}$ then $s(x) = f(x) \cdot \xi$ for some number $f(x)$. Thus we can define the linear functional $\mathrm{Ev}_x : W \to \mathbb{C}$ by
$$\mathrm{Ev}_x(s) = f(x) := s(x)/\xi.$$

In terms of the coordinates on $W^*$ defined by the basis of sections $s_0, ..., s_n \in W$ (which identify $W^*$ with $\mathbb{C}^{n+1}$),
$$\mathrm{Ev}_x = (f_0(x), ..., f_n(x)).$$

## 4.7. Holomorphic sections and projective maps

Now,
$$s = \sum_{i=0}^{n} a_i s_i \in \phi_W(x) \iff \text{Ev}_x(s) = \sum_{i=0}^{n} a_i f_i(x) = s(x)/\xi = 0.$$

Now, we can scale the vector $(f_0(x), ..., f_n(x))$, which is equivalent to scaling the vector $\xi$, and still obtain the same relation. It follows that the line $[f_0(x), ..., f_n(x)]$ annihilates the hyperplane $\phi_W(x)$ and thus corresponds to $\phi_W(x) \in \mathbb{P}(W^*)$. Since the choice of $\xi$ does not change the line $[f_0(x), ..., f_n(x)]$, it is reasonable to write
$$\phi_W(x) = [s_0(x), ..., s_n(x)].$$

### 4.7.3. All holomorphic maps to projective space are projective morphisms.

**PROPOSITION 4.7.1.** *Let $X$ be a Riemann surface and let $\phi : X \to \mathbb{P}_n$ be a holomorphic map. Then there is a holomorphic line bundle $H \to X$ and a subspace $W \subset \Gamma_\mathcal{O}(X, H)$ such that*
$$\mathbf{Bs}(W) = \emptyset \quad \text{and} \quad \phi = \phi_W.$$

**Proof.** Recall that the global sections of the hyperplane section $\mathbb{H} \to \mathbb{P}_n$ are identified with the linear span of any set of homogeneous coordinate functions $z_0, ..., z_n$. We let
$$H := \phi^*\mathbb{H} \quad \text{and} \quad s_j := \phi^*(z_j|_{\phi(X)}),$$
where $z_0, ..., z_n$ are homogeneous coordinates on $\mathbb{P}_n$, and we set
$$W = \text{Span}_\mathbb{C}\{s_0, ..., s_n\}.$$
Then
$$\phi = [s_0, ..., s_n] = \phi_W,$$
as claimed. $\square$

### 4.7.4. Resolving the base locus.
Let $X$ be a Riemann surface and let $H \to X$ be a holomorphic line bundle. Let $W \subset \Gamma_\mathcal{O}(X, H)$. The set $\mathbf{Bs}(W)$ is cut out locally by holomorphic functions. So far we have only taken into account the set, but we can also take into account multiplicities.

**DEFINITION 4.7.2.** *The divisor $\mathscr{B}s(W) = \sum_{x \in X} \mathscr{B}s(W)_x \cdot x$ defined by*
$$\mathscr{B}s(W)_x := \min\{\text{Ord}_x(s) \; ; \; s \in W\}$$
*is called the base divisor.* $\diamond$

Evidently $\mathscr{B}s(W)$ is an effective divisor. Thus there is a holomorphic section $\sigma$ of the line bundle $L = L_{\mathscr{B}s(W)}$ whose zero divisor is $\mathscr{B}s(W)$. By the definition of $\mathscr{B}s(W)$, it follows that every section of $W$ is divisible by $\sigma$:
$$W \ni s = \tilde{s} \cdot \sigma.$$
Evidently $\tilde{s} \in \Gamma_\mathcal{O}(X, H \otimes L^*)$. We have thus proved the following proposition.

PROPOSITION 4.7.3. *For each linear system $W \subset \Gamma_{\mathcal{O}}(X, H)$ there is a holomorphic line bundle $L \to X$, a holomorphic section $\sigma \in \Gamma_{\mathcal{O}}(X, L)$, and a linear system $\tilde{W} \subset \Gamma_{\mathcal{O}}(X, H \otimes L^*)$ such that*

$$\mathbf{Bs}(\tilde{W}) = \emptyset \quad \text{and} \quad W = \sigma \tilde{W}.$$

## 4.8. A finiteness theorem

THEOREM 4.8.1. *Let $X$ be a compact Riemann surface and let $L \to X$ be a holomorphic line bundle. Then the vector space $\Gamma_{\mathcal{O}}(X, L)$ is finite-dimensional.*

**Proof.** Choose a finite cover $\mathcal{U} := \{U_1, ..., U_N\}$ of $X$ by coordinate charts on which $L$ is trivial. Morover, choose the cover so that $z_j : U_j \to \mathbb{D}$ is a homeomorphism onto the unit disk and, with $V_j := z_j^{-1}(\frac{1}{2}\mathbb{D})$, the collection of open sets $\mathcal{V} := \{V_1, ..., V_N\}$ is again an open cover of $X$. We write

$$a_i := z_i^{-1}(0).$$

For each $j = 1, 2, ..., N$, let $e_j \in \Gamma_{\mathcal{O}}(U_j, L)$ be a nowhere-zero section. By shrinking $U_j$ if necessary, we may assume that $e_j$ is bounded away from zero in $U_j$. For each $s \in \Gamma_{\mathcal{O}}(X, L)$, define

$$||s||^{\mathcal{U}} = \max_i \sup_{U_i} |s(x)/e_i(x)|$$

and

$$||s||^{\mathcal{V}} = \max_i \sup_{V_i} |s(x)/e_i(x)|.$$

Clearly

$$||s||^{\mathcal{V}} \leq ||s||^{\mathcal{U}}.$$

Let $g_{ij}$ be the transition functions associated to the above cover $\mathcal{U}$ and local trivializations $\{e_j\}$:

$$g_{ij} := e_i/e_j.$$

Write

$$C := \max_{i,j} \sup_{U_i \cap U_j} |g_{ij}|.$$

By our assumption on the $e_j$, $C < +\infty$.

Now let $x_o \in U_i$. Choose $j$ such that $x_o \in V_j$. Then

$$|(s/e_i)(x_o)| = |g_{ji}(x_o) s/e_j(x_o)| \leq C ||s||^{\mathcal{V}},$$

and thus

$$||s||^{\mathcal{U}} \leq C ||s||^{\mathcal{V}}.$$

Next, fix an integer $k \geq 0$. Let $s \in \Gamma_{\mathcal{O}}(X, L)$ such that $\text{Ord}_{a_i}(s) \geq k$ for each $i = 1, ..., N$. Consider the section $z_i^{-k} s$ on $U_i$ and the functions

$$f_i := z_i^{-k} s/e_i : U_i \to \mathbb{C}.$$

By hypothesis, each of these functions is holomorphic. Then
$$\sup_{V_i} |s/e_i| = \sup_{V_i} |z_i^k f_i| = \tfrac{1}{2^k}\sup_{V_i} |f_i| \leq \tfrac{1}{2^k}\sup_{U_i} |f_i| = \tfrac{1}{2^k}\sup_{U_i} |s/e_i| \leq \tfrac{1}{2^k}\|s\|^{\mathscr{U}}.$$
(The second equality and first inequality follow from the maximum principle.) Taking the maximum over $i$, we thus obtain that any section $s$ for which $\mathrm{Ord}_{a_i}(s) \geq k$ for each $i = 1, ..., N$ satisfies
$$\|s\|^{\mathscr{U}} \leq C\|s\|^{\mathscr{V}} \leq 2^{-k}C\|s\|^{\mathscr{U}}.$$
It follows that as soon as $k > \frac{\log C}{\log 2}$, any such section $s$ must be identically zero.

Finally, consider the map
$$\Gamma_{\mathcal{O}}(X, L) \to \bigoplus_{j=1}^{N} \mathscr{P}_k$$
sending the section $s$ to the $N$-tuple $(P_1, ..., P_N)$, where $\mathscr{P}_k \cong \mathbb{C}^{k+1}$ is the space of polynomials of degree at most $k$, and $P_i$ is the $k^{\text{th}}$ Taylor polynomial of the function $s/e_i$ at the origin and in the coordinate $z_i$. The above calculation shows that this map is injective as soon as $k > \frac{\log C}{\log 2}$. It follows that $\Gamma_{\mathcal{O}}(X, L)$ is finite-dimensional. (The dimension is bounded above by $(\lfloor \frac{\log C}{\log 2} \rfloor + 2)N$.) The proof is complete. □

DEFINITION 4.8.2. On a compact Riemann surface $X$, the number
$$g_{\mathcal{O}}(X) := \dim_{\mathbb{C}} \Gamma_{\mathcal{O}}(X, K_X)$$
is called the arithmetic genus of $X$. ◇

REMARK. As we will see in Chapter 9, the arithmetic genus of a compact Riemann surface $X$ is equal to the geometric genus, i.e., the number of handles of $X$. ◇

## 4.9. Exercises

4.1. Show that if a line bundle $L \to M$ has two sections $s_1$ and $s_2$ such that $s_1(M) \cap s_2(M) = \emptyset$ then $L$ is isomorphic to the trivial bundle.

4.2. Show that the zero section of $L \to M$ and the $\mathbb{C}^*$-action of $M$ are well-defined.

4.3. Show that the line bundles $L \to M$ and $L' \to M$ are isomorphic if and only if for every collection of transition functions $g_{\alpha\beta}$ for $L$ and $g'_{\alpha\beta}$ for $L'$ there is a collection $\{h_\alpha\}$ of zero-free local functions so that $g_{\alpha\beta} = h_\alpha g'_{\alpha\beta} h_\beta^{-1}$.

4.4. Verify the remark preceding Paragraph 4.1.3.

4.5. Check the parenthetical claim at the end of Definition 4.2.5.

4.6. Prove that the addition and scalar multiplication on $T_X^{1,0}$ defined in Paragraph 4.3.2 are well-defined.

4.7. Prove Proposition 4.3.8.

4.8. Let $X$ and $Y$ be compact Riemann surfaces and let $F : X \to Y$ be holomorphic. Show that the degrees of the two divisors $\mathrm{Ram}(F) \in \mathrm{Div}(X)$ and $\mathrm{Branch}(F) \in \mathrm{Div}(Y)$ are equal.

4.9. Show that $L_{D+D'} = L_D \otimes L_{D'}$ and $L_{-D} = L_D^*$.

4.10. Show that $\mathbb{U} = \{(z, \ell) \in \mathbb{C}^{n+1} \times \mathbb{P}_n \; ; \; z \in \ell\}$ is a manifold.

4.11. Show that $\Gamma_{\mathcal{O}}(\mathbb{P}_1, \mathbb{H}^{\otimes m}) \cong \mathrm{Sym}^m(\mathbb{C}^2)$ along the same lines of the alternate proof or Proposition 4.6.1.

4.12. The goal of this exercise is to show that $\mathrm{Sym}^m(\mathbb{C}^{n+1})^* \cong \Gamma_{\mathcal{O}}(\mathbb{P}_n, \mathbb{H}^{\otimes m})$ when $m \geq 2$. We fix homogeneous coordinates $z^0, ..., z^n$ on $\mathbb{P}_n$, which we know to be sections of $H^0(\mathbb{P}_n, \mathbb{H})$.
   (a) Show that $\mathrm{Sym}^m(\mathbb{C}^{n+1})^* \hookrightarrow \Gamma_{\mathcal{O}}(\mathbb{P}_n, \mathbb{H}^{\otimes m})$.
   (b) Show that if $s \in \Gamma_{\mathcal{O}}(\mathbb{P}_n, \mathbb{H}^{\otimes m}) - \{0\}$ then the meromorphic function $f = s/(z^0)^m$ is holomorphic on the chart $U_0 \cong \mathbb{C}^n$.
   (c) Consider the volume form on $U_0 \cong \mathbb{C}^n$ defined by $d\mu = (1+|x^1|^2 + ... + |x^n|^2)^{-(m+n+1)} dV(x)$, where $x^i = z^i/z^0$ are affine coordinates in $U_0$. Show that
   $$\int_{U_0} |f|^2 d\mu < +\infty.$$
   (Hint: Observe that $|s|^2/|z|^{2m}$ is well-defined on $\mathbb{P}_n$.)
   (d) Let $f(x) = \sum_{|\alpha| \geq 0} a_\alpha x^\alpha$ be the power series expansion of $f$. Use orthogonality to show that
   $$\sum_{m \geq 0} \sum_{|\alpha|=m} |a_\alpha|^2 \int_{\mathbb{C}^n} \frac{|x^\alpha|^2 dV(x)}{(1+|x|^2)^{n+m+1}} < +\infty.$$
   (e) Show that the embedding in part (a) is an isomorphism.

4.13. Let $X$ be a compact Riemann surface and let $L \to X$ be a holomorphic line bundle.
   (a) Show that if there exists a section $s \in \Gamma_{\mathcal{O}}(X, L) - \{0\}$ with a zero, then $L$ is not trivial.
   (b) Show that if $\Gamma_{\mathcal{O}}(X, L) \neq \{0\}$ then either $L$ is trivial or $\Gamma_{\mathcal{O}}(X, L^*) = \{0\}$.
   (c) Prove that $\Gamma_{\mathcal{O}}(\mathbb{P}_1, K_{\mathbb{P}_1}) = \{0\}$.

4.14. Let $X$ be a compact Riemann surface and $D$ an effective divisor on $X$ and let $s \in \Gamma_{\mathcal{O}}(X, L_D)$. Consider the subspace $\mathcal{O}_X(-D) := \{f \in \mathcal{M}(X) \; ; \; \mathrm{Ord}(f) \geq -D\} \subset \mathcal{M}(X)$.
   (a) Show that the map $\mathcal{O}_X(-D) \ni f \mapsto f \cdot s \in \Gamma_{\mathcal{M}}(X, L_D)$ is injective.
   (b) Fix a canonical section $s_D$. Under what conditions does the image of the map in part (a) contain $s_D$, and which function is mapped to $s_D$ under these conditions?

*Chapter 5*

# Complex Differential Forms

There are three kinds of differential forms that will be important to us: $(1,0)$-forms, $(0,1)$-forms, and $(1,1)$-forms.

## 5.1. Differential $(1,0)$-forms

DEFINITION 5.1.1. Let $X$ be a Riemann surface. A section $\omega \in \Gamma(X, K_X)$ of the canonical bundle is called a $(1,0)$-form on $X$. ◇

We have already seen that if $(z_\alpha, U_\alpha)$ is a complex chart on $X$, then $dz_\alpha$ is a section of $K_X$ over $U_\alpha$. It follows that in this chart, there is a function $f_\alpha : U_\alpha \to \mathbb{C}$ such that
$$\omega = f_\alpha(z_\alpha)dz_\alpha,$$
and for different charts the functions $f_\alpha$ are related by
$$f_\alpha(z_\alpha) = f_\beta(z_\beta)\frac{dz_\beta}{dz_\alpha}.$$

DEFINITION 5.1.2. If $f_\alpha$ is holomorphic (resp. meromorphic) for all $\alpha$, we say that $\omega$ is a holomorphic (resp. meromorphic) 1-form. That is to say, $\omega$ is a holomorphic (resp. meromorphic) section of the canonical bundle. ◇

EXAMPLE 5.1.3. If $f : X \to \mathbb{C}$ is a complex-valued function, then
$$\omega = \frac{\partial f}{\partial z_\alpha}dz_\alpha \quad \text{on } U_\alpha$$
is a globally-defined $(1,0)$-form. Moreover, if $f$ is holomorphic (resp. meromorphic), then $\omega$ is a holomorphic (resp. meromorphic) 1-form. Indeed, if $z_\alpha$ and $z_\beta$

are two charts, then
$$\frac{\partial f}{\partial z_\alpha}dz_\alpha = \left(\frac{\partial z_\beta}{\partial z_\alpha}\frac{\partial f}{\partial z_\beta}\right)\left(\frac{\partial z_\alpha}{\partial z_\beta}dz_\beta\right) = \frac{\partial f}{\partial z_\beta}dz_\beta,$$
where the last equality follows from the chain rule. ◇

DEFINITION 5.1.4. The $(1,0)$-form $\omega$ of the previous example is denoted $\partial f$. ◇

REMARK. Not every differential $(1,0)$-form is of the form $\partial f$ for some function $f$. For example, a non-trivial holomorphic differential form on a compact Riemann surface cannot be of the form $\partial f$, for then
$$\frac{\partial}{\partial \bar{z}_\alpha}\frac{\partial f}{\partial z_\alpha} = 0,$$
and thus $f$ would be a harmonic function, which must be constant. Thus, for instance, the differential form $dz$ on a torus $\mathbb{C}/L$ (where $z$ is the coordinate on $\mathbb{C}$, which is defined on $\mathbb{C}/L$ only modulo $L$, but whose differential is consequently well-defined) is not the differential of any function.

On the other hand, it is a consequence of the complex conjugate of the forthcoming Theorem 8.1.1 in Chapter 8 that every $(1,0)$-form on an open Riemann surface is of the form $\partial f$ for some function $f$. A local version of this result can be established by taking complex conjugates of the Dolbeault Lemma, proved at the end of the present chapter. ◇

EXAMPLE 5.1.5. The $(1,0)$-forms $\omega$ and $\eta$ on $\mathbb{P}_1$ given by
$$\eta = \begin{cases} dz_0 & \text{on} \quad U_0 \\ -z_1^{-2}dz_1 & \text{on} \quad U_1 \end{cases}$$
and
$$\omega = \begin{cases} z_0^{-1}dz_0 & \text{on} \quad U_0 \\ -z_1^{-1}dz_1 & \text{on} \quad U_1 \end{cases}$$
are meromorphic. ◇

EXAMPLE 5.1.6 (Non-example). There are no non-trivial holomorphic differential $(1,0)$-forms on $\mathbb{P}_1$. Indeed, suppose $\omega$ is a holomorphic differential $(1,0)$-form on $\mathbb{P}_1$. Then on $U_i$ we have $\omega = f_i(z_i)dz_i$ where $f$ is an entire function. Moreover,
$$f_0(z_0) = \frac{dz_1}{dz_0}f_1(z_0^{-1}) = -z_0^{-2}f_1(z_0^{-1}).$$
Thus as $z_0 \to \infty$, $f(z_0)$ has a limit, namely 0. Thus $f$ is holomorphic on the Riemann sphere, hence constant, and thus 0. ◇

DEFINITION 5.1.7. Let $p \in X$ and let $\omega$ be a meromorphic 1-form on $X$, written locally near $p$ as $\omega = fdz$. We define
$$\mathrm{Ord}_p(\omega) := \mathrm{Ord}_p(f)$$
and call $\mathrm{Ord}_p(\omega)$ the order of $\omega$ at $p$. ◇

REMARK. Since $dz$ defines a frame for $K_X$, this definition of Ord is just a special case of the definition already given for meromorphic sections of general line bundles. ◇

## 5.2. $T_X^{*0,1}$ and $(0,1)$-forms

DEFINITION 5.2.1. Let $X$ be a Riemann surface.
(1) The complex line bundle $T_X^{*0,1}$ is the line bundle whose transition functions are the complex conjugates of the transition functions for the canonical bundle.
(2) A section $s \in \Gamma(X, T_X^{*0,1})$ of $T_X^{*0,1}$ is called a $(0,1)$-form. ◇

It follows that if $z$ is a local coordinate on $X$, then $\overline{dz} = d\bar{z}$ is a non-vanishing local section of $T_X^{*0,1}$, and thus every $(0,1)$-form is of the form $f d\bar{z}$ on the domain of the chart $z$.

REMARK. In fact, the operation of complex conjugation sends sections of $K_X$ to sections of $T_X^{*1,0}$. ◇

EXAMPLE 5.2.2. If $f$ is a function on $X$ and $\{(z_\alpha, U_\alpha)\}$ is an atlas, then
$$\omega = \frac{\partial f}{\partial \bar{z}_\alpha} d\bar{z}_\alpha$$
is a well-defined $(0,1)$-form on $X$. ◇

DEFINITION 5.2.3. The $(0,1)$-form of the previous example is denoted $\bar{\partial} f$. ◇

REMARK. Unlike $K_X$, $T_X^{*0,1}$ is not a holomorphic line bundle. Thus we have no notion of holomorphic $(0,1)$-forms. Nevertheless, we will make significant use of $(0,1)$-forms in this book. ◇

## 5.3. $T_X^*$ and 1-forms

DEFINITION 5.3.1. We call an element of $\Gamma(X, K_X) \oplus \Gamma(X, T_X^{*0,1})$ a 1-form. ◇

In other words, a 1-form $\omega$ on a Riemann surface can be written in a chart $(z_\alpha, U_\alpha)$ as
$$\omega = f_\alpha dz_\alpha + g_\alpha d\bar{z}_\alpha.$$

DEFINITION 5.3.2. If $f : X \to \mathbb{C}$ is a function, we define the $(1,0)$-form
$$df := \partial f + \bar{\partial} f,$$
called the full exterior derivative of $f$. ◇

REMARK. If we write $z = x + \sqrt{-1}y$, we find that
$$df = \frac{\partial f}{\partial x}dx + \frac{\partial f}{\partial y}dy,$$
as the reader can verify. ◇

REMARK. Note that 1-forms can be identified with sections of the vector bundle
$$T_X^* \otimes \mathbb{C} = K_X \oplus T_X^{*0,1}.$$
We chose our presentation in order to avoid the use of vector bundles. ◇

REMARK. Note that a holomorphic 1-form on $X$ automatically has no $(0,1)$-component. From here on, the expression *holomorphic 1-form* will be synonymous with the expression *holomorphic $(1,0)$-form*. ◇

## 5.4. $\Lambda_X^{1,1}$ and $(1,1)$-forms

DEFINITION 5.4.1. Let $X$ be a Riemann surface.

(1) The complex line bundle $\Lambda_X^{1,1}$ is by definition
$$\Lambda_X^{1,1} := K_X \otimes T_X^{*0,1}.$$
Therefore the transition functions are the squared moduli of the transition functions of $K_X$.

(2) A section $s \in \Gamma(X, \Lambda_X^{1,1})$ is called a (1,1)-form. ◇

REMARK. Like $T_X^{*0,1}$, $\Lambda_X^{1,1}$ is not a holomorphic line bundle, and thus there is no notion of a holomorphic $(1,1)$-form. ◇

Given a local coordinate $z$ on $X$, we denote a local section of $\Lambda_X^{1,1}$ by
$$f\,dz \wedge d\bar{z}.$$

REMARK. A formal calculation shows that if the coordinate $z = x + \sqrt{-1}y$ is written in terms of real and imaginary parts, then
$$\frac{\sqrt{-1}}{2} dz \wedge d\bar{z} = dx \wedge dy.$$
As our notation suggests, we are planning to integrate (1,1)-forms. ◇

## 5.5. Exterior algebra and calculus

DEFINITION 5.5.1 (Wedge product). If $\omega_1$ and $\omega_2$ are 1-forms, given locally by
$$\omega_1 = f_1 dz + g_1 d\bar{z} \quad \text{and} \quad \omega_2 = f_2 dz + g_2 d\bar{z},$$
then we define the $(1,1)$-form
$$\omega_1 \wedge \omega_2 := (f_1 g_2 - f_2 g_1) dz \wedge d\bar{z},$$
called the wedge product of $\omega_1$ and $\omega_2$. ◇

## 5.5. Exterior algebra and calculus

It is easily seen that $\omega_2 \wedge \omega_1 = -\omega_1 \wedge \omega_2$.

REMARK. The reader can verify that if $\omega_1$ and $\omega_2$ are global 1-forms on $X$, then the expression $\omega_1 \wedge \omega_2$, a priori defined only locally, is actually a global (1,1)-form on $X$. ◇

DEFINITION 5.5.2 (pullback). Let $F : X \to Y$ be a smooth map and let $\{z_\alpha\}$ be an atlas for $Y$. Let $f : Y \to \mathbb{C}$ be a function, let $\omega = g_\alpha dz_\alpha$ be a (1,0)-form, let $\theta = h_\alpha d\bar{z}_\alpha$ be a (0,1)-form, and let $\eta = k_\alpha dz_\alpha \wedge d\bar{z}_\alpha$ be a (1,1)-form. We define

$$\begin{aligned}
F^*f &:= f \circ F, \\
F^*\omega &:= (F^*g_\alpha)d(F^*z_\alpha), \\
F^*\theta &:= (F^*h_\alpha)\overline{d(F^*z_\alpha)}, \\
F^*\eta &:= (F^*k_\alpha)d(F^*z_\alpha) \wedge \overline{d(F^*z^\alpha)}.
\end{aligned}$$

In all cases, we call $F^*\xi$ the pullback of $\xi$ by $F$. ◇

REMARK. Observe that $F^*$ takes functions to functions, 1-forms to 1-forms, and (1,1)-forms to (1,1)-forms. However, the chain rule shows that $F^*$ sends $(1,0)$-forms to $(1,0)$-forms if and only if it sends $(0,1)$-forms to $(0,1)$-forms if and only if $F$ is holomorphic. ◇

Recall that for a function $f$, we defined $\partial f, \bar{\partial} f$, and $df$ as

$$\partial f = \frac{\partial f}{\partial z}dz, \quad \bar{\partial}f = \frac{\partial f}{\partial \bar{z}}d\bar{z}, \quad \text{and} \quad df = \partial f + \bar{\partial} f.$$

We now extend the definition of exterior derivative to forms.

DEFINITION 5.5.3 (Exterior differentiation). Let $X$ be a Riemann surface, let $\omega = gdz + hd\bar{z}$ be a 1-form on $X$, and let $\theta$ be a (1,1)-form on $X$. Then

$$\begin{aligned}
\partial \omega &:= \partial h \wedge d\bar{z} = \frac{\partial h}{\partial z} dz \wedge d\bar{z}, \\
\bar{\partial} \omega &:= \bar{\partial} g \wedge dz = -\frac{\partial g}{\partial \bar{z}} dz \wedge d\bar{z}, \\
d\omega &:= \partial\omega + \bar{\partial}\omega = \left(\frac{\partial h}{\partial z} - \frac{\partial g}{\partial \bar{z}}\right) dz \wedge d\bar{z}, \\
\partial \theta &:= \bar{\partial}\theta := d\theta := 0
\end{aligned}$$

are the various exterior derivatives of the forms in question. ◇

Note that a 1-form $\omega$ is holomorphic if and only if $\omega$ is a $(1,0)$-forms and $\bar{\partial}\omega = 0$. In fact, even more is true.

LEMMA 5.5.4. *If $\omega$ is a holomorphic 1-form on a Riemann surface, then $d\omega = 0$.*

**Proof.** Since $\omega$ is holomorphic, $\omega = gdz$ and $d\omega = -\frac{\partial g}{\partial \bar{z}}dz \wedge d\bar{z} = 0$. □

We leave it to the reader to show that if $F : X \to Y$ is holomorphic then
$$\partial F^* = F^*\partial, \quad \bar{\partial}F^* = F^*\bar{\partial}, \quad \text{and} \quad dF^* = F^*d.$$
(In fact, the third relation holds for all smooth maps.)

LEMMA 5.5.5. *Let $F : X \to Y$ be a holomorphic map and let $\omega$ be a meromorphic 1-form on $Y$. For each $p \in X$,*
$$\mathrm{Ord}_p(F^*\omega) = (1 + \mathrm{Ord}_{F(p)}(\omega))\mathrm{Mult}_p(F) - 1.$$

**Proof.** Choose local coordinates $z$ near $p$ and $w$ near $F(p)$ so that $w = F(z) = z^n$. If the coordinate patches are simply connected, then locally $\omega = w^k e^{f(w)} dw$, and we have
$$F^*\omega = nz^{nk+n-1} e^{f(z^n)} dz.$$
Thus the desired formula holds. □

## 5.6. Integration of forms

### 5.6.1. Integration of 1-forms.

DEFINITION 5.6.1. Let $X$ be a Riemann surface. A *path* on $X$ is a continuous, piecewise smooth function $\gamma : [a, b] \to X$. The *endpoints* of $\gamma$ are $\gamma(a)$ and $\gamma(b)$. We say that $\gamma$ *connects* $\gamma(a)$ and $\gamma(b)$. We say that $\gamma$ is *closed* if $\gamma(a) = \gamma(b)$. We say that $\gamma$ is *simple* if $\gamma$ is injective on $[a, b]$, except possibly if $\gamma(a) = \gamma(b)$. ◇

We shall often abuse notation and refer to the image $\gamma([a, b])$ as a path, or to either $\gamma$ or its image as a *curve* or an *arc*. The latter is mostly used when $\gamma$ is not closed.

An injective piecewise smooth continuous map $\alpha : [a, b] \to [c, d]$ is called a *reparameterization*. We shall also say that $\gamma \circ \alpha^{-1} : [c, d] \to X$ is a reparameterization of $\gamma$. Note that a reparameterization can reverse the orientation of the curve in the surface. The orientation remains the same if $\alpha$ is an increasing function and reverses if $\alpha$ is decreasing.

Let $\gamma : [a, b] \to X$ be a path. Using the compactness of $[a, b]$, it is easily seen that there exist $a = a_0 < a_1 < ... < a_{N-1} < a_N = b$ such that, with
$$\gamma_j = \gamma|_{[a_{j-1}, a_j]} : [a_{j-1}, a_j] \to X, \quad 1 \leq j \leq N,$$

(i) the final point of one path $\gamma_j$ is the initial point of the next path $\gamma_{j+1}$:
$$\gamma_j(a_j) = \gamma_{j+1}(a_j), \quad 1 \leq j \leq N, \quad \text{and}$$

(ii) each path $\gamma_j$ has image contained in a coordinate chart.

We shall refer to this partition of $\gamma$ as a *charted partition*. We are now ready to define integration of a 1-form.

## 5.6. Integration of forms

DEFINITION 5.6.2. Let $X$ be a Riemann surface, $\omega$ a 1-form on $X$, and $\gamma : [a, b] \to X$ a path with charted partition $\gamma_1, ..., \gamma_N$ and coordinates $z_j : U_j \to \mathbb{C}$ so that $\gamma_j([a_{j-1}, a_j]) \subset U_j$. Write $\omega = f_j dz_j + g_j d\bar{z}_j$ on $U_j$. We define
$$\int_\gamma \omega := \sum_{j=1}^N \int_{a_{j-1}}^{a_j} \left\{ f_j(z_j \circ \gamma(t)) \frac{d(z_j \circ \gamma)}{dt} + g_j(z_j \circ \gamma(t)) \overline{\left(\frac{d(z_j \circ \gamma)}{dt}\right)} \right\} dt.$$
The number $\int_\gamma \omega$ is called the integral of $\omega$ over $\gamma$.   ◇

We leave it to the reader to show that this definition is independent of the coordinate charts chosen. Moreover, the following properties are easily checked.

(1) If $\gamma$ is a union of paths $\gamma_1, ..., \gamma_N$, then
$$\int_\gamma \omega = \sum_{j=1}^N \int_{\gamma_j} \omega.$$

(2) The integral is $\mathbb{C}$-linear in $\omega$:
$$\forall c_1, c_2 \in \mathbb{C}, \quad \int_\gamma c_1\omega_1 + c_2\omega_2 = c_1 \int_\gamma \omega_1 + c_2 \int_\gamma \omega_2.$$

(3) If $F : X \to Y$ is a holomorphic map of Riemann surfaces, then
$$\int_{F \circ \gamma} \omega = \int_\gamma F^*\omega.$$

(4) If $f$ is a smooth function defined in a neighborhood of $\gamma([a, b])$ then
$$\int_\gamma df = f(\gamma(b)) - f(\gamma(a)).$$

(5) If $\alpha$ is a reparameterization that preserves the orientation of $\gamma$, then
$$\int_{\gamma \circ \alpha} \omega = \int_\gamma \omega.$$

(6) If $\gamma$ is a path, let $-\gamma$ denote the same path traversed backwards, i.e., with the opposite orientation. Then
$$\int_{-\gamma} \omega = -\int_\gamma \omega.$$

**5.6.2. Integration of $(1,1)$-forms.** Let $X$ be a Riemann surface, $T \subset X$ a triangle and $z : U \to \mathbb{C}$ a chart with $T \subset U$. Given a (1,1)-form $\eta$ on a neighborhood of $U$ in $X$, the restriction of $\eta$ to $U$ is $\eta = f \frac{\sqrt{-1}}{2} dz \wedge d\bar{z}$ for some function $f$ on $U$. We then define
$$\int_T \eta := \int_{z(T)} f \circ z^{-1} dx dy,$$
where the integral on the right is the usual Riemann integral in the plane.

The change of variables formula and the fact that the real Jacobian determinant of the holomorphic transformation $w = \varphi(z)$ is $|\varphi'(z)|^2$ show that the definition of $\int_T \eta$ is independent of the local chart $z$.

Now let $D \subset X$ be an open subset with smooth boundary and compact closure. Cover $D$ with a finite number of closed triangles $T_1, ..., T_N$ whose interiors are disjoint (and whose boundaries need not align in the same way as in a 'triangulation' in the sense of Chapter 3) such that, for some complex charts $z_i : U_i \to \mathbb{C}$, $T_i \subset U_i$ for all $i = 1, ..., N$.

DEFINITION 5.6.3. The quantity
$$\int_D \eta := \sum_{j=1}^{N} \int_{T_j} \eta$$
is called the integral of the $(1, 1)$-form $\eta$ over $D$. ◇

It is easy to verify the following facts.

(1) If $D = D_1 \cup D_2$ with $D_1 \cap D_2 = \emptyset$, then
$$\int_D \eta = \int_{D_1} \eta + \int_{D_2} \eta.$$

(2) For all $c^1, c^2 \in \mathbb{C}$ and all (1,1)-forms $\eta_1, \eta_2$,
$$\int_D c^1 \eta_1 + c^2 \eta_2 = c^1 \int_D \eta_1 + c^2 \int_D \eta_2.$$

(3) If $F : X \to Y$ is an isomorphism of Riemann surfaces, $D \subset X$ is finitely triangulated, and $\eta$ is a 1-form on $F(D)$, then
$$\int_{F(D)} \eta = \int_D F^* \eta.$$

Observe that properties (1), (2), and (3) of integration of $(1, 1)$-forms are the analogs of properties (1), (2), and (3) of integration of 1-forms. Property (5) of integration of 1-forms corresponds to a "change of variables formula", and is accounted for by property (3) of integration of $(1, 1)$-forms.

The analogue of property (4), i.e., the Fundamental Theorem of Calculus, is the well-known *Stokes' Theorem*. In the plane, it is just Green's Theorem, and thus on a surface it can be derived from Green's Theorem by a decomposition of the domain into a union of plane domains. We state Stokes' Theorem in the invariant language of exterior calculus.

THEOREM 5.6.4 (**Stokes' Theorem**). *Let $D$ be a finitely triangulated open subset of a Riemann surface $X$ with piecewise-smooth and co-oriented boundary $\partial D$, and let $\omega$ be a smooth 1-form on a neighborhood of the closure of $D$. Then*
$$\int_{\partial D} \omega = \int_D d\omega.$$

## 5.7. Residues

The notion of the residue of a meromorphic function is not well-defined on a Riemann surface. Indeed, to compute the residue, one needs to integrate along a path, and on a general Riemann surface integration along a path is an operation defined on 1-forms. Because plane domains have a global frame for $K_X$, namely $dz$, the problem in defining the residue of a function disappears in a plane domain.

We have already presented the following definition, which we now restate.

DEFINITION 5.7.1. Let $X$ be a Riemann surface, $p$ a point of $X$, and $\omega$ a meromorphic 1-form on $X$. Let $U$ be a relatively compact open subset of $X$ containing $p$ and having smooth boundary $\partial U$ such that there is no pole of $\omega$ in $\overline{U} - \{p\}$. We define
$$\text{Res}_p(\omega) := \frac{1}{2\pi\sqrt{-1}} \int_{\partial U} \omega,$$
where $\partial U$ is co-oriented: if one walks on the surface forward along $\partial U$, one finds $U$ on the left. ◇

The definition of residue is independent of the domain $U$ chosen as above. To see this independence, we argue as follows. First, let $V \subset\subset U$ be a domain containing $p$. By Stokes' Theorem,
$$\int_{\partial U} \omega - \int_{\partial V} \omega = \int_{U-V} d\omega = 0,$$
where the last equality holds because holomorphic 1-forms on Riemann surfaces are closed. Now if $U_1$ and $U_2$ are domains, we can find $V \subset\subset U_1 \cap U_2$, and thus we conclude that
$$\int_{\partial U_1} \omega = \int_{\partial U_2} \omega.$$
For later purposes it will be useful to have the following proposition.

PROPOSITION 5.7.2. *Let $\omega$ be a meromorphic 1-form on a compact Riemann surface $X$. Then*
$$\sum_{p \in X} \text{Res}_p(\omega) = 0.$$

**Proof.** Let $x_1, ..., x_N$ be the poles of $\omega$, and take disjoint, relatively compact open disks $D_1, ..., D_N$ with $x_i \in D_i$, $1 \leq i \leq N$. Let $U = X - (\overline{D_1 \cup ... \cup D_N})$. Then $\omega$ is holomorphic on $U$, so that
$$0 = -\int_U d\omega = -\int_{\partial U} \omega = \sum_{i=1}^N \int_{\partial D_i} \omega = 2\pi\sqrt{-1} \sum_{p \in X} \text{Res}_p(\omega).$$

The proof is finished. □

We also have the following version of the Argument Principle.

PROPOSITION 5.7.3 (Argument Principle). *For any $f \in \mathscr{M}(X)$,*
$$\mathrm{Res}(df/f) = \mathrm{Ord}(f).$$

The proof is left as an exercise.

## 5.8. Homotopy and homology

We present, in a heuristic manner, some basic ideas in algebraic topology. For the rest of this section we assume the Riemann surface $X$ is connected.

**5.8.1. Homotopy of curves.** Let $\gamma_0, \gamma_1 : [a, b] \to X$ be two paths such that $\gamma_0(a) = \gamma_1(a)$ and $\gamma_0(b) = \gamma_1(b)$. Recall that a homotopy of the paths $\gamma_0$ and $\gamma_1$ is a continuous map $\Gamma : [a, b] \times [0, 1] \to X$ such that for all $t \in [a, b]$
$$\Gamma(t, 0) = \gamma_0(t), \quad \Gamma(t, 1) = \gamma_1(t)$$
and for all $s \in [0, 1]$
$$\Gamma(a, s) = \gamma_0(a) = \gamma_1(a) \quad \text{and} \quad \Gamma(b, s) = \gamma_0(b) = \gamma_1(b).$$
In this case, we say that $\gamma_0$ and $\gamma_1$ are homotopic. It is not hard to construct a homotopy of two paths whose image is the same but whose oriented parameterizations are different. Of course, there are many homotopic curves whose images are different.

**5.8.2. Fundamental group.** Fix a point $p \in X$ and consider the set of all parameterized closed loops in $X$ starting and ending at $p$. Two such loops are said to be equivalent if they are homotopic. The set of equivalence classes is denoted $\pi_1(X, p)$.

The set $\pi_1(X, p)$ carries the structure of a group, defined as follows.
(1) The product $[\gamma_1][\gamma_2]$ of two loops $\gamma_1$ and $\gamma_2$ is the homotopy class of the loop obtained by first following $\gamma_1$ and then following $\gamma_2$.
(2) The inverse of $[\gamma]$ is the homotopy class of the loop obtained by reversing the loop $\gamma$.

It is not hard to show that the product and inverse operations are well-defined.

DEFINITION 5.8.1. The set $\pi_1(X, p)$ together with the multiplication and inverse operations outlined above is called the fundamental group of $(X, p)$. ⋄

REMARK. Suppose $X$ is connected. Let $p, q \in X$ and fix a path $\sigma$ whose initial point is $p$ and whose final point is $q$. Then we can define a map from $F_\sigma : \pi_1(X, q) \to \pi_1(X, p)$ as follows. If $\alpha$ is a loop starting at $q$, then we have a loop starting at $p$, obtained by following $\sigma$, then $\alpha$, then $-\sigma$. The homotopy class of this loop is denoted $F_\sigma[\alpha]$. It is easy to see that $F_{-\sigma} = F_\sigma^{-1}$.

From now on, when $X$ is connected, we will denote $\pi_1(X, p)$ by $\pi_1(X)$. ⋄

## 5.8. Homotopy and homology

**5.8.3. Homology.** We shall pretend that the image $D = \Gamma([a, b] \times [0, 1])$ of our homotopy $\Gamma$ is regular enough to integrate over. (In fact, the image of $D$ need not be so regular; it can be approximated by a domain that is sufficiently regular.) A clever application of Stokes' Theorem gives us the following proposition.

PROPOSITION 5.8.2. *Let $\omega$ be a 1-form such that $d\omega = 0$. If $\gamma_0$ and $\gamma_1$ are homotopic paths, then*

$$\int_{\gamma_0} \omega = \int_{\gamma_1} \omega.$$

REMARK. In particular, the hypothesis of the proposition holds for a holomorphic 1-form. ◇

It follows that a 1-form determines a map

$$[\gamma] \in \pi_1(X) \to \int_\gamma \omega \in \mathbb{C}.$$

It is clear from the properties of integrals that this map is a homomorphism, and thus, since $\mathbb{C}$ is an Abelian group, the kernel of this map must contain the commutator subgroup of $\pi_1(X)$. It can be shown that the intersection of the kernels of the maps obtained by using all closed 1-forms $\omega$ is exactly the commutator subgroup of $\pi_1(X)$.

DEFINITION 5.8.3. The quotient group

$$H_1(X) := \pi_1(X)/[\pi_1(X), \pi_1(X)]$$

is called the *first de Rham homology group* of the surface $X$. (This group can also be realized as a quotient of the Abelian group of closed 1-chains modulo boundaries of 2-chains.) ◇

Let us denote an equivalence class of a closed loop $\gamma$ in $H_1(X)$ by $\{\gamma\}$. Let $X$ be a compact Riemann surface, and let $Z^1(X)$ denoted the set of all smooth, closed 1-forms on $X$. We say that $\omega_1$ and $\omega_2$ are *cohomologous*, and write $\omega_1 \sim \omega_2$, if there is a smooth function $f : X \to \mathbb{C}$ such that

$$\omega_1 - \omega_2 = df.$$

Since $d^2 = 0$, Stokes' Theorem shows us that for any closed loop $\gamma$,

$$\int_\gamma \omega_1 = \int_\gamma \omega_2.$$

(In fact a little more work is needed; we have to perturb $\gamma$ a little to make it a union of pairwise disjoint, simple closed curves.) The quotient space $H^1_{\text{dR}}(X) = Z^1(X)/\sim$ is an Abelian group, called the *first de Rham cohomology group*.

LEMMA 5.8.4. *The bilinear pairing* $\langle\,,\,\rangle : H_1(X) \times H^1_{\mathrm{dR}}(X) \to \mathbb{C}$ *given by*

$$\langle\{\gamma\}, [\omega]\rangle := \int_\gamma \omega$$

*is non-degenerate, and thus* $H^1_{\mathrm{dR}}(X) \cong (H_1(X))^*$.

**Proof.** Fix a smooth 1-form $\omega$ on $X$ and suppose $\int_\gamma \omega = 0$ for all closed loops $\gamma$ in $X$. Then the function

$$F(x) := \int_{x_o}^{x} \omega$$

is well-defined, and $dF = \omega$. It follows that $[\omega] = 0$ in $H^1_{\mathrm{dR}}(X)$, and thus the pairing is non-degenerate. □

## 5.9. Poincaré and Dolbeault Lemmas

THEOREM 5.9.1 (Poincaré Lemma). *Let $X$ be a simply connected Riemann surface and let $\omega$ be a 1-form such that $d\omega = 0$. Then there exists a function $f$ such that $\omega = df$.*

**Proof.** Fix $p \in X$. If $x \in X$, we define

$$f(x) := \int_p^x \omega,$$

where the integral is over any simple arc connecting $p$ to $x$. This integral is well-defined because $d\omega = 0$ and $X$ is simply connected, and again $df = \omega$. □

REMARK. Observe that if $\alpha$ is a (0,1)-form, then $\bar\partial \alpha = 0$. ◇

THEOREM 5.9.2 (Dolbeault Lemma). *Let $\alpha = g dz$ be a $(0,1)$-form on a domain $X \subset \mathbb{C}$. Then for any smoothly bounded open set $U \subset\subset X$ there is a function $f : U \to \mathbb{C}$ such that $\alpha = \bar\partial f$.*

**Proof.** Recall that for a domain $V \subset\subset \mathbb{C}$ with piecewise-smooth boundary and $h \in \mathscr{C}^1(V)$,

$$h(z) = \frac{1}{2\pi\sqrt{-1}} \int_{\partial V} \frac{h(\zeta)}{\zeta - z} d\zeta + \frac{1}{\pi} \iint_V \frac{\partial h}{\partial \bar\zeta} \frac{dA(\zeta)}{\zeta - z}.$$

Let $U' \subset\subset X$ be a smoothly bounded open set such that $U \subset\subset U'$ and choose $\chi \in \mathscr{C}_0^\infty(U')$ such that $\chi|_U \equiv 1$. Defining

$$f_1(z) := \frac{1}{\pi} \iint_{U'} \frac{\chi(\zeta) g(\zeta) dA(\zeta)}{\zeta - z},$$

we have
$$\begin{aligned}
\frac{\partial f_1}{\partial \bar{z}} &= \frac{1}{\pi} \iint_{U'} \chi(\zeta) g(\zeta) \frac{\partial}{\partial \bar{z}} \left( \frac{1}{\zeta - z} \right) dA(\zeta) \\
&= -\frac{1}{\pi} \iint_{U'} \chi(\zeta) g(\zeta) \frac{\partial}{\partial \bar{\zeta}} \left( \frac{1}{\zeta - z} \right) dA(\zeta) \\
&= \frac{1}{\pi} \iint_{U'} \frac{\partial (\chi(\zeta) g(\zeta))}{\partial \bar{\zeta}} \frac{dA(\zeta)}{\zeta - z} \\
&= \frac{1}{2\pi i} \int_{\partial U'} \frac{\chi(\zeta) g(\zeta)}{\zeta - z} d\zeta + \frac{1}{\pi} \iint_{U'} \frac{\partial (\chi(\zeta) g(\zeta))}{\partial \bar{\zeta}} \frac{dA(\zeta)}{\zeta - z} \\
&= \chi(z) g(z).
\end{aligned}$$

Taking $f = f_1|_U$ completes the proof. □

REMARK. In fact it is possible to solve the equation $\bar{\partial} f = \alpha$ for a function $f$ defined on all of $X$. We will show this fact in Chapter 7.   ◇

## 5.10. Dolbeault cohomology

We say that two smooth $(0,1)$-forms $\omega$ and $\omega'$ are equivalent, writing $\omega \sim \omega'$, if there is a smooth function $f$ such that
$$\omega - \omega' = \bar{\partial} f.$$
The (first) Dolbeault cohomology of a compact Riemann surface is the group
$$H^{0,1}_{\bar{\partial}}(X) := \{\text{smooth } (0,1)\text{-forms}\} / \sim .$$
The Dolbeault cohomology group therefore quantifies the obstructions to solving the equation $\bar{\partial} u = \alpha$ for a given $(0,1)$-form $\alpha$: there is a solution $u$ if and only if the Dolbeault cohomology class of $\alpha$ is 0.

As we will see, if $X$ is an open Riemann surface then $H^{0,1}_{\bar{\partial}}(X) = 0$. By contrast, on a compact Riemann surface $X$ the group $H^{0,1}_{\bar{\partial}}(X)$ gives us very interesting information, both topological and complex analytic, about $X$.

THEOREM 5.10.1 (Dolbeault-Serre Isomorphism). *Let $X$ be a compact Riemann surface. Then*
$$H^{0,1}_{\bar{\partial}}(X) \cong \overline{\Gamma_{\mathcal{O}}(K_X)}.$$
*In particular, $H^{0,1}_{\bar{\partial}}(X)$ is finite-dimensional, and the dimension of $H^{0,1}_{\bar{\partial}}(X)$ is $g_{\mathcal{O}}(X)$, the arithmetic genus of $X$.*

Once the isomorphism in Theorem 5.10.1 is established, finite-dimensionality is a consequence of Theorem 4.8.1. Theorem 5.10.1 is proved by considering a partial differential operator on $(0,1)$-forms whose kernel contains exactly one member of any class in $H^{1,0}_{\bar{\partial}}(X)$. This partial differential operator is a natural generalization of the Laplacian to the setting of $(0,1)$-forms on Riemann surfaces with metrics. Theorem 5.10.1 is proved at the end of Chapter 9.

## 5.11. Exercises

5.1. Consider the 2-dimensional real vector space $T^*_{X,p}$, the real cotangent space of a Riemann surface $X$ at $p \in X$. Let $z = x + \sqrt{-1}y$ be a holomorphic local coordinate vanishing at $p$. Consider the real 1-forms $dx(p)$ and $dy(p)$ as elements of $T^*_{X,p}$, and define the linear transformation $J : T^*_{X,p} \to T^*_{X,p}$ by $Jdx = dy$ and $Jdy = -dx$. Compute the eigenvalues of $J$ (which will be complex) and the $J$-eigenspace decomposition of $T^*_{X,p} \otimes_\mathbb{R} \mathbb{C}$.

5.2. Is the line bundle $\Lambda^{1,1}_X$ trivial?

5.3. Show that $F^*$ maps $(1,0)$-forms to $(1,0)$-forms if and only if $F$ is a holomorphic map.

5.4. Show that Definition 5.6.2 of $\int_\gamma \omega$ is independent of the coordinates chosen.

5.5. Verify the properties following Definition 5.6.2.

5.6. Show that the integral $\int_D \eta$ of a $(1,1)$-form over a domain $D$ in a Riemann surface is well-defined.

5.7. Prove Proposition 5.7.3.

5.8. Construct a homotopy of two paths with the same image but different oriented parameterizations.

5.9. Prove Proposition 5.8.2.

5.10. Let $X$ be simply connected and let $\omega$ be a closed 1-form. Show that the function $f$ given by
$$f(x) := \int_p^x \omega,$$
where the integral is over any simple arc connecting $p$ to $x$, is well-defined and satisfies $\omega = df$.

5.11. In the proof of Theorem 5.9.2, why can we assume that $U$ is a simply connected subset of the plane?

5.12. Show that the wedge product of two 1-forms is a globally-defined $(1,1)$-form.

5.13. Show that the pullback $F^*$ by a map $F : X \to Y$ maps global forms on $Y$ to global forms on $X$ and commutes with the exterior derivative $d$. Show further that $F^*$ preserves $(1,0)$-form and $(0,1)$-forms if and only if $F^*$ commutes with $\partial$ and $\bar\partial$ if and only if $F$ is holomorphic.

5.14. Check that Definition 5.6.2 is well-posed, and verify the 6 properties following this definition.

# Chapter 6

# Calculus on Line Bundles

## 6.1. Connections on line bundles

We begin with the definition of $H$-valued $(1,0)$-forms, $(0,1)$-forms, 1-forms, and $(1,1)$-forms.

DEFINITION 6.1.1. Let $H \to X$ be a complex line bundle on a Riemann surface $X$.

(1) An $H$-valued $(1,0)$-form is an element of $\Gamma(X, T_X^{*1,0} \otimes H)$.
(2) An $H$-valued $(0,1)$-form is an element of $\Gamma(X, T_X^{*0,1} \otimes H)$.
(3) An $H$-valued 1-form is an element of the direct sum
$$\Gamma(X, T_X^* \otimes H) := \Gamma(X, T_X^{*1,0} \otimes H) \oplus \Gamma(X, T_X^{*0,1} \otimes H).$$
(4) An $H$-valued $(1,1)$-form is an element of $\Gamma(X, \Lambda_X^{1,1} \otimes H)$.

When the complex line bundle $H$ is not explicitly mentioned, we may refer to *twisted* forms. ◇

Next we pass to the general definition of a connection.

DEFINITION 6.1.2. Let $X$ be a Riemann surface and let $H \to X$ be a complex line bundle. A connection on $X$ is a linear map $\nabla : \Gamma(X, H) \to \Gamma(X, T_X^* \otimes H)$ such that the Leibniz Rule
$$\nabla(fs) = df \otimes s + f \nabla s \quad \text{for all } s \in \Gamma(X, H) \text{ and } f \in \mathscr{C}^1(X)$$
is satisfied. ◇

Let $\xi$ be a frame of $H$ over an open subset $U \subset X$, i.e., a section of $H$ over $U$ such that $\xi(x) \neq 0$ for all $x \in U$. Since $\nabla \xi$ is an $H$-valued 1-form,
$$\nabla \xi = \omega \otimes \xi$$

101

for some differential 1-form $\omega$ that depends on $\xi$.

DEFINITION 6.1.3. The differential 1-form $\omega$ is called the connection form of $\nabla$ in the frame $\xi$. ◇

Any section of $H$ over $U$ is of the form $s = f \cdot \xi$ for some differentiable function $f$. By the Leibniz Rule,
$$\nabla(f \cdot \xi) = df \otimes \xi + f\omega \otimes \xi = df \otimes \xi + \omega \otimes (f \cdot \xi).$$
Thus locally the action of the connection is determined by the connection form.

REMARK. In the literature one often finds the expression $\nabla = d + \omega$, or
$$\nabla s = ds + \omega s.$$
These expressions depend on the choice of frame, but often the frame is not explicitly mentioned. ◇

If we change the frame $\xi$ to another frame $\xi'$, then $\xi' = f\xi$ for some function $f$. Thus
$$\omega' \otimes \xi' = \nabla(\xi') = \nabla(f\xi) = df \otimes \xi + f\omega \otimes \xi = \left(\frac{df}{f} + \omega\right) \otimes \xi',$$
and therefore
$$\omega' = \omega + \frac{df}{f}.$$
In particular, we see that the 1-form $\omega$ is not globally defined.

REMARK (Curvature). For later use, observe that the exterior derivative $d\omega$ is a globally defined 2-form independent of the choice of frame. Indeed, if $\omega = \nabla\xi$, $\xi' = f\xi$, and $\omega' = \nabla\xi'$, then
$$d\omega' = d(\omega + f^{-1}df) = d\omega.$$
The form $d\omega$ is called the *curvature form* of the connection $\nabla$. ◇

REMARK. Interestingly, the difference of two connections for a given line bundle $H \to X$ is also globally defined, so that the space of connections on $X$ is an affine space modeled on the vector space $\Gamma(X, T_X^*)$ of 1-forms on $X$. In general, there is no canonical choice for a special point in this affine space. A geometry, however, often determines a special point in this affine space. This will be the situation for holomorphic line bundles, as we shall soon see. ◇

EXAMPLE 6.1.4. Let $X$ be a Riemann surface.

(1) The exterior derivative $d$ is a connection for the trivial bundle $\mathcal{O} \to X$.

(2) (Non-example) It is mistakenly asserted in a number of sources that the operator

$$\bar{\partial} : \mathscr{C}^\infty(X) = \Gamma(X, \mathcal{O}) \to \Gamma(X, T_X^{*0,1}) \subset \Gamma(X, T_X^*),$$

sending a function $f$ to the $(0,1)$-form $\bar{\partial}f = \frac{\partial f}{\partial \bar{z}} d\bar{z}$, is a connection for the trivial bundle. In fact this is not the case. Indeed,

$$\bar{\partial}(fg) \neq df \otimes g + f\bar{\partial}g,$$

so that the Leibniz Rule is not satisfied. ◇

Line bundles with connections behave rather naturally with respect to operations of multilinear algebra in the sense that if we form a new line bundle from old line bundles by such an operation, and if the old line bundles have connections, we obtain a connection on the new line bundle. We will show how this is done in the three examples of tensor products, complex conjugate bundles, and dual bundles.

Let $H_1$ and $H_2$ be two complex line bundles with connections $\nabla_1$ and $\nabla_2$, respectively. Then

$$\nabla_1 \otimes \nabla_2(\xi_1 \otimes \xi_2) = (\nabla_1 \xi_1) \otimes \xi_2 + \xi_1 \otimes (\nabla_2 \xi_2)$$

defines connections for $H_1 \otimes H_2$ when extended by linearity. The reader can check that the Leibniz Rule holds here, even with the ambiguity $f\xi_1 \otimes \xi_2 = (f\xi_1) \otimes \xi_2 = \xi_1 \otimes (f\xi_2)$.

To a complex line bundle $H \to X$, one can associate its complex conjugate bundle $\overline{H} \to X$. In terms of transition functions, if $g_{\alpha\beta}$ are the transition functions of $H$, then $\overline{g_{\alpha\beta}}$ are the transition functions for $\overline{H}$. Moreover, if $s$ is a section of $H$, then there is an associated section of $\overline{H}$ which we denote $\bar{s}$. The assignment $s \mapsto \bar{s}$ is conjugate-linear. The connection $\nabla^\dagger$ for the complex conjugate bundle $\overline{H}$ is simply defined by

$$\nabla^\dagger \bar{\xi} := \overline{\nabla \xi}.$$

The Leibniz Rule is easy to confirm here, since the exterior derivative $d$ is a real operator.

The connection $\nabla^*$ for the dual $H^*$ of a complex line bundle $H$ is a little more tricky to define. Let $\xi$ be a frame and let $\xi^*$ be its dual frame, so that $\xi^* \otimes \xi$ is the identity map for the (trivial) bundle $\text{End}(H) \cong H^* \otimes H$. (Note that under the identification of $\text{End}(H)$ with the trivial bundle, the identity map is sent to multiplication by 1.) Then

$$\nabla^*(\xi^*) = -\xi^* \otimes \nabla(\xi) \otimes \xi^*$$

defines a connection for $H^*$. (Thinking more formally, since $\xi^* = 1/\xi$, $\nabla^*(\xi^*) = -(1/\xi) \otimes (\nabla \xi) \otimes (1/\xi)$.)

## 6.2. Hermitian metrics and connections

DEFINITION 6.2.1. A Hermitian metric for a line bundle $H \to X$ is a smooth section $h$ of the line bundle $H^* \otimes \overline{H}^* \to X$ such that the function $h : H \otimes \overline{H} \to \mathbb{C}$ defined (with abusive notation) by
$$h(v, \bar{w}) := \langle h, v \otimes \bar{w} \rangle, \quad v, w \in H_x,$$
satisfies $h(v, \bar{w}) = \overline{h(w, \bar{v})}$, $h(v, \bar{v}) \geq 0$, and $h(v, \bar{v}) = 0 \iff v = 0$. ◇

EXAMPLE 6.2.2. Let $H \to X$ be a complex line bundle and $s_1, ..., s_k \in \Gamma(X, H)$ be global sections with no common zeros. If we choose a frame $\xi$, then there are functions $f_1, ..., f_k$ such that $s_i = f_i \xi$. We define
$$h(v, \bar{w}) := \frac{a\bar{b}}{|f_1|^2 + ... + |f_k|^2}, \quad \text{where } v = a\xi \text{ and } w = b\xi.$$
It is clear that if we change frame, the functions $f_i$ and the numbers $a, b$ change in such a way that the value of $h(v, \bar{w})$ remains unchanged. A standard abuse of notation is to write
$$h(v, \bar{w}) = \frac{v\bar{w}}{|s_1|^2 + ... + |s_k|^2}.$$
For example, if $X = \mathbb{P}_1$ and $H = \mathbb{H} \to \mathbb{P}_1$ is the hyperplane line bundle, we can take the sections $z_0, z_1$, where $[z_0, z_1]$ are homogeneous coordinates. (See Section 4.6.) The metric
$$h_{FS}(v, \bar{w}) := \frac{v\bar{w}}{|z_0|^2 + |z_1|^2}$$
is called the Fubini-Study metric. ◇

REMARK. If we take a frame $\xi$ for $H$ over $U$, then the function $h(\xi, \bar{\xi})$ is nowhere vanishing on $U$. We can then define
$$\varphi^{(\xi)} := -\log h(\xi, \bar{\xi}).$$
Thus for any section $s = f\xi$, we have $h(s, \bar{s}) = |f|^2 e^{-\varphi^{(\xi)}}$. A standard abuse of notation is to omit reference to the frame and write
$$h(s, \bar{s}) = |s|^2 e^{-\varphi}, \quad \text{or simply} \quad h = e^{-\varphi}.$$
We shall often employ this abusive notation. ◇

DEFINITION 6.2.3. We say that a line bundle is Hermitian if it has a Hermitian metric. ◇

REMARK. A standard partition of unity argument shows that any complex line bundle has a Hermitian metric. Indeed, if we take a locally finite open cover $\{U_\alpha\}$ of coordinate charts on each of which $H$ is trivial, then we have Hermitian metrics $h_\alpha(a\xi_\alpha, \bar{b}\bar{\xi}_\alpha) = a\bar{b}$. Taking a partition of unity $\{\chi_\alpha\}$ subordinate to this cover, the object
$$h := \sum_\alpha \chi_\alpha h_\alpha$$

is easily seen to be a Hermitian metric. ◇

Since $h(s,\bar{t}) = h \otimes s \otimes \bar{t}$ when we think of a metric $h$ as a section of $H^* \otimes \overline{H^*}$, we compute that
$$d(h(s,\bar{t})) = h(\nabla s, \bar{t}) + h(s, \overline{\nabla t}) + \nabla h(s,t).$$

The $\nabla$ in the third term on the right-hand side is the connection for the line bundle $H^* \otimes \overline{H^*}$ associated to the connection for $H$.

DEFINITION 6.2.4. A connection $\nabla$ for a complex line bundle $H$ is said to be *compatible with a Hermitian metric $h$ for $H$*, or simply a *metric connection for $(H, h)$*, if for any sections $s, t$ and any tangent vector $v$,
$$d(h(s,t))v = h((\nabla s)(v), \bar{t}) + h(s, \overline{(\nabla s)v}).$$

Equivalently, the connection associated to $H^* \otimes \overline{H^*}$ annihilates $h$, i.e., $\nabla h = 0$. ◇

## 6.3. $(1,0)$-connections on holomorphic line bundles

Since $\Gamma(X, T_X^* \otimes H) := \Gamma(X, T_X^{*1,0} \otimes H) \oplus \Gamma(X, T_X^{*0,1} \otimes H)$ by definition, we can decompose a connection into its $(1,0)$- and $(0,1)$-parts:
$$\nabla = \nabla^{1,0} + \nabla^{0,1}.$$

For a general complex line bundle, this splitting is not particularly helpful. However, when the underlying line bundle is holomorphic, the splitting plays a crucial role. The main difference in the setting of holomorphic vector bundles is the ability to define the $\bar{\partial}$-operator for sections of holomorphic line bundles.

DEFINITION 6.3.1. Let $H \to X$ be a holomorphic line bundle. We define $\bar{\partial}$ : $\Gamma(X, H) \to \Gamma(X, T_X^{*0,1} \otimes H)$ as follows. Choose a holomorphic local section $\xi$. Then
$$\bar{\partial}(f \cdot \xi) := \bar{\partial} f \otimes \xi.$$

Note that if $\xi'$ is another holomorphic section, then there is a holomorphic nowhere zero function $g$ such that $\xi' = g\xi$, and thus if $s = f'\xi' = f\xi$, we have $f = f'g$. We obtain
$$\bar{\partial}(f'\xi') := (\bar{\partial}f')\xi' = (\bar{\partial}f')g\xi = (\bar{\partial}(f'g))\xi = (\bar{\partial}f)\xi =: \bar{\partial}(f\xi),$$
which shows that $\bar{\partial}$ is well-defined.

DEFINITION 6.3.2. A connection $\nabla$ for a holomorphic vector bundle $H$ is said to be a $(1,0)$-*connection* if, in some (and hence any) holomorphic frame, the connection form $\omega$ is a $(1,0)$-form. Equivalently, the $(0,1)$-part of $\nabla$ is $\bar{\partial}$. ◇

## 6.4. The Chern connection

The following theorem is sometimes called the Fundamental Theorem of Hermitian Holomorphic Geometry.

THEOREM 6.4.1. *Let $H \to X$ be a holomorphic line bundle with Hermitian metric $h$. Then there is a unique $(1,0)$-connection for $H$ that is compatible with $h$.*

**Proof.** We have

$$\begin{aligned} d(h(s,\bar{t})) = d(s\bar{t}e^{-\varphi}) &= ds\bar{t}e^{-\varphi} + s\overline{dt}e^{-\varphi} + s\bar{t}d(e^{-\varphi}) \\ &= (\partial s - (\partial\varphi)s + \bar{\partial}s)\bar{t}e^{-\varphi} + s\overline{(\partial t - (\partial\varphi)t + \bar{\partial}t)}e^{-\varphi}, \end{aligned}$$

and thus the formula

$$\nabla(f\xi) = (df + f(-\partial\varphi)) \otimes \xi$$

shows the existence of the desired connection. Now suppose we have two such connections, say $\nabla_1$ and $\nabla_2$. Then we have an $H$-valued (1,0)-form $\theta$ such that $\nabla_1 s - \nabla_2 s = \theta \otimes s$. Then

$$\begin{aligned} 0 &= d(h(s,t)) - d(h(s,t)) \\ &= h(\nabla_1 s, \bar{t}) - h(\nabla_2 s, \bar{t}) - (h(\nabla_1 s, \bar{t}) - h(s, \overline{\nabla_2 t})) \\ &= h(\theta s, t) - h(s, \overline{\theta t}). \end{aligned}$$

Defining $\theta^\dagger$ by $h(\theta^\dagger v, \bar{w}) = h(v, \overline{\theta w})$, we have shown that $\theta^\dagger = \theta$. But since $\theta^\dagger$ is a $(0,1)$-form, $\theta = 0$. □

DEFINITION 6.4.2. *On a holomorphic Hermitian line bundle, the unique $(1,0)$-connection compatible with the Hermitian metric is called the **Chern connection**. If the metric is $e^{-\varphi}$ in some holomorphic frame, then the connection form in that frame is $-\partial\varphi$.* ◇

Using the fact that the $(0,1)$-part of the Chern connection is $\bar{\partial}$, note that

$$\partial(s,t) = (\nabla^{1,0}s, t) + (s, \bar{\partial}t).$$

The left side is

$$\partial(s\bar{t}e^{-\varphi}) = (\partial(e^{-\varphi}s))\bar{t} + s\overline{\bar{\partial}t}e^{-\varphi} = (e^{\varphi}\partial(e^{-\varphi}s))\bar{t}e^{-\varphi} + s\overline{\bar{\partial}t}e^{-\varphi}.$$

We thus see that, with respect to a local holomorphic frame $\xi$ such that $s = f\xi$ and $h(\xi, \bar{\xi}) = e^{-\varphi}$,

$$(6.1) \qquad \nabla^{1,0}s = e^{\varphi}\frac{\partial}{\partial z}\left(e^{-\varphi}f\right)dz.$$

## 6.5. Curvature of the Chern connection

Let $h$ be a Hermitian metric for a holomorphic line bundle $H$. For holomorphic frames $\xi$ and $\xi'$, let

$$\varphi^{(\xi)} := -\log h(\xi, \bar{\xi}) \quad \text{and} \quad \varphi^{(\xi')} := -\log h(\xi', \bar{\xi}')$$

be the local potentials of the metric in the two frames. There is a nowhere-zero holomorphic function $g$ such that $\xi' = g\xi$, and thus

$$\varphi^{(\xi')} = \varphi^{(\xi)} - \log |g|^2.$$

Since $\log |g|^2$ is harmonic, we have

$$\partial \bar{\partial} \varphi^{(\xi)} = \partial \bar{\partial} \varphi^{(\xi')}.$$

Thus $\Theta_h$, defined as $\partial \bar{\partial} \varphi^{(\xi)}$ on a neighborhood $U$ where $\xi$ is a frame, is a globally-defined $(1,1)$-form.

DEFINITION 6.5.1. The form $\Theta_h$ is called the Chern connection associated to $h$. ◇

REMARK. We have chosen an ad hoc definition for the curvature of the Chern connection, but to give this definition some additional meaning, we present the following discussion. The Chern connection, being a $(1,0)$-form, can be written as (see (6.1))

$$\nabla = \nabla^{1,0} + \bar{\partial}, \quad \text{where locally} \quad \nabla^{1,0} s = \partial s - (\partial \varphi) s.$$

If we think of $\nabla$ as a 'twisted' version of the exterior derivative, designed to map sections of the line bundle $H$ to $H$-valued 1-forms, we can consider extending this twisted exterior derivative to differential forms with values in $H$. Since we are on a Riemann surface, we only need to know how to apply this twisted exterior derivative to twisted 1-forms, i.e., elements of $\Gamma(X, T_X^{1,0*} \otimes H) \oplus \Gamma(X, T_X^{0,1*} \otimes H)$. The definition of the twisted exterior derivative on tensor products $\alpha \otimes s$, where $\alpha$ is a 1-form and $s$ a section, is

$$\nabla(\alpha \otimes s) := d\alpha \otimes s - \alpha \wedge \nabla s.$$

The minus sign in the second term is the usual one obtained by extending the Leibniz Rule to forms of higher degree. (It is common to use the same $\nabla$ to denote the action on sections of $H$ and on $H$-valued 1-forms, just as in the case of the usual exterior derivative.) For later use, we compute that, for a $(0,1)$-form $\beta$ with values in a line bundle $H$ having Hermitian metric $e^{-\varphi}$ and written locally as $\beta = f d\bar{z}$, $\bar{\partial} \beta = 0$ and thus

$$(6.2) \qquad \nabla^{1,0} \beta = \nabla \beta = -d\bar{z} \wedge \nabla^{1,0} f = e^{\varphi} \frac{\partial}{\partial z}(e^{-\varphi} f) dz \wedge d\bar{z}.$$

The similarity with the exterior derivative ends when we compute two consecutive derivatives; we find that $\nabla \nabla s \neq 0$. In fact, using the local formula $\nabla = d + \theta$,

we have that
$$\nabla\nabla s = \nabla(ds + \theta \otimes s) = d(\theta \otimes s) + \theta \wedge (ds + \theta \otimes s) = (d\theta) \otimes s.$$
Here we have used that $\theta \wedge \theta = 0$.

Observe that $d\theta$ is a 2-form with values in $\text{End}(H)$. But as $\text{End}(H) \cong H \otimes H^* \cong \mathcal{O}$, $d\theta$ is a globally defined 2-form, a fact that can be checked directly by changing frames: indeed, $\theta' = \theta + g^{-1}dg$ implies that $d\theta' = d\theta + d(g^{-1}dg) = d\theta$. Of course, when the connection form is $\theta = -\partial\varphi$, we have $d\theta = -\bar{\partial}\partial\varphi = \partial\bar{\partial}\varphi$.

The failure of the second covariant derivative to vanish means that the order of covariant partial derivatives matters, and therefore suggests that the sections see the space on which they are defined as somehow 'curved'. The curvature operator, which measures this failure of commutativity of mixed partials, is a $0^{\text{th}}$-order differential operator (also called a *multiplier*) with values in $\Lambda^{1,1}_X$. ◇

EXAMPLE 6.5.2. The trivial bundle with the constant metric has zero connection form and zero curvature. However, there are line bundles with non-trivial connections whose curvature is zero, both on open and compact Riemann surfaces, in the latter case as long as the genus is positive (and then the line bundles will be non-trivial). We shall return to this point later in the text. ◇

DEFINITION 6.5.3. We say that a metric $e^{-\varphi}$ has non-negative (resp. positive) curvature if the $(1,1)$-form
$$\frac{\sqrt{-1}}{2\pi}\Theta_h$$
is of the form $f\Omega$ for some non-negative (resp. positive) function $f$ and some area form $\Omega$ on $X$. ◇

REMARK. The factor $\frac{1}{2\pi}$ is not needed in the previous definition, but it does have a special significance that will be revealed shortly. ◇

REMARK. On a holomorphic line bundle, the Chern connection associated to a metric $h = e^{-\varphi}$ has non-negative curvature if and only if the locally defined functions $\varphi$ obtained by using holomorphic frames are subharmonic. ◇

EXAMPLE 6.5.4. If $H \to X$ has global holomorphic sections $s_0, ..., s_k$ whose common zero locus is empty then the metric defined by

(6.3) $$\varphi = \log(|s_0|^2 + ... + |s_k|^2)$$

as in Example 6.2.2 has non-negative curvature. For example, the Fubini-Study metric has non-negative curvature. In fact, the latter metric has positive curvature: suppose we take the local coordinate $\zeta = z_1/z_0$ on $U_0$. Then the Fubini-Study metric is given by the function
$$\varphi = \log(1 + |\zeta|^2),$$

## 6.6. Chern numbers

and so
$$\frac{\sqrt{-1}}{2\pi}\partial\bar{\partial}\varphi = \frac{\sqrt{-1}dz \wedge d\bar{z}}{2\pi(1+|\zeta|^2)^2}$$
is strictly positive on $U_0 = \{\zeta \in \mathbb{C}^n\}$. We note that
$$\frac{\sqrt{-1}}{2\pi}\int_{\mathbb{P}_1} \Theta_h = \int_0^\infty \frac{2rdr}{(1+r^2)^2} = 1.$$
In general, the metric (6.3) has positive curvature if and only if, with
$$V := \text{Span}\{s_0, ..., s_k\} \subset \Gamma_\mathcal{O}(X, H),$$
the induced map $\phi_{|V|} : X \to \mathbb{P}(V^*)$ is an immersion. Indeed, in local coordinates where $s_0 \neq 0$, writing $f_j = s_j/s_0$, $1 \leq j \leq k$, and $f = (f_1, ..., f_k)$, we find that
$$\begin{aligned}\sqrt{-1}\partial\bar{\partial}\log(\sum |s_i|^2) &= \sqrt{-1}\partial\bar{\partial}\log(1+||f||^2) \\ &= \frac{(1+||f||^2)df \wedge \overline{df} - (\bar{f} \cdot df) \wedge (f \cdot \overline{df})}{(1+||f||^2)^2} \\ &\geq \frac{df \wedge \overline{df}}{(1+||f||^2)^2},\end{aligned}$$
where $df \wedge \overline{df} := df_1 \wedge \overline{df_1} + ... + df_k \wedge \overline{df_k}$, and therefore $\sqrt{-1}\partial\bar{\partial}\log(\sum |s_i|^2)$ vanishes if and only if the map $df$ annihilates some tangent vector. ◇

### 6.6. Chern numbers

We begin with the following theorem.

THEOREM 6.6.1. *Let $X$ be a compact Riemann surface and let $h$ be a Hermitian metric for a holomorphic line bundle $H \to X$, and let $\Theta_h$ be the curvature of its Chern connection. Then the number*
$$c(H) := \frac{\sqrt{-1}}{2\pi}\int_X \Theta_h$$
*is independent of the metric $h$.*

**Proof.** If $h$ and $h'$ are two metrics for $H$, then $h/h'$ is a metric for the trivial bundle and thus is a smooth function with no zeros. Denote this function by $e^{-f}$. It follows that
$$\Theta_h - \Theta_{h'} = \sqrt{-1}\partial\bar{\partial}f = d(\sqrt{-1}\bar{\partial}f).$$
Thus by Stokes' Theorem, we see that $c(H)$ is independent of $h$. □

As we have previously stated and shall prove in due course, every holomorphic line bundle $H$ on a Riemann surface has a meromorphic section $s$ that is not identically zero. Using such a section, we can compute the number $c(H)$. To this end, consider the function
$$x \mapsto f(x) = h(s(x), \overline{s(x)})$$

on the set
$$X_s := \{x \in X \; ; \; \mathrm{Ord}_x(s) = 0\}$$
where $s$ has no zeros or poles. Let $X_{s,\varepsilon}$ be the subset of $X$ obtained by removing coordinate disks $|z_j| < \varepsilon$ about the points $x_j$ of $X - X_s$ from $X$. By Stokes' Theorem, we have

$$\begin{aligned}\frac{\sqrt{-1}}{2\pi}\int_{X_{s,\varepsilon}} \partial\bar{\partial}\log h(s,\bar{s}) &= \frac{1}{2\pi}\int_{X_{s,\varepsilon}} dd^c \log h(s,s) \\ &= -\sum_{j=1}^{k} \frac{1}{2\pi}\int_{|z_j|=\varepsilon} d^c(\log|z_j|^{2m_j} - \varphi),\end{aligned}$$

where $m_j = \mathrm{Ord}_{x_j}(s)$. (Recall that $d^c := \frac{\sqrt{-1}}{2}(\bar{\partial} - \partial)$.) Note that the orientation of the inner boundary of $X_{s,\varepsilon}$ is opposite to the orientation of the last integral, and hence the minus sign. A simple calculation shows that

$$\int_{|z|=\varepsilon} d^c \log|z|^2 = 2\pi,$$

and thus we have

(6.4) $$\begin{aligned}\frac{\sqrt{-1}}{2\pi}\int_{X_{s,\varepsilon}} \partial\bar{\partial}\log h(s,\bar{s}) &= -\deg(\mathrm{Ord}(s)) + \frac{1}{2\pi}\int_{|z_j|=\varepsilon} d^c\varphi \\ &= -\deg(\mathrm{Ord}(s)) + \frac{\sqrt{-1}}{2\pi}\int_{X-X_{s,\varepsilon}} \Theta_h,\end{aligned}$$

where the last equality is again by Stokes' Theorem. On the other hand, on $X_{s,\varepsilon}$ we have

$$\sqrt{-1}\partial\bar{\partial}\log h(s,\bar{s}) = \sqrt{-1}\partial\bar{\partial}\varphi - \sqrt{-1}\partial\bar{\partial}\log|s|^2 = \sqrt{-1}\partial\bar{\partial}\varphi,$$

since the log-modulus of a nowhere-zero holomorphic function is harmonic. Thus

$$\frac{\sqrt{-1}}{2\pi}\int_{X_{s,\varepsilon}} \partial\bar{\partial}\log h(s,\bar{s}) = \frac{\sqrt{-1}}{2\pi}\int_{X_{s,\varepsilon}} \Theta_h.$$

Adding $\frac{\sqrt{-1}}{2\pi}\int_{X-X_{s,\varepsilon}} \Theta_h$ to both sides of (6.4), we have the following theorem.

THEOREM 6.6.2. *Let $X$ be a compact Riemann surface, let $H \to X$ be a holomorphic line bundle with Hermitian metric $h$, and let $s \in \Gamma_{\mathscr{M}}(X,H) - \{0\}$. Then the number $c(H)$ is an integer given by the formula*

$$\frac{\sqrt{-1}}{2\pi}\int_X \Theta_h = \deg(\mathrm{Ord}(s)).$$

*In particular, the number $\deg(\mathrm{Ord}(s))$ is independent of the non-identically zero meromorphic section $s$.*

REMARK. The constancy of the function $\Gamma_{\mathcal{M}}(X,H) \ni s \mapsto \deg(\mathrm{Ord}(s))$ was already known to us by the Argument Principle: the quotient of two sections is a meromorphic function, whose order divisor has degree zero. ◇

DEFINITION 6.6.3. The number $c(H)$ is called the Chern number of the holomorphic line bundle $H$. We will also define this number to be the degree of $H$. ◇

REMARK. Chern numbers can be defined for more general connections of complex line bundles, again as the integral of $\frac{\sqrt{-1}}{2\pi}$ times the curvature of any connection of those line bundles. The Chern numbers in this context are also integers and are again independent of the connection, thus agreeing with our definition of Chern numbers in the case of a holomorphic line bundle. ◇

## 6.7. Example: The holomorphic line bundle $T_X^{1,0}$

Let $X$ be a real oriented surface with a Riemannian metric $g$. In view of Theorem 2.2.4, we can choose a complex atlas that makes $X$ a Riemann surface, such that in local coordinates $z = x + \sqrt{-1}y$,

$$g = \frac{e^{-\varphi}}{2}(dx \otimes dx + dy \otimes dy) = e^{-\varphi}\frac{1}{2}dz \otimes d\bar{z}.$$

Notice that this Riemannian metric for $X$ is now a Hermitian metric for $T_X^{1,0}$, in terms of the complex structure provided by Theorem 2.2.4. In the literature, a Hermitian metric for $T_X^{1,0}$ is often referred to as a Hermitian metric for $X$. Equivalently, a Hermitian metric for $X$ is a metric that is invariant under the linear transformation $J : T_X \to T_X$ defined in terms of the holomorphic coordinates $z = x + \sqrt{-1}y$ by

$$J\tfrac{\partial}{\partial x} = \tfrac{\partial}{\partial y} \quad \text{and} \quad J\tfrac{\partial}{\partial y} = -\tfrac{\partial}{\partial x}.$$

As the reader can easily verify, the linear transformation $J$ satisfies $J^2 = -\mathrm{Id}$ and thus has the purely imaginary eigenvalues $\pm\sqrt{-1}$. In fact, $T_X \otimes \mathbb{C}$ decomposes into the two eigenspaces of $J$ as

$$T_X \otimes \mathbb{C} = T_X^{1,0} \oplus T_X^{0,1}.$$

Notice that the function $e^{-\varphi}$ depends on $z$ as follows: if $z'$ is an overlapping local coordinate such that

$$g = e^{-\varphi'}\frac{1}{2}dz' \otimes d\bar{z}',$$

then on the intersection of the coordinate neighborhoods, we must have

$$e^{-\varphi'}\left|\frac{dz'}{dz}\right|^2 = e^{-\varphi}.$$

It follows that the differential $(1,1)$-form

$$\omega_g := \frac{\sqrt{-1}}{2}e^{-\varphi}dz \wedge d\bar{z}$$

is globally defined.

DEFINITION 6.7.1. The form $\omega_g$ is called the metric form, or the area form, associated to $g$. ◇

REMARK. It turns out that the Chern connection of $T_X^{1,0}$ with the Hermitian metric $g$ agrees with the Levi-Čivita connection of the Riemannian metric $g$ on $X$, after we identify $T_X^{1,0}$ with $T_X$ by sending a $(1,0)$-vector to twice its real part.

A Hermitian manifold $X$ of arbitrary dimension whose Levi-Čivita connection agrees with its Chern connection, after the identification of $T_X^{1,0}$ with $T_X$ obtained by sending a section to twice its real part, is called a Kähler manifold. It turns out that being Kähler is equivalent to the property that $d\omega_g = 0$, which trivially holds on Riemann surfaces.

The fact that a Hermitian metric on a Riemann surface is automatically Käher is one of relatively few low-dimensional accidents that account for the extraordinarily rich structure of Riemann surfaces. ◇

## 6.8. Exercises

6.1. Show that, for line bundles $H$, $H_1$, and $H_2$ with connections $\nabla$, $\nabla_1$, and $\nabla_2$, the induced connections $\nabla_1 \otimes \nabla_2$, $\nabla^\dagger$, and $\nabla^*$ for $H_1 \otimes H_2$, $\overline{H}$, and $H^\dagger$, respectively, are indeed connections.

6.2. Graphs of functions on a Riemann surface $X$ yield sections of the trivial bundle on $X$. Show that under this identification, the $\bar{\partial}$-operator associated to the (holomorphic) trivial bundle on a Riemann surface $X$ via Definition 6.3.1 agrees with the usual $\bar{\partial}$-operator on functions.

6.3. Let $g$ be a metric on a Riemann surface $X$ and let $\omega$ be its associated metric form. Show that the map $\Phi_\omega$ sending a $(0,1)$-form $\beta$ to the $(1,0)$-vector field $\xi$ defined by $\omega(\xi, \cdot) = \beta$ is a linear isomorphism. Compute the formula for $\Phi_\omega$ in terms of the local frames $d\bar{z}$ and $\frac{\partial}{\partial z}$ for $T_X^{*0,1}$ and $T_X^{1,0}$, respectively.

6.4. Show that the curvature of the Chern connection is given by the graded commutator
$$\Theta_\varphi f = (\nabla^{1,0}\bar{\partial} + \bar{\partial}\nabla^{1,0})f.$$

6.5. Show that $c(H_1 \otimes H_2) = c(H_1) + c(H_2)$ and $c(H^*) = -c(H)$.

6.6. Compute the Chern numbers of $T_{\mathbb{P}_1}^{1,0} \to \mathbb{P}_1$ and $T_{\mathbb{P}_1}^{*1,0} \to \mathbb{P}_1$.

6.7. Show that if a holomorphic line bundle has a global holomorphic section then its Chern number is non-negative.

6.8. Show that if two complex line bundles are (smoothly) isomorphic then their Chern numbers are equal.

## 6.8. Exercises

6.9. Let $p$ and $q$ be two points on a compact complex torus $E = \mathbb{C}/\Lambda$. Show that the holomorphic line bundle associated to the divisor $p - q$ has Chern number zero but that this line bundle is not trivial.

6.10. The Levi-Čivita connection of a Hermitian Riemann surface $(X, g)$ is the unique connection for $T_X^*$ such that, in any frame, the connection matrix is symmetric. The goal of this exercise is to establish a natural identification of the real cotangent bundle (as a rank-2 real vector bundle) with $T_X^{*1,0}$ (relative to its underlying real structure as a rank-2 vector bundle) and then prove that, via this identification, the Chern connection agrees with the Levi-Čivita connection.

  (a) Fix local holomorphic coordinates $z = x + \sqrt{-1}y$. Show that the fiberwise-linear map sending $dz$ to $\operatorname{Re} dz$ gives a real isomorphism of the real rank-2 vector bundles $T_X^{*1,0}$ and $T_X^*$.
  
  (b) Show that, under the identification of part (a), the dual metric $g^*$ of $g$ for $T_X^*$ induces a Hermitian metric for $T_X^{*1,0}$.
  
  (c) Show that, under the identification of part (a), the Levi-Čivita connection defines a Hermitian connection for $(T_X^{*1,0}, g^*)$.
  
  (d) Show that, under the identification of part (c), the Chern connection and the Levi-Čivita connection of $(T_X^{*1,0}, g^*)$ agree.

*Chapter 7*

# Potential Theory

## 7.1. The Dirichlet Problem and Perron's Method

**7.1.1. Definition of the Dirichlet Problem.** Let $X$ be a Riemann surface. The Dirichlet Problem for an open subset $Y \subset X$ with non-empty boundary $\partial Y$ is the following boundary value problem.

DEFINITION 7.1.1 (Dirichlet Problem). For any continuous $f : \partial Y \to \mathbb{R}$, find $u : \overline{Y} \to \mathbb{R}$ such that $u \in \mathscr{C}(\overline{Y}) \cap \mathscr{C}^2(Y)$, $\Delta u = 0$ on $Y$, and $u|_{\partial Y} = f$. ⋄

**7.1.2. Perron Families.**

DEFINITION 7.1.2. Let $X$ be a Riemann surface. A non-empty family $\mathscr{F}$ of subharmonic functions on $X$ is said to be a Perron family if the following two properties hold.

(P1) For every coordinate disk $D \subset X$ and every $u \in \mathscr{F}$ there exists $v \in \mathscr{F}$ such that $v|_D$ is harmonic and $v \geq u$.

(P2) For every $u_1, u_2 \in \mathscr{F}$ there exists $v \in \mathscr{F}$ such that $v \geq \max\{u_1, u_2\}$. ⋄

The main result about Perron families is the following theorem.

THEOREM 7.1.3. *Let $X$ be a connected Riemann surface. If $\mathscr{F}$ is a Perron family on $X$, then the upper envelope*
$$U = U_{\mathscr{F}} := \sup_{v \in \mathscr{F}} v$$
*is either $\equiv +\infty$ or is harmonic on $X$.*

**Proof.** It suffices to prove the result for disks in $X$, for by connectivity, it is not possible for $U$ to be harmonic on some disk and $\equiv +\infty$ on another disk. Thus we may assume $X = \mathbb{D}$ is the unit disk.

115

Let $\{z_j\} \subset \mathbb{D}$ be a countable dense subset of the unit disk. For each $j$, choose a sequence of subharmonic functions $\{v_{jk}\ ;\ k \geq 1\} \subset \mathscr{F}$ such that
$$U(z_j) = \lim_{k \to \infty} v_{jk}(z_j).$$
Using (P1), choose $v_1 \in \mathscr{F}$ such that $v_1$ is harmonic on $\mathbb{D}$ and $v_1 \geq v_{11}$. If $\{v_1, ..., v_n\} \subset \mathscr{F}$ have been chosen, let $v_{n+1} \in \mathscr{F}$ be harmonic on $\mathbb{D}$ and satisfy
$$v_{n+1} \geq \max\{v_n, v_{m\ell}\ ;\ m \leq n+1,\ \ell \leq n+1\}.$$
It follows that
$$\lim_{n \to \infty} v_n(z_j) = \sup_n v_n(z_j) = U(z_j).$$
Suppose that $U \not\equiv +\infty$. Without loss of generality, we can assume that $U(z_1) < +\infty$. By Harnack's Principle, $v := \lim v_n$ is harmonic on $\mathbb{D}$.

Next, since $U$ is the upper envelope of $\mathscr{F}$, $v_n \leq U$ and thus $v \leq U$. On the other hand, $v \geq w$ on a dense subset for all $w \in \mathscr{F}$. Since for each $w \in \mathscr{F}$ there exists a harmonic (and therefore continuous) $\tilde{w} \in \mathscr{F}$ such that $\tilde{w} \geq w$ locally, $v \geq w$ for all $w \in \mathscr{F}$. It follows that $v \geq U$. Thus $v = U$, and the proof is finished. $\square$

### 7.1.3. Perron's Method.

DEFINITION 7.1.4. *Let $X$ be a Riemann surface and $Y \subset\subset X$ an open subset whose boundary is non-empty. A point $x \in \partial Y$ is said to be regular if there is a neighborhood $U$ of $x$ in $X$ and a continuous function $\beta : \overline{Y} \cap U \to \mathbb{R}$ such that*

(1) $\beta|_{Y \cap U}$ *is subharmonic and*

(2) $0 = \beta(x) > \beta(y)$ *for all $y \in \overline{Y} \cap U - \{x\}$.*

*A function $\beta$ with these two properties is called a barrier at $x$.* ◇

Once we have a regular point, we can refine our choice of barrier. We have the following lemma.

LEMMA 7.1.5. *Let $X$ be a Riemann surface and let $Y \subset\subset X$ be an open subset whose boundary is non-empty. Let $x \in \partial Y$ be a regular point. Fix real numbers $m \leq c$. Then there is a neighborhood $V$ of $x$ in $X$ and a continuous function $v : \overline{Y} \cap V \to \mathbb{R}$ such that*

(1) $v|_{Y \cap V}$ *is subharmonic,*

(2) $c = v(x) \geq v(y)$ *for all $y \in \overline{Y} \cap V$, and*

(3) $v|_{\overline{Y} - V} = m$.

**Proof.** Clearly we may assume $c = 0$. Suppose $U$ is a neighborhood of $x$ and $\beta$ is a barrier at $x$ with respect to this neighborhood $U$. Let $V$ be a neighborhood of $x$

## 7.1. The Dirichlet Problem and Perron's Method

such that $V \subset\subset U$. Then $\sup_{\partial V \cap \overline{Y}} \beta < 0$, and thus there is a constant $a > 0$ such that $a\beta|_{\partial V \cap \overline{Y}} < m$. Then the function

$$v := \begin{cases} \max\{m, a\beta\} & \text{on } V \cap \overline{Y} \\ m & \text{on } \overline{Y} - V \end{cases}$$

satisfies the required conditions. The proof is complete. □

DEFINITION 7.1.6. Let $X$ be a Riemann surface and let $Y \subset\subset X$ be an open subset. Fix a continuous function $f : \partial Y \to \mathbb{R}$. The Perron class $\mathscr{P}_f \subset \mathscr{C}(\overline{Y})$ consists of all functions $u$ that are subharmonic in $Y$ and satisfy $u \leq f$ on $\partial Y$. ◇

The properties of subharmonic functions, together with the continuity of $f$, show that $\mathscr{P}_f$ is a Perron family. It follows from Theorem 7.1.3 that the function $U_f$ defined by

$$U_f(x) := \sup_{v \in \mathscr{P}_f} v(x)$$

is harmonic in $Y$.

We have the following theorem.

THEOREM 7.1.7. *If $x \in \partial Y$ is a regular point, then*

$$\lim_{Y \ni y \to x} U_f(y) = f(x).$$

**Proof.** Fix $\varepsilon > 0$. By continuity of $f$ there exists a relatively compact neighborhood $V$ of $x$ in $X$ with $|f(y) - f(x)| < \varepsilon$ for all $y \in \partial Y \cap V$. Let $a := \min_{\partial Y} f$ and $A := \max_{\partial Y} f$. By Lemma 7.1.5 there is a continuous function $v : \overline{Y} \to \mathbb{R}$ that is subharmonic on $Y$, such that $v(x) = f(x) - \varepsilon$, $v_{\overline{Y} \cap V} \leq f(x) - \varepsilon$, and $v|_{\overline{Y}-V} = a - \varepsilon$. Then $v \in \mathscr{P}_f$ and thus $v \leq U_f$. It follows that

$$\liminf_{Y \ni y \to x} U_f(y) \geq v(x) = f(x) - \varepsilon.$$

Next, Lemma 7.1.5 provides a continuous function $w : \overline{Y} \to \mathbb{R}$ that is subharmonic on $Y$ such that

$$w(x) = -f(x), \quad w|_{\overline{Y} \cap V} \leq -f(x), \quad \text{and} \quad w|_{\overline{Y}-V} = -A.$$

Now, for every $u \in \mathscr{P}_f$ and $y \in \partial Y \cap V$ one has $u(y) \leq f(x) + \varepsilon$. Then $u(y) + w(y) \leq \varepsilon$ for all $y \in \partial Y \cap V$. Furthermore $u(z) + w(z) \leq A - A = 0$ for all $z \in \overline{Y} \cap \partial V$. Applying the Maximum Principle to the subharmonic function $u + w$ on $Y \cap V$, we deduce that $u + w \leq \varepsilon$ on $\overline{Y} \cap V$. Thus $u|_{\overline{Y} \cap V} \leq \varepsilon - w|_{\overline{Y} \cap V}$ for all $u \in \mathscr{P}_f$. Hence

$$\limsup_{Y \ni y \to x} U_f(y) \leq f(x) + \varepsilon.$$

Since $\varepsilon$ was arbitrary, the proof is complete. □

As a corollary, we obtain the following result.

COROLLARY 7.1.8. *Let $X$ be a Riemann surface and let $Y \subset\subset X$ be an open subset each of whose boundary points is regular. Then the Dirichlet Problem has a solution on $Y$.*

We end with a useful sufficient condition for regularity of a boundary point which implies, in particular, that any point at which the boundary is smooth and 1-dimensional is a regular point.

PROPOSITION 7.1.9. *Let $X$ be a Riemann surface, let $Y \subset\subset X$ be an open set, and let $x \in \partial Y$. Suppose there is a coordinate neighborhood $z : U \to \mathbb{C}$ of $x$ such that for some $p \in z(U - \overline{Y})$ the disk $\{\zeta \in z(U) \; ; \; |\zeta - p| < |z(x) - p|\}$ lies in $z(U)$ and does not meet $z(Y \cap U)$. Then $x$ is a regular point.*

**Proof.** The function $\beta(\zeta) := \log \frac{|z(x)-p|}{2} - \log \left|\zeta - \frac{p+z(x)}{2}\right|$ is a barrier at $x$. □

### 7.1.4. Aside: Countable topology of Riemann surfaces.
In the definition of manifold, one has to assume a countable topology. Though we would happily make such an assumption in this book, it turns out that for Riemann surfaces, it is not necessary to do so, as was proved by Radó. The key is the ability to solve the Dirichlet Problem.

THEOREM 7.1.10. *Every Riemann surface has a countable topology.*

**Proof.** Let $U \subset X$ be a coordinate neighborhood. Choose smoothly bounded open disks $D_1, D_2 \subset\subset U$ whose closures are disjoint, and set $Y := X - \overline{D_1 \cup D_2}$. Then $\partial Y$ consists of regular points, and thus there is a continuous function $u : \overline{Y} \to \mathbb{R}$ that is harmonic on $Y$ and satisfies

$$u|_{\partial D_0} \equiv 0 \quad \text{and} \quad u|_{\partial D_1} \equiv 1.$$

The $(1,0)$-form $\omega := \partial u$ is then holomorphic and non-trivial. Consider the universal cover $\pi : \tilde{Y} \to Y$. Let $f$ be a holomorphic function on $\tilde{Y}$ such that $\partial f = \pi^* \omega$. Since the mapping $f : \tilde{Y} \to \mathbb{C}$ is continuous and has discrete fibers, it follows from elementary point set topology that $\tilde{Y}$ has a countable topology, and since the mapping $\pi : \tilde{Y} \to Y$ is continuous and surjective, $Y$ also has countable topology, as desired. □

### 7.1.5. A helpful remark.
In the next paragraph we consider Green's Functions. On a Riemann surface $Y$ with boundary, a Green's Function is a family of solutions of the distributional boundary value problems

$$\sqrt{-1}\partial\bar{\partial} G_x = \delta_x, \quad G_x|_{\partial Y} = 0$$

as $x$ varies over the points of (the interior of) $Y$. Since for a fixed $x$ such a function is harmonic away from $x$, and since the difference of any two solutions (ignoring the boundary condition) is also harmonic, we can expect that the kind of singularity that a Green's Function $G_x$ possesses at $x$ is in some sense uniform.

## 7.1. The Dirichlet Problem and Perron's Method

To see what this singularity is, we work locally. The following lemma, whose proof was already indirectly hinted at on a number of occasions, characterizes the singularity of a Green's Function locally. Recall that with $d^c = \frac{\sqrt{-1}}{2}(\bar\partial - \partial)$, $\sqrt{-1}\partial\bar\partial = dd^c$.

**LEMMA 7.1.11.** *Let $\varphi \in \mathscr{C}_0^\infty(\mathbb{C})$. Then*
$$\int_{\mathbb{C}} \log|z|^2 dd^c \varphi = 2\pi \varphi(0).$$
*That is to say, $dd^c \log|z|^2 = 2\pi \delta_0$ in the sense of currents.*

**Proof.** Let $R > 0$ be such that $\text{Support}(\varphi) \subset \mathbb{D}(0, R)$, and let $\varepsilon \in (0, R)$. Note that since $\log|z|^2$ is locally integrable,
$$\int_{\mathbb{C}} \log|z|^2 dd^c \varphi = \lim_{\varepsilon \to 0} \int_{\mathbb{D}(0,R) - \mathbb{D}(0,\varepsilon)} \log|z|^2 dd^c \varphi.$$

Note that $d^c f \wedge dg = -df \wedge d^c g$. By integration-by-parts, we have

$$\int_{\mathbb{D}(0,R)-\mathbb{D}(0,\varepsilon)} \log|z|^2 dd^c \varphi$$
$$= -\int_{\mathbb{D}(0,R)-\mathbb{D}(0,\varepsilon)} d\log|z|^2 \wedge d^c\varphi + \log \varepsilon^2 \int_{|z|=\varepsilon} d^c \varphi$$
$$= \int_{\mathbb{D}(0,R)-\mathbb{D}(0,\varepsilon)} d^c \log|z|^2 \wedge d\varphi + O(\varepsilon \log \varepsilon^2)$$
$$= -\int_{\mathbb{D}(0,R)-\mathbb{D}(0,\varepsilon)} \varphi dd^c \log|z|^2 + \int_{|z|=\varepsilon} \varphi d^c \log|z|^2 + O(\varepsilon \log \varepsilon^2)$$
$$= \int_0^{2\pi} \varphi(\varepsilon e^{\sqrt{-1}\theta}) d\theta + O(\varepsilon \log \varepsilon^2).$$

The second equality follows because $dd^c \log|z|^2 = 0$ away from 0 and, when $z = \varepsilon e^{\sqrt{-1}\theta}$, $d^c \log|z|^2 = d\theta$. Thus
$$\int_{\mathbb{C}} \log|z|^2 dd^c \varphi = \lim_{\varepsilon \to 0} \int_{\mathbb{D}(0,R)-\mathbb{D}(0,\varepsilon)} \log|z|^2 dd^c \varphi = 2\pi \varphi(0).$$
The proof is complete. □

### 7.1.6. Green's Functions.

**DEFINITION 7.1.12.** Let $X$ be a Riemann surface. A Green's Function on $X$ with singularity at $x \in X$ is a subharmonic function $G_x : X \to [-\infty, 0)$ such that

(G1) $G_x$ is harmonic on $X - \{x\}$,

(G2) if $z$ is any local coordinate in a neighborhood $U$ of $x$ with $z(x) = 0$ then $G_x - \log|z|^2$ is harmonic in $U$, and

(G3) if $H$ is any other subharmonic function satisfying (G1) and (G2) then $G_x \geq H$. ◇

Observe that by the very definition of Green's Function, any Riemann surface that admits a Green's Function must admit a bounded subharmonic function. (In this text we call such surfaces *potential-theoretically hyperbolic*, a notion we will discuss again in Chapter 10.) The amazing fact is that the converse is true.

THEOREM 7.1.13. *If a Riemann surface $X$ admits a bounded non-constant subharmonic function then for any point $x \in X$, $X$ admits a Green's Function with singularity $x$.*

REMARK. Note that no compact Riemann surface admits a Green's Function. Some readers may find this confusing, since there are other notions of Green's Function in the literature such that every compact Hermitian manifold admits such a Green's Function. These other notions arise because Poisson's Equation on a compact manifold can only be solved for data that is orthogonal to the space of harmonic functions; the Green's Function can then be defined on this orthogonal complement. We shall solve such a PDE on compact Riemann surfaces in Chapter 9, but we will not use the term 'Green's Function' for the solution operator. ◇

In order to prove Theorem 7.1.13, we establish the following lemma.

LEMMA 7.1.14. *Let $X$ be a Riemann surface admitting a bounded subharmonic function and let $K$ be a compact subset of $X$ such that $X - K$ is connected and has regular boundary. Then there exists a continuous function $\varphi : \overline{X - K} \to (0, 1]$ that is harmonic and non-constant on $X - K$ and $\equiv 1$ on $\partial K$.*

**Proof.** Let $\psi_o$ be a non-constant negative subharmonic function on $X$ and set $m_o := -\max(\psi_o|_K) > 0$ and $\psi_1 := \psi_o/m_o$. Then $\psi_1$ is negative, subharmonic, non-constant, and $\leq -1$ on $K$. Note that by the Maximum Principle, $\psi_1|_K$ assumes its maximum $-1$ at some point $p \in \partial K$. Also, there exists $q \in X - K$ such that $\psi_1(q) > -1$, for otherwise $p$ is an interior maximum for $\psi_o$, and the Maximum Principle would imply that $\psi_o$ is constant. Let $\psi := \max\{\psi_1, -1\}$. Then $\psi$ is subharmonic on $X$, $-1 \leq \psi < 0$, $\psi(q) < -1$, and $\psi|_K \equiv -1$.

Let $\mathscr{H}$ denote the set of all continuous functions $v$ on $\overline{X - K}$ that are subharmonic on $X - K$ and satisfy the inequality $v + \psi \leq 0$ on $X - K$. The maximum of two functions in $\mathscr{H}$ is clearly in $\mathscr{H}$.

Take a function $u \in \mathscr{H}$ and a disk $D$ in $X - K$. Solve the Dirichlet Problem on $D$ with boundary value $u$ on $\partial D$. Let $w$ be the solution, and set $v = w$ in $D$ and $v = u$ in $X - (K \cup D)$. We have to verify that $v$ is subharmonic. It suffices to verify the sub-mean value property on $\partial D$. But since $w \leq u$ in $D$, we have $v \leq u$, with equality on the boundary of $D$. Thus for a small disk $U$ centered at a point

## 7.1. The Dirichlet Problem and Perron's Method

$x \in \partial D$

$$\fint_U v \, dA - v(x) \leq \fint_U u \, dA - u(x) \leq 0,$$

and the desired sub-mean value inequality now follows. Since the function 0 lies in $\mathcal{H}$, $\mathcal{H}$ is nonempty and therefore a Perron family (on $X - K$). It follows that the function

$$\varphi := \sup_{v \in \mathcal{H}} v$$

is harmonic on $X - K$ and satisfies $0 \leq \varphi \leq 1$ on $X - K$.

It remains to show that $\varphi|_{\partial K} \equiv 1$. Toward this end, choose an open set $U \subset\subset X$ containing $K$ such that $\partial U$ consists of a finite number of smooth curves. (This is easy to do and we shall give a proof in the next section.) Let $v_o$ be the solution of the Dirichlet Problem on $U - K$ for the boundary values 1 on $\partial K$ and 0 and $\partial U$. By the Maximum Principle $0 \leq v_o \leq 1$ on $\overline{U - K}$. Extend $v_o$ to $X - U$ by zero. As before, $v_o$ is subharmonic on $X - K$. We claim that $v_o + \psi \leq 0$ on $X - K$. Indeed, the latter is subharmonic on $X - K$ and $\leq 0$ on $\partial U$ and on $X - U$. Since $\psi \equiv -1$ on $\partial K$, the Maximum Principle implies that $v_o + \psi \leq 0$ on $\overline{U} - K$. Thus $v_o \in \mathcal{H}$.

By the definition of $\varphi$, we have

$$v_o \leq \varphi \leq -\psi.$$

It follows that $\varphi$ is continuous on $\overline{X - K}$ and $\equiv 1$ on $\partial K$, and since $-\psi(q) < 1$, $\varphi$ is non-constant. Moreover, $0 < \varphi < 1$ on $X - K$. The proof is complete. $\square$

REMARK. The function $\varphi$ satisfying the conclusions of Lemma 7.1.14 is a classical object called the harmonic measure of the set $K$. The harmonic measure has many important interpretations and applications that we will not get into in this book. $\diamond$

**Proof of Theorem 7.1.13.** We use Perron's Method. To this end, denote by $\mathscr{G}$ the family of all non-negative functions $u$ on $X$ that have compact support, are subharmonic on $X - \{x\}$, and such that for any local coordinate $z$ near $x$ with $z(x) = 0$, $u + \log |z|$ is subharmonic in some disk $|z| < \varepsilon$.

The function $u_o(z) := -\chi_{\{|z|<1\}} \cdot \log |z|$, where $\chi_A$ denotes the characteristic function of $A$, shows that $\mathscr{G}$ is non-empty. Moreover, $\mathscr{G}$ is a Perron family, as we leave the reader to verify.

We claim next that outside every neighborhood of $x$, $\mathscr{G}$ is uniformly bounded. To this end, fix $r \in (0, 1)$ and let $\varphi_r$ be the function associated to the compact set $\{|z| \leq r\}$ by Lemma 7.1.14. That is to say, $\varphi_r$ is harmonic away from $\{|z| \leq r\}$, where it takes values in $(0, 1)$, and $\varphi_r \equiv 1$ on $|z| = r$. Let $a_r := \max_{|z|=1} \varphi_r \in (0, 1)$.

For $u \in \mathscr{G}$ let $b_r := \max_{|z|=r} u$. Then $u - b_r \varphi_r$ is subharmonic and negative off $|z| \leq r$ and $\leq 0$ off $|z| < r$. In the complement of the (compact) support of $u$, $u - b_r \varphi_r = -b_r \varphi_r \leq 0$. Since the support of $u$ is compact, the Maximum Principle

implies that $u - b_r\varphi_r \leq 0$ on $X - \{|z| \leq 1\}$. But in $\{|z| < 1\}$, $u + \log|z|$ is subharmonic. Thus
$$b_r + \log r = \max_{|z|=r} u + \log r \leq \max_{|z|=1} u \leq \sup_{|z|=1} b_r\varphi_r = a_r b_r.$$
Therefore $b_r \leq \frac{\log(1/r)}{1-a_r} < +\infty$. Since $u$ has compact support, we deduce again that
$$\max_{X-\{|z|<r\}} u = \max_{|z|=r} u \leq \frac{\log(1/r)}{1-a_r}.$$
The function
$$U_{\mathscr{G}} := \sup_{u \in \mathscr{G}} u$$
cannot be identically $+\infty$ and is therefore harmonic on $X - \{x\}$.

Next, we claim that for each $u \in \mathscr{G}$, the subharmonic function $u + \log|z|$ is uniformly bounded near $x$. Indeed,
$$u(z) + \log|z| \leq b_r + \log r \leq \frac{\log(1/r)}{1-a_r} + \log r \leq \frac{a_r \log(1/r)}{1-a_r}.$$

Now let $G_x := -U_{\mathscr{G}}$. Then $G_x$ is harmonic on $X - \{x\}$ and $G_x - \log|z|$ is harmonic on $|z| < 1$. Thus $G_x$ satisfies (G1) and (G2). We need only to verify (G3). To this end, suppose $H$ is a non-positive subharmonic function satisfying (G1) and (G2). Then for each $u \in \mathscr{G}$, $u + H$ is subharmonic in $X$. Since $u$ has compact support, $u + H \leq 0$ away from the support of $u$, and hence by the Maximum Principle, everywhere. But then $H \leq G_x$, which is (G3). The proof is complete. □

**7.1.7. Symmetry of the Green's Function.** Observe that we have now defined a function of two variables: for each $x \in X$ we obtained $G_x$. We define
$$G(x,y) := G_x(y),$$
which is well-defined off the diagonal and furthermore has the property that (in local coordinates)
$$G(x,y) - \log|x-y|$$
is harmonic in $y$ for fixed $x$.

As we already showed, $\sqrt{-1}\partial\bar{\partial}G(x,\cdot) = 2\pi\delta_x$, where $\delta_x$ is the Dirac measure at $x$. The "physical interpretation" of this identity is that $G(x,y)$ represents the potential energy of the electric field induced on $y$ by a charge $\pi/2$ placed at the point $x$. It stands to reason that if we were to place our charge at $y$, we might feel the same effect at $x$. That is to say, we might expect that
$$G(x,y) = G(y,x).$$
We will outline what is involved in proving the result mathematically.

## 7.1. The Dirichlet Problem and Perron's Method

To emphasize dependence on the surface in question, we write $G_X(x,\cdot)$ for the Green's Function on $X$ with singularity at $x$.

Observe that if $Y$ is a relatively compact domain in a Riemann surface $X$ whose boundary is regular for the Dirichlet Problem, then by letting $u$ be the solution to the Dirichlet Problem with boundary value $G_X(x,\cdot)|_{\partial Y}$, we find that

$$G_Y(x,\cdot) = G_X(x,\cdot) - u.$$

Indeed, properties (G1) and (G2) are obvious, and by the Maximum Principle, if $H$ satisfies (G1) and (G2) then $H - (G_X(x,\cdot) - u)$ is harmonic on $Y$ and $\leq 0$ on the boundary, thus $\leq 0$ on $Y$ as desired.

Using Green's Theorem, one can show that, for a smoothly bounded domain $Y$, $G_Y(x,y) = G_Y(y,x)$. Assuming this, we now sketch a proof of the same result for general Riemann surfaces $X$ having non-constant bounded subharmonic functions.

**THEOREM 7.1.15.** *Let $X$ be a Riemann surface admitting a bounded non-constant subharmonic function. Then the Green's Function on $X$ satisfies the symmetry condition*

$$G_X(x,y) = G_X(y,x).$$

**Sketch of proof.** Fix $x \in X$. Let $\mathscr{Y}$ denote the set of relatively compact open subsets $Y$ of $X$ containing $x$ and whose boundary is smooth. For each $Y \in \mathscr{Y}$, let $g_Y$ be the extension to $G_Y$ by zero outside $Y \times Y$. Let

$$\mathscr{F} := \{g_Y \,;\, Y \in \mathscr{Y}\}.$$

Since two smoothly bounded relatively compact domains lie in a single smoothly bounded, relatively compact domain, it is easy to prove that $\mathscr{F}$ is a Perron family. Its upper envelope is easily seen to be the Green's Function $G_X$. It follows that $G_X(x,y) \geq G_Y(x,y) = G_Y(y,x)$. Now fixing $y$ and varying $x$ on the right-hand side, we obtain $G_X(x,y) \geq G_X(y,x)$. Since $x$ and $y$ are arbitrary, they can be interchanged to obtain the opposite inequality. Thus the desired equality. $\square$

**7.1.8. Reproducing formulas.** Using the Green's Function, one can establish a Cauchy-Green-type formula for smoothly bounded (or more generally Dirichlet regular) domains in a Riemann surface.

**PROPOSITION 7.1.16.** *Let $X$ be a Riemann surface and let $Y \subset\subset X$ be an open subset with smooth boundary. Then for any smooth function $f : \overline{Y} \to \mathbb{C}$,*

$$f(x) = \frac{1}{2\pi} \int_{\partial Y} f(y) d_y^c G_Y(x,y) + \frac{1}{2\pi} \int_Y G_Y(x,y) dd^c f(y).$$

*In particular, if $f$ is subharmonic on $X$ and $Y = E(x,r) := \{y \in X \,;\, G_X(x,y) < \log r\}$ then*

$$f(x) \leq \frac{1}{2\pi} \int_{\partial E(x,r)} f(y) d_y^c G_X(x,y),$$

*with equality for all $x$ and $r$ if and only if $f$ is harmonic.*

REMARK. The inequality for subharmonic functions can be rather useful. It will be one of the ingredients we shall use in our proof of regularization of singular Hermitian metrics. ◇

**Proof of Proposition 7.1.16.** This is just an application of the Green-Stokes Formula

$$\int_Y g dd^c f - f dd^c g = \int_{\partial Y} g d^c f - f d^c g$$

to the functions $g = G_Y(x, \cdot)$. One uses $dd^c G_Y(x, \cdot) = 2\pi \delta_x$ and $G_Y(x, \cdot)|_{\partial Y} \equiv 0$. (More precisely, one has to remove a small neighborhood of $x$ and take limits as that neighborhood shrinks to zero. We leave such details to the reader.) □

Now, $d(f \partial g) = \bar{\partial} f \wedge \partial g - f \partial \bar{\partial} g$. By integration and Stokes' Formula, one obtains

$$\int_Y f \sqrt{-1} \partial \bar{\partial} g = -\sqrt{-1} \int_{\partial Y} f \partial g - \sqrt{-1} \int_Y \partial g \wedge \bar{\partial} f.$$

Applying the result with $g = G_Y(x, \cdot)$, we obtain the following Cauchy-Green-type formula.

PROPOSITION 7.1.17. *Let $X$ be a Riemann surface and let $Y \subset\subset X$ be an open subset with smooth boundary. Then for any smooth function $f : \overline{Y} \to \mathbb{C}$,*

$$f(x) = \frac{1}{2\pi\sqrt{-1}} \int_{\partial Y} f(y) \partial_y G_Y(x, y) + \frac{1}{2\pi\sqrt{-1}} \int_Y \partial_y G_Y(x, y) \wedge \bar{\partial} f(y).$$

*In particular, if $f$ is holomorphic then*

$$f(x) = \frac{1}{2\pi\sqrt{-1}} \int_{\partial Y} f(y) \partial G_Y(x, y),$$

*while if $\text{Support}(f) \subset\subset Y$ then*

$$f(x) = \frac{1}{2\pi\sqrt{-1}} \int_Y \partial_y G_Y(x, y) \wedge \bar{\partial} f(y).$$

By similar methods, but slightly modified, we obtain the following solution of the $\bar{\partial}$-equation.

THEOREM 7.1.18. *Let $X$ be a Riemann surface and let $Y \subset\subset X$ be an open set such that $Y - X \neq \emptyset$. Suppose $\alpha$ is a smooth $(0,1)$-form on $X$. Then there is a smooth function $f : X \to \mathbb{C}$ such that $\bar{\partial} f = \alpha$ on $Y$.*

**Proof.** Let $\chi$ be a smooth function with compact support contained in an open set $Z \subset\subset X$ having smooth, 1-dimensional boundary, such that $X - Z \neq \emptyset, \overline{Y} \subset Z$, and $\chi|_Y \equiv 1$. Set

$$f(x) := \frac{1}{2\pi\sqrt{-1}} \int_{y \in Z} \partial_y \left( G_Z(x, y) \wedge \chi(y) \alpha(y) \right).$$

## 7.1. The Dirichlet Problem and Perron's Method

Then
$$\bar{\partial} f(x) := \frac{1}{2\pi\sqrt{-1}} \int_{y \in Z} \partial_y \left( \bar{\partial}_x G_Z(x,y) \wedge \chi(y)\alpha(y) \right).$$

The theorem is proved if we show that for any $(0,1)$-form $\Theta$ with compact support in $Z$,
$$\Theta(x) = \frac{1}{2\pi\sqrt{-1}} \int_Z \partial_y \left( \bar{\partial}_x G_Z(x,y) \otimes \Theta(y) \right).$$

To this end, choose a local coordinate $\zeta$ with $\zeta(x) = 0$, and choose a number $\varepsilon > 0$ so small that the disk $\mathbb{B}(2\varepsilon) := \{|\zeta| < 2\varepsilon\}$ is a coordinate chart. Write $\Theta = h(\zeta) d\bar{\zeta}$ in $\mathbb{B}(2\varepsilon)$. Then

$$\int_{Z-\mathbb{B}(\varepsilon)} \partial_\zeta \left( \sqrt{-1}\Theta(\zeta) \otimes \frac{1}{2\pi} \bar{\partial}_z G_Z(z,\zeta) \right)$$
$$= \frac{-1}{2\pi\sqrt{-1}} \int_{Z-\mathbb{B}_\varepsilon} d\zeta \left( \Theta(\zeta) \otimes \bar{\partial}_z G_Z(z,\zeta) \right)$$
$$= \left( \frac{1}{2\pi\sqrt{-1}} \int_{\partial \mathbb{B}_\varepsilon} h(\zeta) \frac{\partial G_Z(z,\zeta)}{\partial \bar{z}} d\bar{\zeta} \right) d\bar{z},$$

where $z$ is our local coordinate near $x$ with $z(x) = 0$, and we have used the fact that $G_Z(z,\zeta) \equiv 0$ for all $\zeta \in \partial Z$. But
$$\frac{\partial G_Z(z,\zeta)}{\partial \bar{z}} = \frac{1}{\bar{\zeta}} + \frac{\partial H(z,\zeta)}{\partial \bar{z}}$$

for some smooth function $H(z,\zeta)$ that is harmonic in each variable separately, and we have

$$\int_{Z-\mathbb{B}(\varepsilon)} \partial_\zeta \left( \sqrt{-1}\Theta(\zeta) \wedge \frac{1}{2\pi} \bar{\partial}_z G_Z(z,\zeta) \right)$$
$$= \left( \frac{-1}{2\pi\sqrt{-1}} \int_{|\zeta|=\varepsilon} \frac{h(\zeta) d\bar{\zeta}}{\bar{\zeta}} \right) d\bar{z} + O(\varepsilon)$$
$$= \left( \int_0^{2\pi} h(\varepsilon e^{\sqrt{-1}\theta}) \frac{d\theta}{2\pi} \right) d\bar{z} + O(\varepsilon)$$
$$= h(0) d\bar{z} + O(\varepsilon) = \Theta(x) + O(\varepsilon).$$

Letting $\varepsilon \to 0$, we see that
$$\Theta(x) = \int_Z \partial_y \left( \sqrt{-1}\Theta(\zeta) \otimes \frac{1}{2\pi} \bar{\partial}_x G_Z(x,\bar{\zeta}) \right).$$

The proof is complete. □

In fact, given a smooth $(0,1)$-form $\alpha$ on an open Riemann surface $X$, it is possible to find a smooth function $u$ such that $\bar{\partial} u = \alpha$ on all of $X$. The method we will use to prove this fact requires us to establish a certain approximation theorem for holomorphic functions called Runge's Approximation Theorem. We will also need Runge's Theorem for our proof of the Uniformization Theorem. In the next section we state and establish Runge's Theorem.

## 7.2. Approximation on open Riemann surfaces

Harmonic functions are to subharmonic functions as linear functions are to convex functions. Via the Maximum Principle, we can draw from this correspondence a notion of the hull of a subset in a Riemann surface that is analogous to the convex hull of a set in Euclidean space. This notion of convexity is fundamental in the theory of approximation of holomorphic functions on subsets by globally defined holomorphic functions. In this section, we make precise all of these ideas.

In the rest of the section, we always assume that $X$ is an open connected Riemann surface.

### 7.2.1. Holomorphic hulls, Runge domains and regular exhaustion.

DEFINITION 7.2.1. Let $Y \subset X$ be any subset. The holomorphic hull $\widehat{Y}_{\mathcal{O}(X)}$ of $Y$ with respect to $X$ consists of $Y$ together with all of the relatively compact components of $X - Y$ in $X$. An open subset $Y \subset X$ is said to be Runge (in $X$) if $Y = \widehat{Y}_{\mathcal{O}(X)}$. ◇

REMARK. Note that the dependence on $X$ is non-trivial. Indeed, if $S \subset \mathbb{C}$ is the unit circle, then $\widehat{S}_{\mathcal{O}(\mathbb{C})}$ is the unit disk while $\widehat{S}_{\mathcal{O}(\mathbb{C}-\{0\})} = S$. Nevertheless, if $X$ is clear from the context, we sometimes write $\widehat{Y}$ instead of $\widehat{Y}_{\mathcal{O}(X)}$. ◇

LEMMA 7.2.2. *If a subset $Y$ of $X$ is closed (resp. compact) then so is its holomorphic hull $\widehat{Y}_{\mathcal{O}(X)}$.*

**Proof.** The assertion about closedness is easy, and we leave it as an exercise to the interested reader. Suppose $Y$ is compact (and non-empty; otherwise the assertion is trivial). Let $U$ be a relatively compact neighborhood of $Y$, and denote by $U_j$, $j \in J$, the components of $X - Y$.

Observe that each $U_j$ meets $\overline{U}$. Otherwise $\overline{U_j} \subset X - U \subset X - Y$, and since $U_j$ is a connected component of $X - Y$, we would have $U_j = \overline{U_j}$ (the closure in $X$). But this is impossible, since $X$ is connected.

Next, we observe that since the $U_j$ are disjoint and give an open cover of the compact set $\partial U$, only finitely many components $U_j$ meet $\partial U$.

Let $J_o$ be the set of all $j \in J$ for which $U_j$ is relatively compact. Then there are finitely many $j_1, ..., j_m \in J_o$ such that $U_j \cap \partial U \neq \emptyset$, and for all other $j \in J_o$, $U_j \subset U$. It follows that

$$\widehat{Y}_{\mathcal{O}(X)} \subset U \cup U_{j_1} \cup ... \cup U_{j_m}$$

is relatively compact. Since $\widehat{Y}_{\mathcal{O}(X)}$ is closed by the first part of the lemma, the proof is complete. □

## 7.2. Approximation on open Riemann surfaces

As an immediate corollary of Lemma 7.2.2 and the countability of the topology of a Riemann surface, we obtain the following result.

**COROLLARY 7.2.3.** *For an open connected Riemann surface $X$ there is a sequence $K_j \subset X$ of compact subsets, $j = 1, 2, ...$, such that $\widehat{K}_{j,\mathcal{O}(X)} = K_j$, $K_j \subset \text{Interior}(K_{j+1})$, and $\bigcup_j K_j = X$.*

**Proof.** Since $X$ has countable topology, there are compact subsets $L_1 \subset L_2 \subset ...$ with $\bigcup_j L_j = X$. Let $K_0 = \emptyset$ and $K_1 = \widehat{L}_1$. Suppose $K_m, K_{m-1}, m \geq 1$, have been chosen so that $\widehat{K}_m = K_m$ and $K_{m-1} \subset \text{Interior}(K_m)$. Let $M$ be a compact subset whose interior contains $L_m \cup K_m$. Setting $K_{m+1} = \widehat{M}$ completes the proof. $\square$

**LEMMA 7.2.4.** *Let $K_1, K_2$ be compact subsets of a Riemann surface such that $K_1 \subset \text{Interior}(K_2)$ and $\widehat{K}_2 = K_2$. Then there is an open subset $Y \subset X$ such that $\widehat{Y}_{\mathcal{O}(X)} = Y$ and $K_1 \subset Y \subset K_2$. Moreover, $Y$ can be chosen so that $\partial Y$ consists only of regular points for the Dirichlet problem.*

**Proof.** Cover $\partial K_2$ by a finite number of closed disks $D_1, ..., D_N$ such that $D_i \cap K_1 = \emptyset$ and set $Y = K_2 - (D_1 \cup ... \cup D_N)$. Then $Y$ is open and $K_1 \subset Y \subset K_2$. We need to show that $\widehat{Y} = Y$. To this end, let $U_j, j \in J$, be the components of $X - K_2$, none of which, by assumption, is relatively compact. Every $D_i$ meets at least one $U_j$. Since every connected component of $X - Y$ meets some $D_i$, $U_j$ cannot be relatively compact. Thus $\widehat{Y} = Y$ as claimed.

By slightly deforming the disks $D_i$, we can also guarantee that $\partial Y$ is smooth, and thus regular for the Dirichlet Problem. $\square$

**LEMMA 7.2.5.** *Let $Y$ be a Runge open subset of $X$. Then every connected component of $Y$ is also Runge.*

**Proof.** Every connected component of $Y$ is clearly open. Let $Z = X - Y$. Then $Z$ is closed, as are all of its components.

First, observe that for every component $U_Y$ of $Y$, $\overline{U_Y}$ meets $Z$. Otherwise $U_Y$ has no boundary points, which is a contradiction to the connectivity of $X$.

Next, let $C$ be a connected component of $X - U_Y$. We claim that $C$ meets $Z$. Indeed, $C$ is closed and meets $\partial U_Y$, and the latter is non-empty and hence in $Z$.

Finally, since $C$ is closed, any component of $Z$ that meets $C$ is in fact contained in $C$. Since no such component is relatively compact, $C$ is not relatively compact. The proof is complete. $\square$

**THEOREM 7.2.6.** *On an open connected Riemann surface $X$ there exist relatively compact Runge open subsets $Y_1 \subset\subset Y_2 \subset\subset ...$ so that $\bigcup_j Y_j = X$ and each $\partial Y_j$ is smooth.*

**Proof.** Fix a sequence of compact sets $K_j$ as in Corollary 7.2.3. By Lemma 7.2.4 we can find relatively compact Runge open sets $Z_j$ with regular (and even smooth) boundary such that $K_j \subset Z_j \subset \text{Interior}(K_{j+1})$ for all $j \geq 1$. We let $Y_j$ be the component of $Z_j$ containing $K_j$. The proof is finished. □

DEFINITION 7.2.7. *A collection of relatively compact sets $\{Y_j\}$ satisfying the conclusions of Theorem 7.2.6 is called a normal exhaustion of $X$.* ◇

Finally, in the proof of the Uniformization Theorem we will need the following lemma.

LEMMA 7.2.8. *Let $X$ be a simply connected open Riemann surface. If $Y \subset X$ is Runge, then $Y$ is simply connected.*

**Proof.** Assume $\pi_1(Y) \neq \{1\}$. Let $\gamma : [0,1] \to Y$ be a Jordan curve in $Y$ whose homotopy class is non-zero in $\pi_1(Y)$. (The existence of such a curve can be obtained, for example, by choosing a complete Riemannian metric on $Y$ and taking a distance-minimizing closed geodesic.) The Jordan curve separates $X$ into two components $U$ and $V$, one of which, say $U$, is relatively compact. Indeed, $X$ is simply connected, so the curve is contractible in $X$. But since $[\gamma] \in \pi_1(Y) - \{0\}$, some component of $X - Y$ is contained in $U$, and thus is relatively compact. Thus $Y$ is not Runge. □

### 7.2.2. The Runge Theorem of Behnke-Stein.

THEOREM 7.2.9 (Behnke-Stein Runge Theorem). *Let $X$ be an open Riemann surface and let $Y \subset\subset X$ be a Runge open set. Then $\mathcal{O}(X)|_Y$ is dense in $\mathcal{O}(Y)$ in the topology of uniform convergence on compact sets.*

**Proof.** Fix a relatively compact open set $Z \subset\subset X$ with $Y \subset\subset Z$. We begin by showing that every holomorphic function on $Y$ can be approximated uniformly by functions in $Z$. In view of the Hahn-Banach Theorem, it suffices to show that every continuous linear functional on $\mathscr{C}(X)$ (with the locally uniform topology) that annihilates $\mathcal{O}(Z)$ also annihilates $\mathcal{O}(Y)$. By the Riesz Representation Theorem, every such linear functional is represented by integration against a compactly supported real-valued measure $d\mu$ on $Y$. Let $K := \text{Supp}(d\mu)$.

Fix a measure $d\mu$ whose associated linear functional annihilates $\mathcal{O}(Z)$. Associate to $d\mu$ the linear functional $S$ on $(0,1)$-forms with compact support in $Z$ defined by

$$S(\theta) := \int_Y f\, d\mu, \quad \text{where} \quad \bar{\partial} f = \theta \text{ in } Z.$$

Note that $S$ is well-defined since if $\bar{\partial} f = \bar{\partial} \tilde{f}$, then $f - \tilde{f}$ is holomorphic on $Z$ and thus by hypothesis,

$$\int_Y (f - \tilde{f})\, d\mu = 0.$$

## 7.2. Approximation on open Riemann surfaces

Now, by the proof of Theorem 7.1.18, the function
$$f(x) := \frac{1}{2\pi\sqrt{-1}} \int_Z \partial_y \left( G_Z(x,y)\theta(y) \right)$$
solves the equation $\bar{\partial} f = \theta$ on $Z$. Therefore, with
$$\sigma(y) := \frac{1}{2\pi\sqrt{-1}} \int_Y G_Z(x,y) d\mu(x),$$
we have
$$S(\theta) = \frac{1}{2\pi\sqrt{-1}} \int_Z \partial(\sigma\theta).$$
Observe that $\sigma$ is harmonic on $Z - K$ and vanishes on $\partial Z$. Thus $\sigma$ must be zero on every unbounded component of $Z - K$ in $Z$. It follows that $\sigma$ is zero on $Z - \widehat{K}_{\mathcal{O}(Z)}$.

Now let $f \in \mathcal{O}(Y)$. Since $Y$ is Runge, $\widehat{K}_{\mathcal{O}(Z)} \subset\subset Y$. Thus there is a function $h \in \mathcal{C}_o^\infty(Z)$ such that $h = f$ on $\widehat{K}_{\mathcal{O}(Z)}$. It follows that $\mathrm{Supp}(\bar{\partial} h) \subset Z - \widehat{K}_{\mathcal{O}(Z)}$, and since $\sigma$ vanishes on $Z - \widehat{K}_{\mathcal{O}(Z)}$, we have that $\sigma \bar{\partial} h = 0$. Therefore
$$\int_Y f d\mu = \int_Y h d\mu = \int_Z \partial(\sigma \bar{\partial} h) = 0.$$
Thus we have proved that $\mathcal{O}(Z)|_Y$ is dense in $\mathcal{O}(Y)$.

Finally, fix $f \in \mathcal{O}(Y)$ and $\varepsilon > 0$. Let $Y = Y_0 \subset\subset Y_1 \subset\subset Y_2 \subset\subset \ldots \subset X$ be a normal exhaustion. By what we have just proved, $\mathcal{O}(Y_i)|_{Y_{i-1}}$ is dense in $\mathcal{O}(Y_{i-1})$ for all $i = 1, 2, \ldots$. Let $f_0 := f$ and, assuming $f_0, \ldots, f_{i-1}$ have been defined, choose $f_i \in \mathcal{O}(Y_i)$ such that
$$\sup_{Y_{i-1}} |f_i - f_{i-1}| < 2^{-i}\varepsilon.$$
Define
$$g := f + \sum_{j=1}^\infty f_j - f_{j-1}.$$
The reader can check that $g$ is well-defined and converges locally uniformly on $X$ and is therefore holomorphic. In particular, $g$ converges uniformly on $Y$ and
$$\sup_Y |f - g| \leq \sum_{j=1}^\infty \sup_Y |f_j - f_{j-1}| < \varepsilon.$$
The proof is complete. □

**7.2.3. Function-theoretic description of hulls.** The Maximum Principle suggests a connection between the holomorphic hull of a compact set $K$ and $\mathcal{O}(X)|_K$. The following proposition shows that indeed there is such a connection, in which the Behnke-Stein Theorem plays a key role.

PROPOSITION 7.2.10. *Let $X$ be an open Riemann surface and let $K \subset X$ be a compact set. Then*

$$\widehat{K}_{\mathcal{O}(X)} = \left\{ x \in X \; ; \; |f(x)| \leq \sup_K |f| \text{ for all } f \in \mathcal{O}(X) \right\}.$$

**Proof.** Let $x$ be a point in some relatively compact component $U \subset\subset X - K$. Then $\partial U \subset K$. By the Maximum Principle, for any $f \in \mathcal{O}(X)$, $|f(x)| \leq \sup_{\partial U} |f| \leq \sup_K |f|$.

Conversely, suppose $x \notin \widehat{K}_{\mathcal{O}(X)}$. Let $V$ be an unbounded component of $X - K$ containing $x$. Take a small disk $\Delta \subset V$ centered at $x$. Then

$$\widehat{(K \cup \Delta)}_{\mathcal{O}(X)} = \widehat{K}_{\mathcal{O}(X)} \cup \Delta \quad \text{and} \quad \widehat{K}_{\mathcal{O}(X)} \cap \Delta = \emptyset.$$

By Theorem 7.2.9 the function

$$f(z) := \begin{cases} 0, & z \in \widehat{K}_{\mathcal{O}(X)} \\ 1, & z \in \Delta \end{cases}$$

can be approximated by global holomorphic functions on $X$. It follows that there is a function $F \in \mathcal{O}(X)$ such that

$$|F(x)| > 1/2 > \sup_K |F|.$$

Thus $x \notin \{y \in X \; ; \; |f(y)| \leq \sup_K |f| \text{ for all } f \in \mathcal{O}(X)\}$. The proof is complete. □

## 7.3. Exercises

7.1. Show that $\mathscr{P}_f$ as in Definition 7.1.6 is a Perron family.

7.2. Show that the family $\mathscr{G}$ in the proof of Theorem 7.1.13 is Perron.

7.3. Let $X$ be a compact Riemann surface with smooth, non-empty boundary of dimension 1. Assume that $X$ admits a bounded non-constant subharmonic function and thus has a Green's Function. (This is in fact the case.) Show that $G_X(x,y) = G_X(y,x)$.

7.4. Show that the Dirichlet Problem cannot be solved in general on $\mathbb{D} - \{0\}$.

7.5. Give the details of the proof of Proposition 7.1.16.

7.6. Does $\mathbb{C} - \{0,1\}$ have a Green's Function?

7.7. Compute the Green's Function for the upper half-plane.

7.8. Prove that the holomorphic hull of a closed set is closed.

7.9. Show that the function $g$ defined at the end of the proof of Theorem 7.2.9 is well-defined and holomorphic on $X$.

7.10. Use Green's Functions to give an integral formula for the Dirichlet Problem on a compact Riemann surface with piecewise-smooth boundary.

7.11. Show that given any closed discrete sequence $\{x_j\}$ in an open Riemann surface $X$ there is a function $f \in \mathcal{O}(X)$ such that
$$\lim_{j \to \infty} |f(x_j)| = +\infty.$$

7.12. With the previous exercise in mind, let $\{z_j\} \subset \mathbb{D}$ be a closed discrete subset of the unit disk $\mathbb{D}$ whose closure in $\overline{\mathbb{D}}$ contains $\partial \mathbb{D}$. According to Exercise 7.11, there is a holomorphic function that blows up along $\{z_j\}$. Can it happen that for every $x \in \partial \mathbb{D}$,
$$\lim_{\mathbb{D} \ni z \to x} |f(z)| = +\infty?$$

*Chapter 8*

# Solving $\bar{\partial}$ for Smooth Data

We begin this chapter by establishing the fact, stated at the end of Chapter 5, that given any smooth $(0,1)$-forms $\alpha$ on an open Riemann surface, the equation $\bar{\partial}u = \alpha$ can always be solved.

## 8.1. The basic result

Using the technique of the Runge Approximation Theorem, we can now extend Theorem 7.1.18 to the following result.

THEOREM 8.1.1. *Let $X$ be an open Riemann surface. For any smooth $(0,1)$-form $\theta$ on $X$ there is a smooth function $f : X \to \mathbb{C}$ such that $\bar{\partial}f = \theta$.*

**Proof.** Let $Y_0 \subset\subset Y_1 \subset\subset \ldots \subset X$ be a normal exhaustion of $X$. We shall construct a sequence of functions $f_n : Y_n \to \mathbb{C}$ for $n = 0, 1, 2, \ldots$ such that

$$\bar{\partial}f_n = \theta \text{ on } Y_{n-1} \quad \text{and} \quad \sup_{Y_{n-1}} |f_{n+1} - f_n| < \tfrac{1}{2^n}, \quad n \geq 1.$$

The choice of $f_0$ does not matter, but we take it to be $0$ just to fix a choice. Take $f_1$ to be any solution of $\bar{\partial}f_1 = \theta$ on $Y_0$. Such $f_1$ exists by Theorem 7.1.18. Now suppose $f_1, \ldots, f_k$ have been chosen. Let $u_{k+1} : Y_{k+1} \to \mathbb{C}$ be a smooth function solving $\bar{\partial}u_{k+1} = \theta$ on $Y_k$. Then $u_{k+1} - f_k$ is holomorphic on $Y_{k-1}$ and thus there is a holomorphic function $g_k \in \mathcal{O}(X)$ that is within $2^{-k}$ of $u_{k+1} - f_k$ on $Y_{k-1}$. Let $f_{k+1} := u_{k+1} - g_k$. Then

$$\bar{\partial}f_{k+1} = \theta \text{ on } Y_k \quad \text{and} \quad \sup_{Y_{k-1}} |f_{k+1} - f_k| = \sup_{Y_{k-1}} |u_{k+1} - f_k - g_k| < 2^{-k}.$$

Now $f_n = f_0 + \sum_{j=1}^{n} f_j - f_{j-1}$. Note that on $Y_{k-1}$, $k < n$, $f_n = f_k + h_{k,n}$ for some holomorphic function $h_{k,n}$ whose sup norm on $Y_{k-1}$ is at most $2^{1-k}$. It follows that $f_n \to f$ for some smooth function $f$ in $\mathscr{C}^1$-norm, and thus

$$\bar\partial f = \lim \bar\partial f_n = \lim \theta = \theta.$$

The proof is complete. □

## 8.2. Triviality of holomorphic line bundles

Let $L \to X$ be a holomorphic line bundle. Choose a locally trivial open cover $\{U_i\}$ such that each $U_j$ and each $U_k \cap U_\ell$ is connected and simply connected and such that $L|_{U_j}$ is trivial. For each $i$, let $e_i \in \Gamma_{\mathcal{O}}(U_i, L)$ be a nowhere-zero section, and define the holomorphic functions

$$h_{ij} := e_i/e_j \quad \text{on} \quad U_i \cap U_j.$$

Then $h_{ij}$ is a nowhere-zero holomorphic function, and thus there are holomorphic functions $g_{ij}$ such that $h_{ij} = e^{g_{ij}}$.

Let $\{\chi_j\}$ be a partition of unity subordinate to the open cover $\{U_j\}$. Consider the smooth functions

$$\tilde g_i := \sum_j \chi_j g_{ij} \quad \text{on} \quad U_i.$$

Then

$$\tilde g_i - \tilde g_j = \sum_k \chi_k(g_{ik} - g_{jk}) = g_{ij}.$$

Then $\bar\partial \tilde g_i = \bar\partial \tilde g_j$ on $U_i \cap U_j$, so there is a globally defined smooth $(0,1)$-form $\gamma$ such that $\bar\partial \gamma = \bar\partial \tilde g_j$ on $U_j$. Let $h$ be a smooth function satisfying $\bar\partial h = \gamma$, and set

$$g_i := \tilde g_i - h.$$

Then $g_i \in \mathcal{O}(U_i)$ and $g_i - g_j = g_{ij}$. Define

$$s_i := e^{-g_i} e_i.$$

Then on $U_i \cap U_j$ one has

$$s_i = e^{-g_i} e_i = e^{-g_{ij}} e^{-g_j} e_i = e^{-g_j} e_j = s_j.$$

Hence $s = s_i$ on $U_i$ is a globally defined holomorphic section of $L$, and evidently $s$ has no zeros. We have therefore proved the following theorem.

THEOREM 8.2.1. *Every holomorphic line bundle on an open Riemann surface is trivial.*

## 8.3. The Weierstrass Product Theorem

Let $X$ be a Riemann surface and $D$ a divisor on $X$. We seek to find out when $D = \mathrm{Ord}(f)$ for some $f \in \mathscr{M}(X)$. If $X$ is a compact Riemann surface, a necessary condition is that $\deg(D) = 0$, but we already know that this condition is not sufficient unless $X = \mathbb{P}_1$. The sufficient condition for surfaces of positive genus will be addressed in Chapter 14; it is the content of Abel's Theorem.

In this section we solve the problem for open Riemann surfaces. Unlike compact Riemann surfaces, there is no obstruction in the open case. In fact, as a corollary of Theorem 8.2.1, we obtain the following generalization of the famous Weierstrass Product Theorem.

THEOREM 8.3.1. *Let $D$ be a divisor on an open Riemann surface. Then there is a function $f \in \mathscr{M}(X)$ such that $\mathrm{Ord}(f) = D$.*

**Proof.** Let $D$ be a divisor, and let $L$ be the line bundle associated to the divisor $D$. The line bundle $L$ has a canonical meromorphic section $s_D$ such that

$$\mathrm{Ord}(s_D) = D.$$

But by Theorem 8.2.1, there is a nowhere-zero section $s \in \Gamma_{\mathcal{O}}(X, L)$. Thus the meromorphic function $f := s_D/s$ satisfies

$$\mathrm{Ord}(f) = \mathrm{Ord}(s_D) - \mathrm{Ord}(s) = \mathrm{Ord}(s_D) = D.$$

This completes the proof. □

## 8.4. Meromorphic functions as quotients

Let $X$ be a Riemann surface. If $f$ is a meromorphic function on $X$, then locally $f = g/h$ for holomorphic functions $g$ and $h$. However, in general this is not the case globally. Indeed, on a compact Riemann surface there are no non-constant holomorphic functions. But as we shall see in Chapter 12, there are always many meromorphic functions.

Amazingly, on a Riemann surface the only obstruction to writing a meromorphic function as a quotient of holomorphic functions globally is compactness. In fact, we have the following result.

THEOREM 8.4.1. *Let $X$ be an open Riemann surface and let $f \in \mathscr{M}(X)$. Then there are holomorphic functions $g, h : X \to \mathbb{C}$ such that $f = g/h$.*

**Proof.** Let $D := (f)_\infty$ be the divisor of poles of $f$. Then $D$ defines a holomorphic line bundle $L$ and a canonical section $s$ such that

$$\mathrm{Ord}(s) = D.$$

Since $D$ is effective, $s$ is holomorphic.

Since $X$ is open, we have, by Theorem 8.2.1, a section $t \in \Gamma_{\mathcal{O}}(X, L)$ with no zeros. Let $h = s/t$. Then $h$ is a holomorphic function, and

$$\text{Ord}(fh) = \text{Ord}(f) + \text{Ord}(s) = (f)_0 - (f)_\infty + (f)_\infty = (f)_0$$

is effective. It follows that $g = fh$ is a holomorphic function. The proof is complete. $\square$

## 8.5. The Mittag-Leffler Problem

In this section we consider the problem of specifying the singular parts of a meromorphic function.

On a compact Riemann surface, the singular structure of a meromorphic function is not arbitrary. By contrast, on an open Riemann surface we have total freedom to prescribe the singular structure of a meromorphic function.

**8.5.1. Principal parts and the Mittag-Leffler Problem.** We begin by defining the notion of a singular structure, which is classically called a *principal part*.

DEFINITION 8.5.1. Let $X$ be a Riemann surface. A principal part is a collection $\{(U_j, f_j) \,;\, j \in J\}$ such that

(1) $\mathscr{U} = \{U_j \,;\, j \in J\}$ is a locally finite open cover,
(2) $f_j \in \mathscr{M}(U_j)$ for all $j \in J$, and
(3) $f_i - f_j \in \mathcal{O}(U_i \cap U_j)$ for all $i, j \in J$. $\diamond$

EXAMPLE 8.5.2. Let $f$ be a meromorphic function on $X$. Choose a locally finite open cover $\mathscr{U} := \{U_j \,;\, j \in J\}$ and holomorphic functions $g_j \in \mathcal{O}(U_j)$. Then the functions $f_j := f|_{U_j} - g_j$, together with $\mathscr{U}$, form a principal part. $\diamond$

Thus the principal part in the above example is determined by a global meromorphic function.

DEFINITION 8.5.3. A principal part $\{(U_j, f_j) \,;\, j \in J\}$ on a Riemann surface $X$ is said to be *trivial*, or may also be called *the principal part of a meromorphic function*, if there is a meromorphic function $f \in \mathscr{M}(X)$ and holomorphic functions $g_j \in \mathcal{O}(U_j)$, $j \in J$, such that $f_j = f|_{U_j} - g_j$ for all $j \in J$. $\diamond$

DEFINITION 8.5.4. The Mittag-Leffler Problem is to characterize the set of trivial principal parts. $\diamond$

**8.5.2. Solution on open Riemann surfaces.**

THEOREM 8.5.5. *On an open Riemann surface $X$, every principal part is trivial.*

## 8.5. The Mittag-Leffler Problem

**Proof.** Let $g_{ij} := f_i - f_j \in \mathcal{O}(U_i \cap U_j)$. Choose a partition of unity $\{\chi_j\}$ subordinate to the open cover $\mathscr{U}$, and define the $(0,1)$-form $\alpha_j$ on $U_j$ by

$$\alpha_j := \bar{\partial}\Big(\sum_i \chi_i g_{ij}\Big).$$

Observe that

$$\alpha_j - \alpha_k = \bar{\partial}\Big(\sum_i \chi_i (g_{ij} - g_{ik})\Big) = \bar{\partial}\Big(g_{kj} \sum_i \chi_i\Big) = \bar{\partial} g_{kj} = 0.$$

It follows that the $(0,1)$-form $\alpha$ defined to be $\alpha_j$ in $U_j$ is smooth. By Theorem 8.1.1 there is a smooth function $h$ such that $\bar{\partial} h = \alpha$.

Consider the function

$$g_j := -h + \sum_i \chi_i g_{ij}.$$

Then $\bar{\partial} g_j = \alpha_j - \alpha = 0$, so that $g_j \in \mathcal{O}(U_j)$. We also have

$$g_j - g_k = \sum_i \chi_i (g_{ij} - g_{ik}) = g_{kj} = f_k - f_j.$$

It follows that $f_j + g_j = f_k + g_k$, and thus we have a global meromorphic function $f \in \mathscr{M}(X)$ defined to be $f_j + g_j$ in $U_j$. Evidently $f$ shows that $\{(U_j, f_j)\}$ is trivial. $\square$

**8.5.3. Solution on compact Riemann surfaces.** We turn now to the case of compact Riemann surfaces.

EXAMPLE 8.5.6. On a compact Riemann surface there may be principal parts that are not the principal parts of meromorphic functions. For example, on a complex torus one can take a single open coordinate chart $U$ and one point $p \in U$ and consider the function $f_U(z) = (z-p)^{-1}$. Then the principal part $\{(U, f_U), (V, 1)\}$, where $V$ is any open subset of the torus containing the complement of $U$, is not the principal part of a meromorphic function. $\diamond$

Up to a certain point, the method of proof for open Riemann surfaces can be carried over to compact Riemann surfaces. The proof breaks down exactly at the point where we try to solve the equation $\bar{\partial} h = \alpha$; this equation cannot be solved on a general compact Riemann surface.

Let us look again at the form $\alpha$ constructed in the proof of Theorem 8.5.5. To emphasize the dependence of the choice of partition of unity, we write $\alpha = \alpha^\chi$. Suppose we choose another partition of unity $\tilde{\chi}$. Then we have

$$\sum_i (\chi_i - \tilde{\chi}_i)(g_{ij} - g_{ik}) = g_{kj} \sum_i (\chi_i - \tilde{\chi}_i) = g_{kj}(1-1) = 0.$$

Thus we can define the global function $\psi_{\chi,\tilde{\chi}}$ by
$$\psi_{\chi,\tilde{\chi}} = \sum_i (\chi_i - \tilde{\chi}_i) g_{ij} \quad \text{on} \quad U_j,$$
and we find that
$$\alpha^\chi - \alpha^{\tilde{\chi}} = \bar{\partial} \psi_{\chi,\tilde{\chi}}.$$
It follows that the existence of a function $h$ satisfying the equation $\bar{\partial} h = \alpha$ is independent of the choice of partition of unity.

The proof of Theorem 8.5.5 is easily modified to obtain one part of the following result. (Recall the definition of Dolbeault cohomology discussed in Section 5.10.)

THEOREM 8.5.7. *Let $P := \{(U_j, f_j)\}$ be a principal part on a Riemann surface $X$ and let $a_P$ be the $\bar{\partial}$-cohomology class of the form $\alpha$ associated to $P$ by the construction in the proof of Theorem 8.5.5. Then $P$ is the principal part of a meromorphic function if and only if $a_P = 0$.*

**Proof.** We leave it to the reader to confirm that the *if* direction has already been proved, and we prove only the converse. Suppose, then, that we have a meromorphic function $f$ and holomorphic functions $g_j$ such that $f = f_j + g_j$ on $U_j$. Then
$$\alpha = \bar{\partial} \sum_i \chi_i (g_j - g_i) = -\sum_i g_i \bar{\partial} \chi_i - g_j \bar{\partial}(1) = -\sum_i g_i \bar{\partial} \chi_i = \bar{\partial} h,$$
where $h = -\sum \chi_i g_i$. Thus $a_P = 0$ by definition. The proof is complete. □

REMARK. We can take one more formal step. Roughly speaking, two principal parts are said to be equivalent if their difference is the principal part of a meromorphic function. More precisely, we say $(\{U_j, f_j\})$ is equivalent to $(\{V_i, g_i\})$ if the principal part $\{(U_j \cap V_i, f_j - g_i)\}$ is the principal part of a meromorphic function.

Our results show that the map sending the equivalence class $[P]$ of a principal part $P$ to the $\bar{\partial}$-cohomology class $\alpha_P$ is 1-1. It turns out that $[P] \to a_P$ is also surjective onto $H^1_{\bar{\partial}}(X)$. In other words, every element of $H^1_{\bar{\partial}}(X)$ is of the form $a_P$ for some principal part $P$. ◇

**8.5.4. Principal parts of meromorphic 1-forms.** Suppose $\{U_i\}_{i \in \mathcal{I}}$ is an open cover of $X$ and $\eta_i$ is a meromorphic 1-form on $U_i$. (We will say $\{\eta_i\}$ is subordinate to $\{U_i\}$.) Assume that $\eta_i - \eta_j$ is holomorphic on $U_i \cap U_j$. Then at any point $p \in U_i \cap U_j$, the residue of $\eta_i$ and that of $\eta_j$ agree. Thus we can define
$$\operatorname{Res}_p(\{\eta_i\}_{i \in I}) = \operatorname{Res}_p(\eta_i), \quad p \in U_i.$$
We can then define a divisor
$$\operatorname{Res}(\{\eta_i\}) := \sum_{p \in X} \operatorname{Res}_p(\{\eta_i\}) \cdot p.$$
With this notation, we have the following theorem.

## 8.5. The Mittag-Leffler Problem

THEOREM 8.5.8 (Mittag-Leffler Theorem for meromorphic 1-forms). *Let $X$ be a compact Riemann surface, let $\{U_j\}$ be an open cover of $X$, and let $\{\eta_j\}$ be a collection of meromorphic 1-forms subordinate to $\{U_j\}$ such that $\eta_i - \eta_j$ is holomorphic on $U_i \cap U_j$. Then*

$$\deg(\mathrm{Res}(\{\eta_j\})) := \sum_{x \in X} \mathrm{Res}_x(\{\eta_i\}) = 0$$

*if and only if there is a meromorphic 1-form $\eta$ on $X$ such that $\eta - \eta_j$ is holomorphic on $U_j$.*

**Proof.** One direction is easy. We already know that for a global meromorphic 1-form $\eta$, $\deg \mathrm{Res}(\eta) = 0$. Since $\eta - \eta_j$ is holomorphic on $U_j$ for all $j$,

$$\mathrm{Res}(\eta) = \mathrm{Res}(\{\eta_j\}).$$

We now establish the converse. Let $x_1, ..., x_N$ be the set of poles of $\{\eta_i\}$. Choose a partition of unity $\{\chi_i\}$ subordinate to $\{U_i\}$, and let

$$\tilde{\eta}_i := \sum_j \chi_j(\eta_i - \eta_j) \quad \text{on} \quad U_j.$$

Then each $\tilde{\eta}_j$ is smooth. Moreover,

$$\tilde{\eta}_i - \tilde{\eta}_j = \eta_i - \eta_j$$

is a holomorphic 1-form on $U_i \cap U_j$, and thus

$$\bar{\partial}\tilde{\eta}_i = \bar{\partial}\tilde{\eta}_j \quad \text{on} \quad U_i \cap U_j.$$

It follows that the $(1,1)$-form $\alpha$ defined by

$$\alpha := \bar{\partial}\tilde{\eta}_i \quad \text{on} \quad U_i$$

is globally defined.

We claim that

$$\frac{1}{2\pi\sqrt{-1}} \int_X \alpha = \deg(\mathrm{Res}(\{\eta_i\})).$$

To see this claim, let $D_{j,\varepsilon}$ be small coordinate disks of radius $\varepsilon$, each containing exactly one pole $x_j$, $j = 1, ..., N$, of $\{\eta_i\}$. Observe that each $\eta_j$ is holomorphic on $U_j - \{x_1, ..., x_N\}$, and thus since $\eta_i - \eta_j = \tilde{\eta}_i - \tilde{\eta}_j$, the form $\beta := \eta_j - \tilde{\eta}_j$ is smooth on $X - \{x_1, ..., x_N\}$ where it satisfies the equation $d\beta = -\alpha$. Since $\alpha$ is smooth, we have

$$\begin{aligned}
\int_X \alpha &= \lim_{\varepsilon \to 0} \int_{X - \bigcup_j D_{j,\varepsilon}} \alpha \\
&= -\lim_{\varepsilon \to 0} \int_{X - \bigcup_j D_{j,\varepsilon}} d\beta \\
&= \lim_{\varepsilon \to 0} \sum_{j=1}^N \int_{\partial D_{j,\varepsilon}} \beta \\
&= 2\pi\sqrt{-1} \sum_j \operatorname{Res}_{x_j}(\{\eta_{i(j)}\}) - \sum_j \lim_{\varepsilon \to 0} \int_{\partial D_{j,\varepsilon}} \tilde{\eta}_{i(j)} \\
&= 2\pi\sqrt{-1} \sum_j \operatorname{Res}_{x_j}(\{\eta_{i(j)}\}),
\end{aligned}$$

where $i(j)$ is any integer such that $D_{j,\varepsilon} \subset U_{i(j)}$. The last equality holds since $\tilde{\eta}_i$ are smooth forms. Thus we have the stated equality.

In particular, if $\deg(\operatorname{Res}(\{\eta_i\})) = 0$, then $\int_X \alpha = 0$. As we will show in Chapter 9 (Corollary 9.3.3), the vanishing of $\int_X \alpha$ implies that there is a function $f$ such that $\alpha = \partial\bar{\partial} f$, and thus $\alpha = \bar{\partial}\mu$ with $\mu = -\partial f$. Let $\xi_i := \tilde{\eta}_i - \mu$. Then $\bar{\partial}\xi_i = 0$, so $\xi_i$ are holomorphic, and

$$\xi_i - \xi_j = \tilde{\eta}_i - \tilde{\eta}_j = \eta_{ij} = \eta_i - \eta_j.$$

It follows that $\eta := \xi_i - \eta_i$ on $U_i$ defines a global meromorphic 1-form having the same polar structure as $\{\eta_i\}$. This completes the proof. □

## 8.6. The Poisson Equation on open Riemann surfaces

In this section we solve Poisson's Equation with smooth forcing term on an open Riemann surface. Although there are many applications of the solvability of this equation, we shall not pursue them here.

### 8.6.1. Pfluger's Harmonic Runge Theorem. 
The method used to prove Runge's Approximation Theorem for holomorphic functions can be applied to harmonic functions.

8.6.1.1. *Local solvability.* The starting point is the following lemma.

LEMMA 8.6.1. *Let $X$ be a Riemann surface, let $Y \subset\subset X$ be an open subset such that $X - Y \neq \emptyset$, and let $\omega$ be a $(1,1)$-form with compact support in $Y$. Then there is a function $f$ on a neighborhood of $\overline{Y}$ such that*

$$\sqrt{-1}\partial\bar{\partial} f = \omega.$$

## 8.6. The Poisson Equation on open Riemann surfaces

**Proof.** Let $Y' \subset\subset X$ be a smoothly bounded open subset such that $\overline{Y} \subset Y'$ and $X - Y' \neq \emptyset$. Let $G_{Y'}$ be the Green's Function of $Y'$. One simply takes

$$f(x) := \frac{1}{2\pi} \int_{Y'} G_{Y'}(x,y)\omega(y).$$

The proof is finished. □

**8.6.1.2.** *Pfluger's Theorem.* Recall that $H(X)$ denotes the set of harmonic functions on a Riemann surface $X$.

**THEOREM 8.6.2.** *Let $X$ be an open connected Riemann surface and let $Y \subset\subset X$ be a Runge open set. Then $H(X)|_Y$ is dense in $H(Y)$ in the topology of uniform convergence on compact sets.*

**Proof.** The approach is the same as in the proof of Runge's Theorem 7.2.9. Fix a relatively compact open set $Z \subset\subset X$ with $Y \subset\subset Z$. We begin by showing that every harmonic function on $Y$ can be approximated uniformly by harmonic functions in $Z$. As before, the Hahn-Banach Theorem will achieve our aim if we show that every linear functional that annihilates $H(Z)$ also annihilates $H(Y)$. By the Riesz Representation Theorem every such linear functional is represented by a compactly supported real-valued measure $d\mu$ on $Y$. Let $K := \text{Supp}(d\mu)$.

We associate to the measure $d\mu$ the following linear functional on $(1,1)$-forms $\theta$ compactly supported on $Z$:

$$S(\theta) := \int_Y f \, d\mu, \quad \text{where} \quad \sqrt{-1}\partial\bar{\partial}f = \theta \text{ in } Z.$$

Note that $S$ is well-defined since if $\sqrt{-1}\partial\bar{\partial}f = \sqrt{-1}\partial\bar{\partial}g$, then $f - g$ is harmonic on $Z$ and thus by hypothesis,

$$\int_Y (f-g) \, d\mu = 0.$$

Now, by the proof of Lemma 8.6.1, the function

$$f(x) := \frac{1}{2\pi} \int_Z G_Z(x,y)\theta(y)$$

solves the equation $\sqrt{-1}\partial\bar{\partial}f = \theta$ on $Z$. Therefore, with

$$\sigma(y) := \frac{1}{2\pi} \int_Y G_Z(x,y) \, d\mu(x),$$

we have

$$S(\theta) = \frac{1}{2\pi} \int_Z \sigma\theta.$$

Observe that $\sigma$ is harmonic on $Z - K$ and vanishes on $\partial Z$. By the Maximum Principle, $\sigma$ must be zero on every unbounded component of $Z - K$ in $Z$. It follows that $\sigma$ is zero on $Z - \widehat{K}_{\mathcal{O}(Z)}$.

Now let $f \in H(Y)$. Since $Y$ is Runge, $\widehat{K}_{\mathcal{O}(Z)} \subset\subset Y$. Thus there is a function $g \in \mathscr{C}_o^\infty(Z)$ such that $g = f$ on $\widehat{K}_{\mathcal{O}(Z)}$. Then

$$\int_Y f d\mu = \int_Y g d\mu = \int_Z \sigma \sqrt{-1}\partial\bar{\partial}g = 0,$$

where the last equality holds since $\text{Supp}(\sqrt{-1}\partial\bar{\partial}g) \subset Z - \widehat{K}_{\mathcal{O}(Z)}$, and hence $\sigma|_{\text{Supp}(\sqrt{-1}\partial\bar{\partial}g)} \equiv 0$. Thus we have proved that $H(Z)|_Y$ is dense in $H(Y)$.

Finally, fix $f \in H(Y)$ and $\varepsilon > 0$. Let $Y = Y_0 \subset\subset Y_1 \subset\subset Y_2 \subset\subset \ldots \subset X$ be a normal exhaustion. By what we have just proved, $H(Y_i)|_{Y_{i-1}}$ is dense in $H(Y_{i-1})$ for all $i = 1, 2, \ldots$. Let $f_0 := f$ and define inductively $f_i \in H(Y_i)$ such that

$$\sup_{Y_{i-1}} |f_i - f_{i-1}| < 2^{-i}\varepsilon.$$

Define

$$g := f + \sum_{j=1}^{\infty} f_j - f_{j-1}.$$

Then $g$ converges uniformly on $Y$ and

$$\sup_Y |f - g| \leq \sum_{j=1}^{\infty} \sup_Y |f_j - f_{j-1}| < \varepsilon.$$

Moreover, $g$ converges uniformly on each $Y_j$, and therefore $g \in H(X)$. The proof is complete. □

**8.6.2. Solving $\partial\bar{\partial}$ with smooth data on open Riemann surfaces.** We can now use the analogue of the proof of Theorem 8.1.1 to solve Poisson's equation with smooth data.

THEOREM 8.6.3. *Let $\omega$ be a smooth $(1,1)$-form on an open Riemann surface $X$. Then there is a smooth function $\psi : X \to \mathbb{C}$ such that*

$$\sqrt{-1}\partial\bar{\partial}\psi = \omega.$$

**Proof.** Let $Y_0 \subset\subset Y_1 \subset\subset \ldots \subset X$ be a normal exhaustion of $X$. We shall construct a sequence of functions $f_n : Y_n \to \mathbb{C}$ for $n = 1, 2, \ldots$ such that

$$\sqrt{-1}\partial\bar{\partial}f_n = \omega \text{ on } Y_{n-1} \quad \text{and} \quad \sup_{Y_{n-1}} |f_{n+1} - f_n| < \tfrac{1}{2^{n-1}}, \quad n \geq 1.$$

Take $f_1$ to be any solution of $\sqrt{-1}\partial\bar{\partial}f_1 = \omega$ on $Y_0$. Such $f_1$ exists by Lemma 8.6.1. Now suppose $f_1, \ldots, f_k$ have been chosen. Let $u_{k+1} : Y_{k+1} \to \mathbb{C}$ be a smooth function solving $\sqrt{-1}\partial\bar{\partial}u_{k+1} = \omega$ on $Y_{k+1}$. Then $u_{k+1} - f_k$ is harmonic on $Y_{k-1}$ and thus there is a harmonic function $g_k \in H(X)$ that is within $2^{-k}$ of $u_{k+1} - f_k$ on $Y_{k-1}$. Let $f_{k+1} := u_{k+1} - g_k$. Then

$$\sqrt{-1}\partial\bar{\partial}f_{k+1} = \omega \text{ on } Y_k \quad \text{and} \quad \sup_{Y_{k-1}} |f_{k+1} - f_k| = \sup_{Y_{k-1}} |u_{k+1} - f_k - g_k| < 2^{-k}.$$

Note that $f_n = f_0 + \sum_{j=1}^n f_j - f_{j-1}$ and that, on $Y_{k-1}$, $k < n$, $f_n = f_k + h_{k,n}$ for some harmonic function $h_{k,n}$ on $Y_{k-1}$ whose sup norm on $Y_{k-1}$ is at most $2^{-k}$. It follows that $f_n$ converges to some smooth function $f$ locally in $\mathscr{C}^2$-norm, and thus
$$\sqrt{-1}\partial\bar{\partial} f = \lim \sqrt{-1}\partial\bar{\partial} f_n = \lim \omega = \omega.$$
The proof is complete. □

In the next chapter we will establish an analogue of Theorem 8.6.3 for compact Riemann surfaces.

## 8.7. Exercises

8.1. Let $X$ be an open Riemann surface. Suppose $\theta \in \Gamma(X, T_X^{*0,1})$ has compact support. Can a solution $u$ of the equation $\bar{\partial} u = \theta$ have compact support?

8.2. In the notation of Theorem 8.5.7, show that if the cohomology class $a_P$ of a principal part $P$ is zero then $P$ is trivial.

8.3. If $f : \mathbb{D} \to \mathbb{C}$ is a function on the unit disk, find a solution to the equation $\Delta u = f$ satisfying $u|_{\partial \mathbb{D}} \equiv 0$.

8.4. If $x$ is a point in a compact Riemann surface $X$, show that there is no locally integrable function $G_x$ such that $\Delta G_x = \frac{\pi}{2}\delta_x$.

8.5. Show that if $x$ and $y$ are two distinct points of $\mathbb{P}_1$ then there is a locally integrable function $H$ such that $\Delta H = \frac{\pi}{2}\delta_x - \frac{\pi}{2}\delta_y$. Is there a function $F$ such that $\Delta F = \frac{\pi}{2}\delta_x - \frac{\pi}{4}\delta_y$? What about $\frac{\pi}{2}\delta_x - \frac{\pi}{4}\delta_y - \frac{\pi}{4}\delta_z$?

8.6. Let $X$ be a compact Riemann surface of genus 1 and let $x, y \in X$ be two distinct points. Show that there is a locally integrable function $H$ such that $\Delta H = \frac{\pi}{2}\delta_x - \frac{\pi}{2}\delta_y$.

8.7. Prove the assertion made at the end of Example 8.5.6, namely that the given principal part is not the principal part of a meromorphic function.

# Chapter 9

# Harmonic Forms

## 9.1. The definition and basic properties of harmonic forms

Let $X$ be a compact Riemann surface and $\Omega$ a strictly positive $(1,1)$-form on $X$. We will consider the algebra differential forms on $X$. The goal is to find canonical representatives in cohomology classes of differential forms, specifically by solving a certain extremal problem defined with regard to $\Omega$.

### 9.1.1. The Hodge star.
We begin by defining the Hodge $\star$ operator on each member of such a triple. For a function $f$ we set

$$\star f := \bar{f}\Omega.$$

For a 1-form $\alpha = \alpha^{1,0} + \alpha^{0,1}$ we set

$$\star \alpha := \frac{1}{\sqrt{-1}} \left( \overline{\alpha^{0,1}} - \overline{\alpha^{1,0}} \right).$$

For a $(1,1)$-form $\omega = f\Omega$ we set

$$\star \omega := \bar{f}.$$

Observe that $\star$ maps functions to $(1,1)$-forms and vice versa, and that it maps $(1,0)$-forms to $(0,1)$-forms and vice versa. Moreover, on functions and $(1,1)$-forms $\star\star = \text{Id}$ while on 1-forms, $\star\star = -\text{Id}$.

### 9.1.2. Inner products.
Given two functions $f, g : X \to \mathbb{C}$, we define

$$(f, g) := \int_X f \star g.$$

Given two 1-forms $\alpha$ and $\beta$ on $X$, we define

$$(\alpha, \beta) := \int_X \alpha \wedge \star \beta.$$

145

Given two $(1,1)$-forms $\omega$ and $\theta$, we define
$$(\omega, \theta) := \int_X \omega \star \theta.$$
We declare
$$((f, \alpha, \omega), (g, \beta, \theta)) := (f, g) + (\alpha, \beta) + (\omega, \theta).$$

DEFINITION 9.1.1. $\mathscr{E}(X, \Omega)$ denotes the inner product space of all differential forms (which can be graded, and so thought of as triples of functions, 1-forms, and $(1,1)$-forms), together with the above inner product. When we work with the various components, we will write
$$\mathscr{E}_0(X, \Omega), \quad \mathscr{E}_1(X, \Omega), \quad \text{and} \quad \mathscr{E}_{1,1}(X, \Omega).$$
Thus
$$\mathscr{E}(X, \Omega) = \mathscr{E}_0(X, \Omega) \oplus \mathscr{E}_1(X, \Omega) \oplus \mathscr{E}_{1,1}(X, \Omega).$$
The Hilbert space closures of these inner product spaces are denoted
$$L_0^2(X, \Omega), \quad L_1^2(X, \Omega), \quad L_{1,1}^2(X, \Omega)$$
and
$$L^2(X, \Omega) = L_0^2(X, \Omega) \oplus L_1^2(X, \Omega) \oplus L_{1,1}^2(X, \Omega),$$
respectively. ◇

**9.1.3. The formal adjoint of $d$.** The formal adjoint of the exterior derivative operator $d$ is the operator $d^*$ defined to be zero on functions, and defined on 1-forms (resp. $(1,1)$-forms) $\alpha$ by the relations
$$(d^*\alpha, f) = (\alpha, df) \quad \text{for all } f \in \mathscr{E}_0(X) \text{ (resp. } \in \mathscr{E}_1(X)).$$

Let us calculate the operators $d^*$.

(1) (1-forms) We have
$$(d^*\alpha, f) = \overline{\int_X \alpha \wedge \star df} = \overline{\int_X df \wedge \star \alpha} = -\overline{\int_X f d(\star \alpha)}$$
$$= -\overline{\int_X f \star (\star d \star \alpha)} = -\int_X (\star d \star \alpha) \star f = (-\star d \star \alpha, f).$$

(2) $((1,1)$-forms)
$$(d^*\omega, \beta) = \overline{\int_X \omega \star d\beta} = \overline{\int_X d\beta \star \omega} = \overline{\int_X \beta \wedge d(\star \omega)}$$
$$= -\overline{\int_X \beta \wedge \star(\star d \star \omega)} = -\int_X (\star d \star \omega) \wedge \star \beta = (-\star d \star \omega, \beta).$$

Thus in all cases, we have
$$d^* = -\star d \star.$$

## 9.1.4. Laplace-Beltrami operator and harmonic forms.

DEFINITION 9.1.2. *The Laplace-Beltrami operator associated to* $(X, \Omega)$ *is*
$$\Delta := dd^* + d^*d.$$
*An element in the kernel of* $\Delta$ *is called a Harmonic form.* ◇

The term *harmonic* may be a little confusing, since we already have a notion of harmonic functions. At the very least, these two notions should agree if we are going to use the same name for them. To do the local calculations, it is convenient to introduce the local (real valued) function $h$ associated to the local coordinate function $z$ by the relation

$$h(z)\Omega = \frac{\sqrt{-1}}{2} dz \wedge d\bar{z}.$$

Now, for a function $f$ we have

$$\begin{aligned}\Delta f &= -\star d \star df = \star d\sqrt{-1}(\overline{\bar{\partial}f} - \overline{\partial f}) \\ &= -2\star\sqrt{-1}\partial\bar{\partial}\bar{f} = -2\star\overline{f_{z\bar{z}}}\sqrt{-1}dz \wedge d\bar{z} = -4h(z)f_{z\bar{z}}.\end{aligned}$$

Therefore harmonic functions are harmonic in the previous sense of the word.

Next, for a $(1,1)$-form $\omega = f\Omega$, we have

$$\Delta\omega = -d(\star d\bar{f}) = -d\star(\overline{\partial f} + \overline{\bar{\partial}f}) = \sqrt{-1}d(\partial f - \bar{\partial}f) = -2\sqrt{-1}\partial\bar{\partial}f = (\Delta f)\Omega,$$

so that a form $\omega = f\Omega$ is harmonic if and only if the function $f$ is harmonic. Thus the map

$$f \mapsto f\Omega$$

gives a $1-1$ correspondence between harmonic $(1,1)$-forms and harmonic functions.

PROPOSITION 9.1.3. *Global harmonic* $(1,1)$-*forms on a compact connected Riemann surface* $(X, \Omega)$ *are constant multiples of* $\Omega$ *and conversely.*

**Proof.** There are no non-constant harmonic functions on a compact connected Riemann surface, and the proposition follows because $f\Omega$ is harmonic if and only if $f$ is harmonic. □

Finally, we turn to 1-forms. If $\alpha = fdz + gd\bar{z}$, then

$$\begin{aligned}&-(dd^* + d^*d)\alpha \\ &= d(\star d\sqrt{-1}(\bar{f}d\bar{z} - \bar{g}dz)) - \star d \star \sqrt{-1}(g_z - f_{\bar{z}})2h\Omega \\ &= 2d\star((\overline{f_{\bar{z}}} + \overline{g_z})h\Omega) + 2\star d(\sqrt{-1}h\overline{(g_z - f_{\bar{z}})}) \\ &= 2d(h(f_{\bar{z}} + g_z)) + 2\star\sqrt{-1}(\partial(\overline{h(g_z - f_{\bar{z}})})) + 2\bar{\partial}(\overline{h(g_z - f_{\bar{z}})}) \\ &= 2(\partial + \bar{\partial})(h(f_{\bar{z}} + g_z)) + 2(\bar{\partial}(h(g_z - f_{\bar{z}})) - \partial(h(g_z - f_{\bar{z}})) \\ &= 4(\bar{\partial}(hg_z) + \partial(hf_{\bar{z}})) = (hf_{z\bar{z}} + h_z f_{\bar{z}})dz + (hg_{z\bar{z}} + h_{\bar{z}}g_z)d\bar{z}.\end{aligned}$$

REMARK. Given a coordinate system $z$, let us denote by $H^{(z)}$ the real-valued function satisfying
$$H^{(z)}\sqrt{-1}dz \wedge d\bar{z} = \Omega.$$
One can always choose a coordinate system $\zeta$ about a point $o$ such that $H^{(\zeta)}(o) = 1$ and $H^{(\zeta)}_\zeta(o) = 0$. To see this, start with any coordinate system $z$, and write $H(z) = a + bz + \overline{bz} + O(|z|^2)$. Note that $a \in \mathbb{R}$. By scaling $z$, we can assume $a = 1$. Now take $z = \zeta - \frac{b}{2}\zeta^2$. Then $dz = (1-b\zeta)d\zeta$, so
$$\begin{aligned} H^{(z)}\sqrt{-1}dz \wedge d\bar{z} &= (1 + b\zeta + \overline{b\zeta} + ...)(1-b\zeta)(1-\overline{b\zeta})d\zeta \wedge d\bar{\zeta} \\ &= (1 + O(|\zeta|^2))\sqrt{-1}d\zeta \wedge d\bar{\zeta}. \end{aligned}$$
A coordinate system in which $\Omega = (1 + O(|\zeta|^2))\sqrt{-1}d\zeta \wedge d\bar{\zeta}$ is called a normal coordinate system. A 1-form is harmonic if and only if its local component functions are harmonic to second order in any normal coordinate system. ◇

**9.1.5. Regularity for the Laplace-Beltrami operator.** With the formulas for the Laplacian in hand, we can now easily prove the following result.

PROPOSITION 9.1.4 (Regularity for $\Delta$). *Let $\Omega$ be a smooth area form on a Riemann surface $X$, and let $\xi \in \mathscr{E}(X, \Omega)$ be a smooth element. Then any weak solution $\eta$ of the equation $\Delta \eta = \xi$ is also smooth.*

**Proof.** Evidently the result is local. From the formulas for the Laplacian on 0-forms, 1-forms, and $(1,1)$-forms, it suffices to prove the following results: given any smooth function $h$, any solutions $f$ and $g$ for the equations $f_z = h$ and $g_{\bar{z}} = h$, respectively, must be smooth. Since $\overline{f_z} = f_{\bar{z}}$, it suffices to prove the regularity of solutions to the second of these two equations. Now, any two solutions differ by a (weak) holomorphic function, and since a holomorphic function is harmonic, it is smooth by our work in Paragraph 1.7.6 of Chapter 1. Thus it suffices to prove that there is one smooth solution. But such a smooth solution is easily obtained from the Cauchy-Green Formula; see, for example, the proof of Theorem 5.9.2. □

REMARK. Because the Laplace operator is so symmetric on a Riemann surface, we can obtain its regularity by *ad hoc* means. In Chapter 11 we will consider a partial differential operator for which regularity is more involved and follows more closely the standard approach to elliptic regularity. ◇

**9.1.6. Notation for harmonic forms.** For ease of reading, we omit the notational dependence on $\Omega$. Let us write
$$\mathscr{H}(X) := \{(f, \alpha, \omega) \,;\, \Delta f = 0, \Delta \alpha = 0, \text{ and } \Delta \omega = 0\}.$$
We have the decomposition $\mathscr{H}(X) = \mathscr{H}_0(X) \oplus \mathscr{H}_1(X) \oplus \mathscr{H}_{(1,1)}(X)$, with the obvious meaning for the factors.

REMARK. Previously we denoted by $H(X)$ the harmonic functions on a Riemann surface. The new notation $\mathscr{H}_0(X)$ refers to $L^2$-structure being present. ◇

## 9.2. Harmonic forms and cohomology

In this section we tie the harmonic forms on a compact Riemann surface to the topology of that surface.

### 9.2.1. De Rham cohomology.
Recall that the de Rham cohomology group is

$$H^1_{\mathrm{dR}}(X) := \frac{\mathrm{Kernel}(d : \Gamma(X, \Lambda^1_X) \to \Gamma(X, \Lambda^{1,1}_X))}{d\mathscr{C}^\infty(X)}.$$

THEOREM 9.2.1. *Each class $[\alpha] \in H^1_{\mathrm{dR}}(X)$ has exactly one harmonic 1-form.*

**Proof.** Since a form $\alpha$ is harmonic if and only if

$$d\alpha = 0 \quad \text{and} \quad d^*\alpha = 0,$$

every harmonic 1-form is closed. Next, we define the functional

$$N(\beta) := ||\beta||^2$$

on smooth 1-forms $\beta$. Among all smooth $\beta \in [\alpha]$, we shall try to minimize $N(\beta)$. Note that if we do find a minimizer $\beta$ of $N$, then for all $\varepsilon > 0$

$$0 \leq N(\beta + \varepsilon df) - N(\beta) = 2\mathrm{Re}\,(\beta, df)\varepsilon + O(\varepsilon^2)$$

and thus applying this estimate to real- and imaginary-valued $f$, we find that

$$d^*\beta = 0.$$

Since $d\beta = 0$ by hypothesis, the form $\beta$ is harmonic and thus smooth.

Next, suppose a minimizer exists. Then it is unique. Indeed, if $\beta_1$ and $\beta_2$ are minimizers, then $\beta_2 = \beta_1 + df$ and we have

$$||\beta_2||^2 = ||\beta_1||^2 + ||df||^2 + 2\mathrm{Re}\,(\beta_1, df) = ||\beta_1||^2 + ||df||^2 \geq ||\beta_1||^2,$$

so equality holds if and only if $df = 0$.

The above calculations show that the minimizer of $N$ is orthogonal to the collection of exact 1-forms. However, we cannot yet guarantee that this minimizer exists, since the inner product space $([\alpha], ||\cdot||)$ is not necessarily closed. This problem is easily surmounted by simply seeking the minimizer $\beta$ in the Hilbert space closure $\mathfrak{H}$ of $([\alpha], ||\cdot||)$. Such a minimizer must be orthogonal to the subspace of all exact 1-forms and thus orthogonal to its closure. Thus the minimizer is harmonic in the sense of currents. By Proposition 9.1.4 our minimizer is smooth, and the proof is complete. □

### 9.2.2. Arithmetic and geometric genus.

THEOREM 9.2.2. *Let $X$ be a compact Riemann surface. Then we have the orthogonal decomposition*

(9.1) $$\mathcal{H}_1(X) = \Gamma_\mathcal{O}(X, K_X) \oplus \overline{\Gamma_\mathcal{O}(X, K_X)}.$$

*In particular, $\mathcal{H}(X)$ is finite-dimensional.*

**Proof.** We know that $\mathcal{H}_0(X)$ and $\mathcal{H}_{(1,1)}(X)$ are both 1-dimensional. Since, by Theorem 4.8.1, $\Gamma_\mathcal{O}(X, K_X)$ is finite-dimensional, the right-hand side of (9.1) is finite-dimensional, and in fact has dimension $2g_{\mathcal{O}(X)}$, where $g_{\mathcal{O}(X)}$ is the arithmetic genus of $X$.

We now turn to the proof of (9.1). It is clear that if $\alpha$ and $\beta$ are holomorphic 1-forms, then $\alpha + \bar{\beta}$ is harmonic. Moreover, the orthogonality of the two summands in the right-hand side of the decomposition is clear from the definition of the $\star$ operator.

Conversely, suppose $\sigma \in \mathcal{H}_1(X)$. First,
$$(\Delta\sigma, \sigma) = |d\sigma|^2 + |d^*\sigma|^2,$$
and thus harmonic forms are precisely those forms that are closed and annihilated by $d^*$. Next, let $\sigma := \sigma_{1,0} + \sigma_{0,1}$ be the unique decomposition into $(1,0)$- and $(0,1)$-forms. The equation $d\sigma = 0$ means
$$\bar\partial\sigma_{1,0} + \partial\sigma_{0,1} = 0,$$
while the equation $d^*\sigma = 0$ means
$$\partial\overline{\sigma_{1,0}} = \bar\partial\overline{\sigma_{0,1}}.$$
But together these two equations imply that
$$\bar\partial\sigma_{1,0} = \partial\sigma_{0,1} = 0,$$
which simply says that $\sigma_{1,0}$ and $\overline{\sigma_{0,1}}$ are holomorphic. The proof is complete. $\square$

COROLLARY 9.2.3. *The arithmetic genus $g_\mathcal{O}(X)$ of a compact Riemann surface $X$ agrees with the geometric genus of $X$, i.e., the number of handles of $X$.*

**Proof.** Every handle on $X$ yields exactly two homology classes of closed loops in $X$. Thus we have $\dim_\mathbb{R} H_1(X) = 2g$. By Lemma 5.8.4, $\dim_\mathbb{R} H^1_{dR}(X) = 2g$. By Theorem 9.2.1, $\dim_\mathbb{R} \mathcal{H}_1(X) = 2g$. But by Theorem 9.2.2, $\dim_\mathbb{R} \mathcal{H}_1(X) = 2g_\mathcal{O}(X)$. It follows that $g = g_\mathcal{O}(X)$, and the proof is complete. $\square$

REMARK. Alternatively, as we will see in the next section, finite-dimensionality of $\mathcal{H}(X)$ can be deduced, after some work, from the Spectral Theorem for compact self-adjoint operators. This finite-dimensionality is a natural part of the process of solving Poisson's Equation on the compact Riemann surface $X$. The fact that

$\mathscr{H}_1(X) = \Gamma_{\mathcal{O}}(X, K_X) \otimes \overline{\Gamma_{\mathcal{O}}(X, K_X)}$ is of course formal, so we will have obtained a second proof of the finite-dimensionality of $\Gamma_{\mathcal{O}}(X, K_X)$.

In fact, the two proofs of finite-dimensionality are related, though the second proof looks rather more sophisticated. The relationship involves notions from PDE: the Laplace operator is an example of an *elliptic* operator, and elliptic operators support a maximum principle. Another example of an elliptic operator is the operator $\frac{\partial}{\partial \bar{z}}$. The main thrust of the proof of Theorem 4.8.1 is the use of the Maximum Principle to give a comparison of two norms on a compact Riemann surface.

There is an analogue of the Hodge Theorem for $H$-valued forms, and the reader is asked to work it out in the exercises for this section in the case that $H$ is a holomorphic line bundle. Given this result, one can use the Spectral Theorem again to note that the twisted harmonic forms make up a finite-dimensional vector space, and therefore $\Gamma_{\mathcal{O}}(X, H \otimes K_X)$ is finite-dimensional. Letting $H = L \otimes T_X^{1,0}$ gives a PDE proof of Theorem 4.8.1. ⋄

## 9.3. The Hodge decomposition of $\mathscr{E}(X)$

### 9.3.1. The main theorem.

THEOREM 9.3.1. *There is an orthogonal decomposition*

(9.2) $$\mathscr{E}(X) = \Delta(\mathscr{E}(X)) \oplus \mathscr{H}(X).$$

DEFINITION 9.3.2. The decomposition (9.2) is called the *Hodge decomposition*. ⋄

The proof of Theorem 9.3.1 has two parts. The first part, which is rather easy, shows that $\mathscr{H}(X)$ is the orthogonal complement of the closure of $\Delta(\mathscr{E}(X))$ in $L^2(X)$. This is done in Lemma 9.3.4 below. The second and most complicated part of the proof is to show that $\Delta(\mathscr{E}(X))$ is closed in $\mathscr{E}(X)$. In view of Proposition 9.1.4, it suffices to find weak solutions of $\Delta \theta = u$ when $u$ is smooth. The idea for solving the latter equation is to show that the operator $I + \Delta$ has a compact inverse, and then use the Spectral Theorem for compact self-adjoint operators to solve $\Delta$ on the orthogonal complement of its kernel. In establishing the compactness, we need a famous result of Rellich, which we prove below.

Before proving Theorem 9.3.1, we state one important corollary.

COROLLARY 9.3.3. *Let $f : X \to \mathbb{C}$ be a smooth function on a compact Riemann surface. Then the equation $\Delta u = f$ has a solution if and only if*

$$\int_X f\Omega = 0.$$

*Equivalently, the equation $\sqrt{-1}\partial\bar{\partial} f = \omega$ has a solution if and only if $\int_X \omega = 0$.*

**Proof.** Since every harmonic function on $X$ is constant and every harmonic $(1,1)$-form is a constant multiple of $\Omega$, a function $f$ (resp. $(1,1)$-form $\omega$) is orthogonal to the harmonic functions (resp. $(1,1)$-forms) if and only if $\int_X f\Omega = 0$ (resp. $\int_X \omega = 0$). The proof is complete. □

REMARK. Note the contrast between Corollary 9.3.3 and Theorem 8.6.3, the latter of which states that the equation $\sqrt{-1}\partial\bar{\partial}u = \omega$ always has a solution on an open Riemann surface. ⋄

### 9.3.2. Obstructions from the kernel of $\Delta$.

LEMMA 9.3.4. *Let* $\mathfrak{H}(X) := \mathscr{E}(X) \ominus \Delta(\mathscr{E}(X))$. *Then*

$$\mathfrak{H}(X) = \mathscr{H}(X).$$

**Proof.** Consider the equation

$$(\eta, \Delta x) = (\Delta \eta, x).$$

If $\eta \in \mathscr{H}(X)$, then the right-hand side is zero, which means that $\eta \perp \Delta(\mathscr{E}(X))$, i.e., $\eta \in \mathfrak{H}(X)$. On the other hand, if $\eta \perp \Delta(\mathscr{E}(X))$ then the left-hand side is zero, and thus $\Delta\eta \perp \mathscr{E}(X)$. Since smooth objects are dense in our Hilbert space, $\Delta\eta = 0$. □

### 9.3.3. The strategy for computing $\Delta(\mathscr{E}(X))$.

By Lemma 9.3.4, $\Delta(\mathscr{E}(X))$ is densely contained in the closed subspace $V := \mathscr{H}(X)^\perp \cap \mathscr{E}(X)$. Our next goal is to show that for every $u \in V$ there exists $\theta$ in $L^2(X)$ such that $\Delta\theta = u$ in the weak sense, i.e., for all smooth $\xi$,

$$(\theta, \Delta\xi) = (u, \xi).$$

Unfortunately, although we are working in the complement of the kernel of $\Delta$ and we know that

$$(\Delta\xi, \xi) = ||d\xi||^2 + ||d^*\xi||^2 \geq 0,$$

at this point we do not have any positive lower bounds on the smallest non-zero eigenvalue of $\Delta$. (If we had such estimates, we could solve the equation $\Delta\theta = u$ with estimates, and we shall take that approach in Chapter 11, where we solve a related equation under additional assumptions.)

The trick to obtaining such a positive lower bound is to consider the operator $I + \Delta$. As we will show, $(I + \Delta)^{-1}$ is compact, and thus we can get some crucial information on its eigenvalues from the Spectral Theorem. Of course, this method will not give us an estimate for the smallest positive eigenvalue, but that is not a problem for us now.

## 9.3. The Hodge decomposition of $\mathscr{E}(X)$

**9.3.4. A compact subspace of $L^2(X)$.** To prove the compactness of $(I+\Delta)^{-1}$, we bring into the picture another Hilbert space $H$, defined as the closure of the set of smooth forms $\xi$ with respect to the norm

$$||\xi||_H := \sqrt{||\xi||_{L^2}^2 + ||d\xi||_{L^2}^2 + ||d^*\xi||_{L^2}^2}.$$

Since

$$||\cdot||_{L^2} \leq ||\cdot||_H,$$

we see that the inclusion $H \hookrightarrow L^2(X)$ is bounded, but in fact more is true.

LEMMA 9.3.5 (Rellich's Compactness Lemma). *The inclusion $H \hookrightarrow L^2(X)$ is a compact operator. That is to say, for each sequence $\{\xi_j\} \subset L^2(X)$ such that*

$$\sup_j ||\xi_j||_H < +\infty$$

*there is a subsequence $\{\xi_{j_k}\}$ converging in $L^2(X)$.*

REMARK. For $x, y \in \mathbb{R}^2$, $|x| \leq |x-y| + |y|$ and thus $|x|^2 \leq 2(|x-y|^2 + |y|^2)$. It follows that

$$1 + |x|^2 \leq 1 + 2|x-y|^2 + 2|y|^2 \leq 2(1+|y|^2)(1+|x-y|^2),$$

and thus we have the inequality

$$\frac{1+|x|^2}{1+|y|^2} \leq 2(1+|x-y|^2), \tag{9.3}$$

which will be useful to us shortly. $\diamond$

**Proof of Rellich's Lemma.** Since the smooth forms are dense, we may assume that the sequence $\{\xi_j\}$ consists of smooth forms. Let $\{U_a\}$ be a finite cover of $X$ by coordinate charts, with subordinate partition of unity $\{\varphi_a\}$. It suffices to extract a convergent subsequence from $\{\varphi_a \xi_j\}$ for each $a$. We fix $a$ and sometimes do not refer to it in the argument. Since we are working locally, there are vector-valued functions $F_j$ (with compact support) such that

$$\int_{\mathbb{R}^2} |F_j|^2 dA = ||\varphi_a \xi_j||_{L^2}^2.$$

(Note that $F_j$ also carries the information of the metric $\Omega$.) A calculation in local coordinates shows that

$$||\varphi_a \xi_j||_H^2 \sim \int_{\mathbb{R}^2} (|F_j|^2 + |DF_j|^2) dA.$$

We work with the Fourier transform

$$\hat{F}_j(t) := \int_{\mathbb{R}^2} e^{-\sqrt{-1}\langle t, x\rangle} F_j(x) dA(x).$$

Letting $\varphi \in \mathscr{C}^\infty(X)$ be a smooth function supported in the coordinate neighborhood in question and identically 1 on the support of $F_j$, we have $F_j = \varphi F_j$, and by thus by (9.3) and the product formula for Fourier transforms we have

$$\sqrt{1+|t|^2}|\hat{F}_j(t)| \lesssim \int_{\mathbb{R}^2} \sqrt{1+|s|^2}|\hat{F}_j(s)|\sqrt{1+|t-s|^2}|\hat{\varphi}(t-s)|dA(s)$$

and

$$\sqrt{1+|t|^2}|D\hat{F}_j(t)| \leq \int_{\mathbb{R}^2} |\hat{F}_j(s)|\sqrt{1+|t|^2}|D\hat{\varphi}(t-s)|dA(s)$$
$$\lesssim \int_{\mathbb{R}^2} \sqrt{1+|s|^2}|\hat{F}_j(s)|\sqrt{1+|t-s|^2}|\widehat{D\varphi}(t-s)|dA(s).$$

It follows that

$$(1+|t|^2)(|\hat{F}_j(t)|^2 + |D\hat{F}_j(t)|^2) \lesssim \int_{\mathbb{R}^2}(1+|s|^2)|\hat{F}_j(s)|^2 dA(s)$$
$$= \int_{\mathbb{R}^2}(|F_j|^2 + |DF_j|^2)dA$$
$$\sim \|\varphi_a \xi_j\|_H^2,$$

where the inequality in the first line follows from the Cauchy-Schwartz Inequality, standard properties of the Fourier transform, and the compact support of $\varphi$, and the equality in the second line follows from the Plancherel Theorem and standard properties of Fourier transforms. These uniform $\mathscr{C}^1$-bounds on the Fourier transforms of $F_j$ combine with the Arzela-Ascoli Compactness Theorem to produce a locally uniformly convergent subsequence $\{\hat{F}_{j_k}\}$. Moreover, by Plancherel's Theorem we have

$$\|F_{j_k} - F_{j_\ell}\|^2 = \int_{\mathbb{R}^2} |\hat{F}_{j_k} - \hat{F}_{j_\ell}|^2 dA$$
$$\leq \int_{|t|<R} |\hat{F}_{j_k}(t) - \hat{F}_{j_\ell}(t)|^2 dA(t)$$
$$+ \frac{1}{1+R^2}\int_{|t|>R}(1+|t|^2)|\hat{F}_{j_k}(t) - \hat{F}_{j_\ell}(t)|^2 dA(t)$$
$$\leq \int_{|t|<R} |\hat{F}_{j_k}(t) - \hat{F}_{j_\ell}(t)|^2 dA(t) + \frac{\|\varphi_a \xi_{j_k}\|_H^2 + \|\varphi_a \xi_{j_\ell}\|_H^2}{1+R^2}.$$

The first term of the last line is small for all $k, \ell >> 0$ by the locally uniform convergence of $\{\hat{F}_{j_k}\}$, while the second term can be made as small as we like by choosing $R$ sufficiently large. Thus $\{F_{j_k}\}$ is a Cauchy sequence, and the proof is complete. $\square$

REMARK. In the proof of Rellich's Lemma, we used a number of properties of the Fourier transform which, though elementary, may not be as familiar to some students. We have suppressed the technicalities in order to emphasize the central points of the proof.

## 9.3. The Hodge decomposition of $\mathscr{E}(X)$

In Chapter 11 we will need the Fourier transform again, to establish the regularity of a certain differential operator that arises in our proof of the Korn-Lichtenstein Theorem. This regularity is also standard, but this time we have provided more details, including a more thorough discussion of Fourier transforms. The reader interested in learning more about the Fourier transform is invited to look ahead at Paragraph 11.4.8. ◇

**9.3.5. The Spectral Theorem for Compact Self-adjoint Operators.** In the proof of the Hodge Decomposition Theorem we will make use of the following result from functional analysis.

THEOREM 9.3.6 (Spectral Theorem for Compact Self-adjoint Operators). *Let $\mathcal{H}$ be a separable Hilbert space and let $T : \mathcal{H} \to \mathcal{H}$ be an injective, compact, self-adjoint operator. Then there is a discrete sequence of positive numbers $\lambda_m$ such that if $\{\lambda_m\}$ is finite then $\mathcal{H}$ is finite-dimensional, and otherwise $\lambda_m \searrow 0$. Moreover, the vector spaces*

$$\mathcal{H}^m(T) := \{\theta \in \mathcal{H} \; ; \; T\theta = \lambda_m \theta\}$$

*are finite-dimensional, and we have the orthogonal Hilbert space decomposition*

$$\mathcal{H} = \bigoplus_{m=1}^{\infty} \mathcal{H}_m(T).$$

A proof of the Spectral Theorem can be found in [**Rudin-1991**].

**9.3.6. Completion of the proof of Theorem 9.3.1.** By Lemma 9.3.4 it suffices to show that given $u \perp \mathscr{H}(X)$ there exists $\theta \in \mathscr{E}(X)$ such that $\Delta\theta = u$. By Proposition 9.1.4 it suffices to find a weak solution, which is what we now do.

LEMMA 9.3.7. *For each $\theta \in L^2(X)$ there exists $u \in H$ such that $||u||_H \leq ||\theta||_{L^2}$ and $(I + \Delta)u = \theta$.*

**Proof.** Consider the quadratic form $Q(\theta, \theta) := ||\theta||_H^2$ defined on $H$. (Note that on smooth forms $\theta$, $Q(\theta, \theta) = ((I + \Delta)\theta, \theta)_{L^2}$.) Clearly $Q(\theta, \theta) \geq ||\theta||_{L^2}^2$ for all $\theta \in H$. Now, the inequality

$$|(\theta, \eta)_{L^2}|^2 \leq ||\theta||_{L^2}^2 ||\eta||_{L^2}^2 \leq ||\theta||_{L^2}^2 Q(\eta, \eta)$$

shows that, for $\theta \in L^2(X)$, the conjugate-linear functional

$$\eta \mapsto (\theta, \eta)_{L^2}$$

is continuous on $H$. By the Riesz Representation Theorem there is an element $T\theta \in H$ such that

(9.4) $\qquad ||T\theta||_H \leq ||\theta||_{L^2} \qquad$ and $\qquad Q(T\theta, \eta) = (\theta, \eta)_{L^2}.$

If $\eta$ is smooth, then $Q(T\theta, \eta) = (T\theta, (I + \Delta)\eta)_{L^2}$, which, combined with (9.4), means that $u := T\theta$ is a weak solution of $(I + \Delta)u = \theta$. This completes the proof. □

LEMMA 9.3.8. *Let $\iota : H \hookrightarrow L^2(X)$ be the inclusion. The operator $G := \iota T : L^2(X) \to L^2(X)$ is compact.*

**Proof.** By (9.4), the operator $T : L^2(X) \to H$ is bounded and injective. The lemma thus follows from Rellich's Lemma and the fact that the composition of a compact operator and a bounded operator is compact. □

**End of the proof of Theorem 9.3.1.** The estimate (9.4) shows that the eigenvalues of $G$ are no more than 1. Moreover, since

$$(\theta, T\eta) = Q(T\theta, T\eta) = \overline{Q(T\eta, T\theta)} = \overline{(\eta, T\theta)} = (T\theta, \eta),$$

the operator $T$ is self-adjoint. Since $T$ is also injective, its eigenvalues are all positive. Therefore $G$ is self-adjoint and all of its eigenvalues lie in $(0, 1]$. By the Spectral Theorem there is a discrete set of positive numbers $1 = \lambda_1 > \lambda_2 > ....$ such that $\lambda_m \to 0$ and the eigenspaces

$$\mathscr{K}_m(X) := \{\theta \in \mathscr{E}(X) \, ; \, G\theta = \lambda_m \theta\}$$

are finite-dimensional and provide a decomposition of $L^2(X)$. Observe that for any $\theta \in \mathscr{K}_m(X)$,

$$\Delta\theta = (\lambda_m^{-1} - 1)\theta = \frac{1 - \lambda_m}{\lambda_m}\theta.$$

In particular, $\mathscr{K}_1(X) = \mathscr{H}(X)$.

Let us fix orthonormal bases

$$\{\theta_j^{(m)} \, ; \, 1 \leq j \leq N_m\} \subset \mathscr{K}_m(X), \quad m = 1, 2, ....$$

Let $u \in V$. Decompose $u$ orthogonally in $L^2(X)$ as

$$u = \sum_{m=2}^{\infty} \sum_{j=1}^{N_m} c_{jm} \theta_j^{(m)}.$$

(There is no $m = 1$ term because $u \perp \mathscr{H}(X)$.) Let

$$\theta := \sum_{m=2}^{\infty} \sum_{j=1}^{N_m} \frac{\lambda_m c_{jm}}{1 - \lambda_m} \theta_j^{(m)}.$$

Then the inequality

$$\frac{|\lambda_m|}{|1 - \lambda_m|} \leq \frac{\lambda_2}{1 - \lambda_2}, \quad m \geq 2,$$

shows that $\theta$ is convergent and

$$\|\theta\|_{L^2} \leq \frac{\lambda_2}{1 - \lambda_2} \|u\|_{L^2}.$$

Next,

$$\Delta\theta = \sum_{m=2}^{\infty} \sum_{j=1}^{N_m} \frac{\lambda_m c_{jm}}{1 - \lambda_m} \Delta\theta_j^{(m)} = \sum_{m=2}^{\infty} \sum_{j=1}^{N_m} c_{jm} \theta_j^{(m)} = u,$$

and thus we have found our weak solution $\theta$. The proof of Theorem 9.3.1 is complete. $\square$

REMARK. Since $\mathcal{K}_1(X) = \mathcal{H}(X)$, the elements of $\mathcal{K}_1(X)$ are smooth. It turns out that all of the vectors in each $\mathcal{K}_m(X)$ are smooth—a fact we will not use. $\diamond$

## 9.4. Existence of positive line bundles

In this section we prove, in several ways, that every Riemann surface admits a line bundle that has a metric of positive curvature. In the next chapter we will give yet another proof using the Uniformization Theorem. The existence of positive line bundles is fundamental in Chapter 12, where we prove that every compact Riemann surface embeds in $\mathbb{P}_3$ and that every open Riemann surface embeds in $\mathbb{C}^3$.

### 9.4.1. Compact Riemann surfaces I: Using Hodge Theory.

PROPOSITION 9.4.1. *Let $X$ be a compact Riemann surface. Then there is a holomorphic line bundle $L \to X$ and a smooth Hermitian metric $e^{-\varphi}$ for $L$ such that $\sqrt{-1}\partial\bar{\partial}\varphi$ is a strictly positive $(1,1)$-form.*

In fact, we will prove the following, more general result.

PROPOSITION 9.4.2. *Let $L \to X$ be a holomorphic line bundle and let $\Omega$ be a $(1,1)$-form such that $\int_X \Omega = c(L)$. Then there is a smooth metric $e^{-\varphi}$ for $L$ such that*
$$\frac{\sqrt{-1}}{2\pi}\partial\bar{\partial}\varphi = \Omega.$$

**Proof.** Fix any smooth metric $e^{-\varphi_o}$ for $L$. By Theorem 6.6.1,
$$\frac{\sqrt{-1}}{2\pi}\int_X \partial\bar{\partial}\varphi_o = c(L).$$
It follows that the $(1,1)$-form $\eta := \frac{\sqrt{-1}}{2\pi}\partial\bar{\partial}\varphi_o - \Omega$ satisfies $\int_X \eta = 0$. By Corollary 9.3.3 there is a smooth function $u : X \to \mathbb{R}$ such that
$$\frac{\sqrt{-1}}{2\pi}\partial\bar{\partial}u = \eta.$$
The metric $e^{-\varphi} := e^{u-\varphi_o}$ therefore has curvature $\partial\bar{\partial}\varphi$ satisfying
$$\frac{\sqrt{-1}}{2\pi}\partial\bar{\partial}\varphi = \Omega,$$
as desired. $\square$

COROLLARY 9.4.3. *If $L \to X$ is a holomorphic line bundle such that $c(L) > 0$ then $L$ admits a metric whose curvature is an area form.*

EXAMPLE 9.4.4. Let $X$ be a compact Riemann surface, and let $p \in X$ be a point. Then the line bundle $L \to X$ associated to the divisor $1 \cdot p$ admits a metric $e^{-\varphi}$ such that $\omega := \frac{\sqrt{-1}}{2\pi}\partial\bar{\partial}\varphi$ is a probability measure with smooth, nowhere-zero density. ⋄

**9.4.2. Compact Riemann surfaces II: An ad hoc technique.** Our construction of positive line bundles on compact Riemann surfaces uses the very powerful Hodge Theorem. If one wants to avoid using the Hodge Theorem, there are a number of other ways to proceed. In this section, we take a particular ad hoc approach that provides us with a less precise conclusion.

One might try to construct a metric of positive curvature as follows. Let $\omega$ be a strictly positive $(1,1)$-form on $X$ such that

$$\int_X \omega = 2\pi.$$

Take a cover $\{U_i\}$ of $X$ by connected, simply connected coordinate neighborhoods (with coordinate functions $z_j$) such that the intersections $U_{ij} := U_i \cap U_j$ and $U_{ijk} = U_i \cap U_j \cap U_k$ are also simply connected.

On each $U_i$, we have

$$\omega = e^{-\psi_i(z_i)}\sqrt{-1}dz_i \wedge d\bar{z}_i.$$

We define the function

$$\varphi_i(z_i) := \frac{1}{2\pi}\int_{U_i} \log|z_i - \zeta|^2 e^{-\psi_i(\zeta)}\sqrt{-1}d\zeta \wedge d\bar{\zeta}.$$

Note that

$$\sqrt{-1}\partial\bar{\partial}\varphi_i = \omega$$

is independent of $i$ and, in particular, that $\varphi_i - \varphi_j$ is harmonic. It follows by the simple connectivity of $U_{ij}$ that there is a holomorphic function $g_{ij}$ such that

$$\varphi_i - \varphi_j = \log|g_{ij}|^2.$$

In fact, while the squared magnitude of $g_{ij}$ is

$$|g_{ij}|^2 = e^{\varphi_j - \varphi_i},$$

the argument of $g_{ij}$ can be defined by

$$\arg g_{ij} = 2\int_{v_{ij}}^{z} d^c(\varphi_i - \varphi_j),$$

where $v_{ij} \in U_{ij}$ is a chosen point. (By convention we choose $v_{ji} = v_{ij}$.) The choice of the curve connecting $v_{ij}$ to $z$ is not important, except to declare that the curve remain in $U_{ij}$ which is simply connected (hence the independence on the curve). So let us fix one such curve, $\Gamma_{ij}(z)$. We remark only that if we change the point $v_{ij}$, then the function $\arg g_{ij}$ changes by an additive constant that depends only on $i$ and $j$.

## 9.4. Existence of positive line bundles

Now, since we have taken

$$g_{ij} := \exp\left\{\varphi_i - \varphi_j + 2\sqrt{-1}\int_{v_{ij}}^z d^c\varphi_i - d^c\varphi_j\right\},$$

we have

$$g_{ij}g_{jk}g_{ki}$$
$$= \exp\left(2\sqrt{-1}\left\{\int_{v_{ij}}^z d^c\varphi_i - d^c\varphi_j + \int_{v_{jk}}^z d^c\varphi_j - d^c\varphi_k + \int_{v_{ki}}^z d^c\varphi_k - d^c\varphi_i\right\}\right).$$

We note that in fact $g_{ij}g_{jk}g_{ki}$ is constant. Indeed, it is holomorphic and of constant magnitude 1. It follows that

$$g_{ij}g_{jk}g_{ki} = e^{\sqrt{-1}\theta_{ijk}}$$

for some collection $\{\theta_{ijk}\}_{i,j,k\in A} \subset \mathbb{R}$ that is skew symmetric in the indices $ijk$.

REMARK. Our next goal would be to show that one can find collections of numbers $\{\eta_{ij} = -\eta_{ji}\}_{i,j\in A} \subset \mathbb{R}$ and $\{m_{ijk} = \mathrm{sgn}(\sigma)m_{\sigma(ijk)}\}_{i,j,k\in A} \subset \mathbb{Z}$ such that

(9.5) $$\theta_{ijk} = \eta_{ij} + \eta_{jk} + \eta_{ki} + 2\pi m_{ijk}.$$

While (9.5) does hold, I could not come up with a direct elementary proof. ◇

Nevertheless, we can still get somewhere. After a slight perturbation of the points $v_{ij}$, $v_{ik}$ and $v_{jk}$, we may assume that the numbers $\theta_{ijk}$ are a rational multiple of $2\pi$. (This is a bit of a bold assertion, and relies heavily on the fact that we have finitely many open sets $U_i$ in our cover.) We take an integer $m$ so that $m\theta_{ijk}$ is an integer multiple of $2\pi$ for all $i,j,k$. Setting

$$\tilde{g}_{ij} = g_{ij}^m,$$

we find that

$$\tilde{g}_{ij}\tilde{g}_{jk}\tilde{g}_{ki} = e^{\sqrt{-1}m\theta_{ijk}} = 1.$$

Thus $\{\tilde{g}_{ij}\}$ satisfies the cocycle condition, and hence defines a line bundle $H \to X$. Also, since

$$e^{-\varphi_i}|g_{ij}|^2 = e^{-\varphi_j},$$

we see that $e^{-\varphi} := \{e^{-m\varphi_i}\}$ defines a smooth Hermitian metric for $H$. Moreover, by construction the curvature current of $e^{-\varphi}$ is

$$\sqrt{-1}\partial\bar{\partial}\varphi = m\omega,$$

which is strictly positive. Of course, our Hodge-theoretic method tells us that we can find a holomorphic line bundle with a metric whose curvature is $\omega$ (see Example 9.4.4).

**9.4.3. Open Riemann surfaces.** In this paragraph we prove that the trivial bundle of an open Riemann surface admits a metric of strictly positive curvature. Of course, since we are dealing with the trivial bundle, our claim is that on an open Riemann surface there is a strictly subharmonic function. (A function $u$ is said to be strictly subharmonic if $\Delta u > 0$ in the sense of distributions, i.e., for any positive smooth function $f$ with compact support, $\int u \Delta f > 0$.) In fact, we will prove more. Recall that a continuous map $f : A \to B$ of topological spaces is said to be proper if for each compact subset $K \subset B$, $f^{-1}(K)$ is compact in $A$.

DEFINITION 9.4.5. For $a \in \mathbb{R} \cup \{\infty\}$, a proper map $f : A \to (-\infty, a)$ is called an exhaustion of $A$.

THEOREM 9.4.6 (Behnke-Stein). *An open Riemann surface has a strictly subharmonic exhaustion.*

**Proof.** Let $Y_1 \subset\subset Y_2 \subset\subset \ldots \subset X$ be a normal exhaustion. Choose open sets $U_1 \subset U_2 \subset \ldots$ such that $\overline{Y_j} \subset U_j \subset\subset Y_{j+1}$ for all $j \geq 1$. For each $j$, choose holomorphic functions $f_{j1}, \ldots, f_{jN_j}$ such that

(1) $\sup_{Y_j} |f_{jk}| < 1$,

(2) for each $z \in Y_{j+2} - U_j$, $\max_k |f_{jk}(z)| > 1$, and

(3) for each $z \in Y_j$ there exists $k$ such that $f_{jk}(z) \neq 0$ and $df_{jk}(z) \neq 0$.

Properties (1), (2), and (3) are easily achieved using the Runge Approximation Theorem. By taking powers of the $f_{jk}$ if necessary, we can guarantee that

$$\sum_{k=1}^{N_j} |f_{jk}(z)|^2 < 2^{-j}, \quad z \in Y_j, \quad \text{and} \quad \sum_{k=1}^{N_j} |f_{jk}(z)|^2 > j, \quad z \in Y_{j+2} - U_j.$$

It follows that the function

$$\varphi := \sum_{j=1}^{\infty} \sum_{k=1}^{N_j} |f_{jk}|^2$$

converges to a smooth function (in fact, real analytic) on $X$ and satisfies

$$\varphi > j \quad \text{on} \quad X - U_j.$$

Moreover, for any tangent vector $\xi \in T_{X,z}$,

$$\partial \bar{\partial} \varphi(z)(\xi, \bar{\xi}) = \sum_{j,k} |\partial f_{jk}(z)(\xi)|^2 > 0$$

by property (3) above. Thus $\varphi$ is strictly subharmonic. The proof is finished. $\square$

## 9.5. Proof of the Dolbeault-Serre isomorphism

In this brief section we prove Theorem 5.10.1.

Let $\alpha$ be a $d$-closed $(0,1)$-form on a compact Riemann surface. Then
$$d^*\alpha = \star d(\sqrt{-1}\,\overline{\alpha}) = \star(\sqrt{-1}\,\overline{d\alpha}) = 0,$$
and thus $\alpha$ is harmonic.

Next we claim that in every Dolbeault class there is a $d$-closed form. To this end, let $\alpha$ be a smooth $(0,1)$-form. Suppose $\alpha$ is not $d$-closed. We seek a $d$-closed $(0,1)$-form $\beta$ and a smooth function $f$ such that
$$\beta - \alpha = \bar{\partial} f.$$

Now, consider the $(1,1)$-form
$$d\alpha = \partial \alpha.$$

By Stokes' Theorem,
$$\int_X d\alpha = 0.$$

It follows from Corollary 9.3.3 that there is a function $f$ such that
$$-\partial\bar{\partial} f = \partial \alpha.$$

Letting
$$\beta := \alpha + \bar{\partial} f,$$
we have
$$d\beta = \partial \beta = \partial \alpha + \partial\bar{\partial} f = 0.$$

To complete the proof of Theorem 5.10.1, we must show that every Dolbeault class contains exactly one $d$-closed form. But if $\alpha_1$ and $\alpha_2$ are $d$-closed 1-forms with $\alpha_2 - \alpha_1 = \bar{\partial} f$, then
$$0 = d(\alpha_2 - \alpha_1) = \partial\bar{\partial} f.$$

Thus $f$ is harmonic, hence constant, so $\alpha_1 = \alpha_2$. The proof of Theorem 5.10.1 is now complete. □

## 9.6. Exercises

To end this chapter, we give a few long and somewhat tough exercises that work out the twisted version of the Hodge Theorem.

> 9.1. (Hilbert spaces) Let $X$ be a compact Riemann surface and let $H \to X$ be a holomorphic line bundle. Let $h$ be a Hermitian metric for $H$ and let $\omega$ be a Hermitian metric for $X$ (i.e. a Hermitian metric for the line bundle $T_X^{1,0} \to X$).

(a) Show that for two sections $s, t \in \Gamma(X, T_X^{*0,1} \otimes H)$ written locally as $s = \alpha_\xi \otimes \xi$ and $t = \beta_\xi \otimes \xi$ where $\xi$ is a frame of $H$ and $\alpha, \beta$ are $(0, 1)$-forms, the expression
$$\langle s, t \rangle := \frac{1}{2\sqrt{-1}} \alpha_\xi \wedge \overline{\beta_\xi} e^{-\varphi(\xi)}$$
defines a smooth complex measure on $X$ such that $\langle s, s \rangle$ is a positive measure.

(b) Define the corresponding measures associated to $\Gamma(X, T_X^{*1,0} \otimes H)$, $\Gamma(X, H)$, and $\Gamma(X, \Lambda_X^{1,1} \otimes H)$. (Observe that while for twisted 1-forms there is no need for the metric $\omega$ in the definition of these measures, this is not so for sections or twisted $(1,1)$-forms.)

Let us define the Hilbert spaces $L_{p,q}^2(\omega, h)$, $p, q \in \{0, 1\}$, to be the Hilbert space closures of the set of smooth $(p, q)$-forms with respect to the inner products
$$(s, t) := \int_X \langle s, t \rangle .$$

9.2. ($\bar{\partial}^*$ and $\square$) Recall that we have operators $\bar{\partial}_{0,0} : \Gamma(X, H) \to \Gamma(X, T_X^{*0,1} \otimes H)$ and $\bar{\partial}_{0,1} : \Gamma(X, T_X^{*0,1} \otimes H) \to \Gamma(X, \Lambda_X^{1,1} \otimes H)$, which are both denoted $\bar{\partial}$ when the degree of the forms is clear from the context. We denote by $\omega = e^{-\psi}\sqrt{-1}dz \wedge d\bar{z}$ the local expression for the area form. The point of this exercise is to compute the so-called *formal adjoint* $\bar{\partial}_{0,1}^*$ of the operator $\bar{\partial}_{0,1}$, defined by
$$((\bar{\partial}_{0,0})^* s, f) = (s, \bar{\partial}_{0,0} f),$$
and to define and compute the associated Laplace operator.

(a) Show that $(\bar{\partial}_{0,0})^*(fd\bar{z} \otimes \xi) = \left[-e^{\varphi+\psi}\frac{\partial}{\partial z}(e^{-\varphi}f)\right]\xi$, and confirm directly that the right hand side is globally defined when $fd\bar{z} \otimes \xi$ is the local expression of a global section, $e^{-\varphi} = h(\xi, \bar{\xi})$, and $e^{-\psi}\sqrt{-1}dz \wedge d\bar{z} = \omega$.

(b) Show that the operator
$$\square_{0,1} := \bar{\partial}_{0,0}(\bar{\partial}_{0,0})^* : \Gamma(X, T_X^{*0,1} \otimes H) \to \Gamma(X, T_X^{*0,1} \otimes H)$$
is given locally by
$$\square_{0,1}(fd\bar{z} \otimes \xi) = \frac{\partial}{\partial \bar{z}}\left(e^\psi\left(-e^\varphi\frac{\partial}{\partial z}(e^{-\varphi}f)\right)\right) d\bar{z} \otimes \xi$$
and confirm directly that the right side of this identity is globally defined when the twisted form on the left side is the local expression of a global twisted form.

(c) Compute the corresponding operator $\square_{0,0} : \Gamma(X, H) \to \Gamma(X, H)$.

## 9.6. Exercises

9.3. $((\nabla^{1,0})^*$ and $\Box)$ Recall that the Chern connection $\nabla = \nabla^{1,0} + \bar{\partial}$ acts on $H$-valued $(0,1)$-forms by
$$\nabla^{1,0}(f d\bar{z} \otimes \xi) = (\partial f - f \partial \varphi) \wedge d\bar{z} \otimes \xi,$$
to produce an $H$-valued $(1,1)$-form. (See formula (6.1) in Chapter 6.)
  (a) Show that $\Box = (\nabla^{1,0})^* \nabla^{1,0}$.
  (b) Show that $\Box = \frac{1}{2}\Delta$, where
$$\Delta = DD^* + D^*D : \Gamma(X, T_X^{*0,1} \otimes H) \to \Gamma(X, T_X^{*0,1} \otimes H).$$
(For the interested reader, the Laplace-type operators $\bar{\partial}\bar{\partial}^* + \bar{\partial}^*\bar{\partial}$ and $\nabla^{1,0}(\nabla^{1,0})^* + (\nabla^{1,0})^*\nabla^{1,0}$, when acting on $H$-valued $(0,1)$-forms, only agree when the complex dimension of $X$ is 1, or when $H$ is trivial and $X$ is Kähler. In general the difference between these two operators is a $0^{\text{th}}$ order operator related to the curvature of $H$ and the metric of $X$. It is always the case that
$$\Delta = \nabla^{1,0}(\nabla^{1,0})^* + (\nabla^{1,0})^*\nabla^{1,0} + \Box.$$
In Chapter 11, we will look at another identity for the operator $\Box$ that is especially useful when the curvature $\partial\bar{\partial}\varphi$ of $e^{-\varphi}$ is positive.)

9.4. Show that if $\Box \theta$ is smooth then $\theta$ is smooth.

9.5. (Twisted Sobolev and Rellich) Let us define the inner product $(\cdot,\cdot)_1$ on $\Gamma(X, t_X^{*1,0} \otimes H)$ by
$$(\theta, \eta)_1 := (\Delta\theta + \theta, \eta) = (\bar{\partial}^*\theta, \bar{\partial}^*\eta) + (\theta, \eta),$$
and the Hilbert space $W^{2,1}$ to be the Hilbert space closure of $\Gamma(X, t_X^{*1,0} \otimes H)$ with respect to $(\cdot,\cdot)_1$.
  (a) Show that $W^{1,2} \subset L^2_{0,1}(\omega, h)$.
  (b) Show that the inclusion $W^{1,2} \hookrightarrow L^2_{0,1}(\omega, h)$ is compact. (This exercise requires some estimates that we omitted in the notes even in the untwisted case. By doing the exercise, you fill in the details of the untwisted case as well.)

9.6. Show that for each $\theta \in L^2_{0,1}(\omega, h)$ there exists $u \in W^{1,2}$ such that $\Box u + u = \theta$.

9.7. Use the Spectral Theorem for Compact Self-adjoint Operators to show that there is a basis for $L^2(\omega, h)$ consisting of smooth eigensections of $\Box$.

9.8. Prove that for any $\theta \in L^2_{0,1}(\omega, h)$, the equation $\Box u = \theta$ has a solution if and only if $\theta \perp \text{Ker } \Box$.

9.9. Show that if $\theta \perp \text{Ker } \Box$ and $u$ is the solution of $\Box u = \theta$ having smallest possible norm, then
$$||u|| \le \tfrac{1}{\lambda_+}||\theta||$$

where $\lambda_+ > 0$ is the smallest positive eigenvalue of $\Box$, i.e., the smallest eigenvalue of $\Box|_{\text{Ker}\,\Box^\perp}$.

9.10. Formulate and prove an analogue of Theorem 9.3.1 for the twisted setting.

Chapter 10

# Uniformization

## 10.1. Automorphisms of the complex plane, projective line, and unit disk

**10.1.1. Aut($\mathbb{C}$).** We begin with the following result.

PROPOSITION 10.1.1. *Any injective holomorphic map $f : \mathbb{C} \to \mathbb{C}$ is affine linear.*

**Proof.** If the singularity at $\infty$ is a pole, then $f$ extends to a rational map of $\mathbb{P}_1$ to itself, and being holomorphic, $f$ must be a polynomial. By the Fundamental Theorem of Algebra, $f$ is affine linear.

We claim that $f$ cannot have an essential singularity. Indeed, if it does, then the image of any of the sets $A_R := \{|z| > R\}$ is open (since holomorphic functions are open maps) and dense by the remark following Theorem 1.4.5. Fix $z_o \in \mathbb{C}$ and let $a = f(z_o)$. Then the image under $f$ of some neighborhood $|z - z_o| < \varepsilon$ is a neighborhood of $a$ in $\mathbb{C}$. Take $A_R$ with $R > \varepsilon + |z_o|$. Since $f(A_R)$ is open and dense, it meets $f(\{|z - z_o| < \varepsilon\})$. It follows that $f$ cannot be injective. The proof is complete. □

COROLLARY 10.1.2. *Every automorphism $f : \mathbb{C} \to \mathbb{C}$ is affine linear.*

**10.1.2. Aut($\mathbb{P}_1$).**

LEMMA 10.1.3. *For each triple $\{P, Q, R\}$ of distinct points in $\mathbb{P}_1$ there is an automorphism $f : \mathbb{P}_1 \to \mathbb{P}_1$ such that $f(P) = [1, 0]$, $f(Q) = [1, 1]$ and $f(R) = [0, 1]$.*

**Proof.** Let us take affine coordinates $\zeta = w/z$ for the points $[z, w]$ of $\mathbb{P}_1$, so that $[1, 0]$ is the point at $\infty$. Then evidently

(10.1) $$f(\zeta) = \frac{(\zeta - \zeta(P))(\zeta(Q) - \zeta(R))}{(\zeta - \zeta(R))(\zeta(Q) - \zeta(P))}$$

has the desired property. Moreover, since $f$ is non-constant and of the form $\frac{a\zeta+b}{c\zeta+d}$, $f$ is invertible (with the inverse being given by $\frac{d\zeta-b}{a-c\zeta}$). □

COROLLARY 10.1.4. *Any automorphism of $\mathbb{P}_1$ is of the form* (10.1).

**Proof.** Let $h \in \mathrm{Aut}(\mathbb{P}_1)$ and let $P$, $Q$ and $R$ be the images of $[1,0]$, $[1,1]$ and $[0,1]$ respectively. By Lemma 10.1.3 there exists $f \in \mathrm{Aut}(\mathbb{P}_1)$ mapping $P$, $Q$ and $R$ to $[1,0]$, $[1,1]$ and $[0,1]$ respectively. Thus $f \circ h$ fixes each of the points $[1,0]$, $[1,1]$ and $[0,1]$, and so it suffices to show that if $g \in \mathrm{Aut}(\mathbb{P}_1)$ fixes $[1,0]$, $[1,1]$ and $[0,1]$ then $g$ is the identity. To this end, since $g$ fixes a point in $\mathbb{P}_1$, it is a holomorphic diffeomorphism of the complement $\mathbb{C}$, where it fixes two distinct points. But an affine transformation of $\mathbb{C}$ fixing two distinct points is the identity. □

### 10.1.3. Aut($\mathbb{D}$).

THEOREM 10.1.5. *Every automorphism $f : \mathbb{D} \to \mathbb{D}$ is of the form*

$$f(z) = e^{\sqrt{-1}\theta} \frac{b-z}{1-\bar{b}z}$$

*for some $\theta \in [0, 2\pi)$ and $b \in \mathbb{D}$.*

**Proof.** Fix $f \in \mathrm{Aut}(\mathbb{D})$, and write $a = f(0)$. Consider the map $g(w) := \frac{a-w}{1-\bar{a}w}$, which maps $a$ to the origin and satisfies

$$1 - |g(w)|^2 = \frac{(1-|w|^2)(1-|a|^2)}{|1-\bar{a}w|^2}.$$

It follows that $g$ fixes both the unit disk and its boundary, and since $g(g(w)) = w$, $g$ is invertible. Now, $g \circ f$ is an automorphism of the unit disk fixing the origin. By the Schwarz Lemma, $g \circ f(z) = e^{\sqrt{-1}\theta} z$ for some $\theta \in \mathbb{R}$. Thus with $b = e^{-\sqrt{-1}\theta} a$, $f$ has the desired form. □

## 10.2. A review of covering spaces

**10.2.1. Definition.** Let $V$ be a connected topological space.

DEFINITION 10.2.1. A covering space of $V$ is a space $U$ together with a continuous map $F : U \to V$ such that for each $v \in V$ there is a neighborhood $W$ of $v$ in $V$ such that $F^{-1}(W)$ is a disjoint union of sets $U_\alpha, \alpha \in A$, each of which is homeomorphic, via $F$, to $W$. ◇

REMARK. Note that if $F : U \to V$ is a covering space, then $F$ is a surjective local homeomorphism. However, the converse is not true. There are many examples of local homeomorphisms that are not uniform, as a covering map is. In fact, there are so many examples that triples $(F, U, V)$ with $F : U \to V$ a surjective local homeomorphism go by a special name: we call $U$ a sheaf over $V$. ◇

**10.2.2. Curve lifting property.** Covering spaces have a curve lifting property: given a path $\gamma : [0,1] \to V$ and a point $p \in F^{-1}(\gamma(0))$, there is a unique path $\tilde{\gamma} : [0,1] \to U$ such that $\tilde{\gamma}(0) = p$ and $F \circ \tilde{\gamma} = \gamma$. The existence of the lift is established by successively lifting neighboring small pieces of the curve using the local isomorphism property of the covering map.

Using the curve lifting property, one can show that, more generally, any map $g : S \to U$ from a simply connected space has a lift $\tilde{g} : S \to V$, i.e., $g = F \circ \tilde{g}$. Even more generally, a map $h : W \to U$ has a lift to $V$ if and only if $h_*\pi_1(W) < F_*\pi_1(U)$.

Two covers $F_1 : U_1 \to V$ and $F_2 : U_2 \to V$ are said to be isomorphic if there is an isomorphism (in whatever category) $G : U_1 \to U_2$ such that $F_1 = F_2 \circ G$.

**10.2.3. Universal cover.** There exists a universal covering, i.e., a simply connected space $U_0$ and a covering map $F_0 : U_0 \to V$. Moreover, $U_0$ is unique up to isomorphism. The word *universal* refers to the following universal property: if $F : U \to V$ is any other cover, then $F$ factors through $F_0$ uniquely, in the sense that there is a unique cover $G : U_0 \to U$ such that $F_0 = F \circ G$.

The fundamental group $\pi_1(V)$ acts on the universal cover $F_0 : U_0 \to V$ as follows. Fix a point $p \in U_0$ and let $q = F_0(p)$. For any loop $\gamma$ such that $[\gamma] \in \pi_1(V, q)$, let $\tilde{\gamma}$ be the unique lift of $\gamma$ to $U_0$ starting at $p$. We define

$$[\gamma]p := \tilde{\gamma}(1).$$

It is not hard to see that the orbit space $U_0/\pi_1(V)$ is naturally isomorphic to the original space $V$.

**10.2.4. Intermediate covers.** Given any subgroup $H < \pi_1(V)$, the action defined above on $U_0$, when restricted to $H$, defines an action of $H$ on the universal cover. The orbit space $U_0/H$ is itself a covering space of $V$, and two such covers are isomorphic if and only if the corresponding subgroups are conjugate. Thus there is a 1-1 correspondence between isomorphism classes of connected coverings and conjugacy classes of subgroups of $\pi_1(V)$. To see how the reverse of this correspondence works, let $F : U \to V$ be a covering, and fix a point $p \in F^{-1}(V)$. We define $H := \{[\gamma] \in \pi_1(V, F(p))\,;\, [\gamma]p = p\} = (\pi_1(V, F(p)))_p$ to be the stabilizer at $p$. Since the orbit of $\pi_1(V)$ is the whole fiber, different choices of $p$ will yield different but conjugate subgroups.

DEFINITION 10.2.2. The degree of a covering $F : U \to V$ is the index $[\pi_1(V) : H]$ of the subgroup $H < \pi_1(V, q)$ corresponding to the cover $F$. ◇

EXAMPLE 10.2.3. (1) Consider the torus $X = \mathbb{C}/L$ obtained from $\mathbb{C}$ after taking a quotient by a lattice. The natural quotient map $\pi : \mathbb{C} \to X$ is the universal cover. The fundamental group of $X$ is the free Abelian group on two generators, and it is isomorphic to the lattice $L$. (The generating loops are the two (non-parallel) edges of a fundamental parallelogram.)

Any proper subgroup of $L$ is isomorphic either to a (sub)lattice $L'$ or to $\mathbb{Z}$. The corresponding quotients are tori $\mathbb{C}/L'$ or $\mathbb{C}^*$. The degree of the cover of the former is just the degree of the induced map $F : \mathbb{C}/L' \to \mathbb{C}/L$ and the degree of the cover $\mathbb{C}^* \to X$ is infinite.

(2) Consider the upper half-plane $\mathbf{H} = \{z \in \mathbb{C} \; ; \; \operatorname{Im} z > 0\}$. The map $z \mapsto e^{2\pi i z}$ maps $\mathbf{H}$ surjectively onto the punctured unit disk, and it is easy to check that this map is a covering map. This exponential map is the projection for the quotient by the group of translations by (real) integers. Any subgroup is cyclic and generated by some translation $z \mapsto z + N$. The resulting quotient is still a punctured disk, and the quotient map is $z \mapsto e^{2\pi\sqrt{-1}z/N}$. Thus this punctured disk is just an $N$-fold cover of the original punctured disk. Indeed, $(e^{2\pi\sqrt{-1}z/N})^N = e^{2\pi i z}$. ◇

## 10.3. The Uniformization Theorem

### 10.3.1. Statement.

THEOREM 10.3.1 (Uniformization). *Let $X$ be a connected Riemann surface. The universal cover of $X$ is exactly one of the following Riemann surfaces:*

(1) *the Riemann sphere $\mathbb{P}_1 = \mathbb{C} \cup \{\infty\}$,*

(2) *the complex plane $\mathbb{C}$,*

(3) *the unit disk $\mathbb{D}$.*

Before turning to the proof, we will discuss some of the implications of the Uniformization Theorem.

### 10.3.2. Potential-theoretic classification of Riemann surfaces.
There are several ways to classify Riemann surfaces. For compact Riemann surfaces it is natural to use the genus of the surface as an invariant, as we shall see later on. For more general Riemann surfaces, one of the earliest classifications of Riemann surfaces is in terms of their subharmonic functions.

DEFINITION 10.3.2. A Riemann surface is said to be

(1) potential-theoretically hyperbolic if it has a non-constant bounded subharmonic function,

(2) potential-theoretically elliptic if it is compact,

(3) potential-theoretically parabolic otherwise. ◇

REMARK. We call this classification potential-theoretic because the condition of having a bounded subharmonic function is equivalent to the existence of a Green's Function. ◇

## 10.3. The Uniformization Theorem

The Riemann sphere is clearly elliptic. The unit disk is hyperbolic, as the function $|z|^2$ shows. On the other hand, the complex plane is parabolic. This parabolicity can be shown by a Liouville-type theorem for subharmonic functions, but we will show it in a slightly different way.

PROPOSITION 10.3.3. *Let $Y$ be a compact Riemann surface, and fix $k$ distinct points $p_1, ..., p_k \in Y$. Then the surface $X = Y - \{p_1, ..., p_k\}$ is potential-theoretically parabolic.*

**Proof.** Suppose $f$ is a bounded subharmonic function on $X$. Let $\tilde{f} : Y \to \mathbb{R} \cup \{-\infty\}$ be an extension of $f$ defined by

$$\tilde{f}(p_k) := \limsup_{z \to p_k} f(z).$$

Observe that $\tilde{f}$ has the same bound as $f$.

Since $\Delta f \geq 0$ in the sense of distributions on $X$ and since $\tilde{f}$ is bounded and upper semi-continuous on $Y$, $\Delta \tilde{f} \geq 0$ in the sense of distributions on $Y$. But then $\tilde{f}$ is subharmonic on $Y$, hence constant. Thus $f$ is constant, and hence $X$ is potential-theoretically parabolic. $\square$

**10.3.3. Riemann surfaces covered by $\mathbb{P}_1$.** Observe that since $\mathbb{P}_1$ is compact, any surface covered by $\mathbb{P}_1$ is also compact. It follows from the Riemann-Hurwitz Formula that the only Riemann surface covered by $\mathbb{P}_1$ is itself $\mathbb{P}_1$.

We could also argue as follows. Suppose $\mathbb{P}_1$ covers a compact Riemann surface $X$. Then the arithmetic genus of $X$ must be zero, for otherwise we could pull back a holomorphic form from $X$ to $\mathbb{P}_1$, and the latter has no holomorphic forms. But then it follows that $X$ is simply connected. Thus the covering map is an isomorphism.

**10.3.4. Riemann surfaces covered by $\mathbb{C}$.** Observe that if a Riemann surface $X$ is hyperbolic, it cannot be covered by $\mathbb{C}$. Indeed, we could then pull back a bounded subharmonic function by the covering map and obtain a contradiction. Thus any Riemann surface covered by $\mathbb{C}$ is either parabolic or elliptic. The converse, however, is not true. In fact, we have the following result.

PROPOSITION 10.3.4. *The only Riemann surfaces covered by $\mathbb{C}$ are*

(1) $\mathbb{C}$,

(2) $\mathbb{C} - \{0\}$, *and*

(3) *the complex tori $\mathbb{C}/\Lambda$.*

**Proof.** Let $X$ be covered by $\mathbb{C}$. Then the group $\Gamma \subset \operatorname{Aut}(\mathbb{C})$ of deck transformations is a discrete subgroup of $\mathbb{C}$. It follows that if the group is non-trivial then it is generated by either one or two independent translations.

In the latter case the deck group is a lattice, and we obtain a torus. On the other hand, if the group is generated by one translation, then we can conjugate the group to the group of translation $\{z \mapsto z+n \; ; \; n \in \mathbb{Z}\}$. Then the map $\pi : \mathbb{C} \to \mathbb{C} - \{0\}$ given by $\pi(z) = e^{2\pi\sqrt{-1}z}$ is the covering map. The proof is complete. $\square$

**10.3.5. Poincaré classification.** Thus we come to another type of classification of Riemann surfaces.

DEFINITION 10.3.5. A Riemann surface is said to be

(1) Poincaré hyperbolic if it is covered by the unit disk,
(2) Poincaré parabolic if it is covered by the complex plane, and
(3) Poincaré elliptic if it is $\mathbb{P}_1$. $\diamond$

In fact, the Poincaré classification is geometric. We begin with the following proposition.

PROPOSITION 10.3.6. *If a Riemann surface is Poincaré hyperbolic (resp. parabolic, elliptic) then its holomorphic tangent bundle has a metric of constant negative (resp. zero, constant positive) curvature.*

Let us start with the unit disk. On it, one has the metric $g_\mathbb{D}$, the so-called Poincaré metric, defined by

$$g_\mathbb{D} := \frac{dz \cdot d\bar{z}}{(1-|z|^2)^2}.$$

The curvature of $g_\mathbb{D}$ is

$$\begin{aligned}
\Theta_{g_\mathbb{D}} &:= \sqrt{-1}\partial\bar{\partial}\log(1-|z|^2)^2 \\
&= \sqrt{-1}\partial\left(\frac{-2z d\bar{z}}{1-|z|^2}\right) \\
&= -2\sqrt{-1}\left(\frac{dz \wedge d\bar{z}}{1-|z|^2} - \frac{|z|^2 dz \wedge d\bar{z}}{(1-|z|^2)^2}\right) \\
&= -4 \cdot \frac{\sqrt{-1}}{2}\omega_\mathbb{D},
\end{aligned}$$

where $\omega_\mathbb{D}$ is the area $(1,1)$-form associated to $g_\mathbb{D}$. Thus $\mathbb{D}$ has a metric of constant negative curvature.

PROPOSITION 10.3.7. *Every automorphism of $\mathbb{D}$ is an isometry of $g_\mathbb{D}$.*

**Proof.** Fix $f \in \operatorname{Aut}(\mathbb{D})$. Recall that $f$ is the composition of a rotation and the map $g(w) := \frac{a-w}{1-\bar{a}w}$ for some $a$. As we already observed, $1-|g(w)|^2 = \frac{(1-|w|^2)(1-|a|^2)}{|1-\bar{a}w|^2}$. Moreover,

$$g'(w) = -\frac{1-|a|^2}{(1-\bar{a}w)^2},$$

## 10.3. The Uniformization Theorem

and thus
$$\frac{|g'(w)|^2}{(1-|g(w)|^2)^2} = \frac{(1-|a|^2)^2}{(1-|g(w)|^2)^2|1-\bar{a}w|^4} = \frac{1}{(1-|w|^2)^2}.$$

Thus $g$ is an isometry of the Poincaré metric. Since a rotation clearly preserves the Poincaré metric, the proof is complete. □

Now, since the deck group of any Riemann surface $X$ covered by the disk is a group of automorphisms of the unit disk and hence by Proposition 10.3.7 a group of isometries of the Poincaré metric, the latter metric descends to a metric of $g_X$, also called the Poincaré metric. The definition of $g_X$ is as follows. Let $\pi : \mathbb{D} \to X$ denote the covering map and fix $v \in T^{1,0}_{X,x}$. Since $\pi$ is locally invertible, we can select a local inverse $\pi^{-1}$ defined on a small neighborhood of $x$. Then we define
$$g_X(v,\bar{v}) = \frac{|d\pi^{-1}(v)|^2}{(1-|\pi^{-1}(x)|^2)^2}.$$

By covering theory, any two branches of $\pi^{-1}$ are conjugate by a deck transformation, which is an automorphism of $\mathbb{D}$ and hence an isometry of $g_\mathbb{D}$. Thus $g_X$ is well-defined.

It is now clear that
$$-\sqrt{-1}\partial\bar{\partial}\log g_X = -4\omega_X$$
where $\omega_X$ is the (1,1)-form associated to $g_X$. Indeed,
$$\begin{aligned}-\pi^*\sqrt{-1}\partial\bar{\partial}\log g_X &= -\sqrt{-1}\partial\bar{\partial}\log \pi^* g_X \\ &= -\sqrt{-1}\partial\bar{\partial}\log g_\mathbb{D} = -4\omega_\mathbb{D} = -4\pi^*\omega_X,\end{aligned}$$
and since $\pi$ is locally invertible, thus an immersion, we have our claim.

Next we turn to the complex plane. The Euclidean metric $|dz|^2$ has zero curvature. Moreover, since the deck group of any Riemann surface covered by the plane consists of translations, which are isometries of the Euclidean metric, we can push the Euclidean metric down to any such surface.

Finally, we deal with the Riemann sphere. Consider the open cover $U_0, U_1$ by affine charts. Then $z_0 = \frac{1}{z_1}$ in $U_0 \cap U_1$. We define the metric
$$g_{FS} = \frac{|dz_j|^2}{(1+|z_j|^2)^2} \quad \text{on } U_j, \quad j = 0, 1.$$

Observe that since $dz_0 = -(z_1)^{-2}dz_1$,
$$\frac{|dz_0|^2}{(1+|z_0|^2)^2} = \frac{1}{|z_1|^4}\frac{|dz_1|^2}{(1+|z_1|^{-2})^2} = \frac{|dz_1|^2}{(1+|z_1|^2)^2}.$$

Thus $g_{FS}$ is a well-defined metric for the holomorphic tangent bundle on $\mathbb{P}_1$. This metric is called the Fubini-Study metric.

The curvature of the Fubini-Study metric is

$$\sqrt{-1}\partial\bar{\partial}\log(1+|z_j|^2) = \sqrt{-1}\partial\left(\frac{z_j d\bar{z}_j}{1+|z_j|^2}\right) = \sqrt{-1}\frac{dz_j \wedge d\bar{z}_j}{(1+|z_j|^2)^2} = 2\omega_{FS},$$

and thus we have proved Proposition 10.3.6. □

REMARK. On $\mathbb{P}_1$, the hyperplane line bundle $\mathbb{H} \to \mathbb{P}_1$ and the tangent bundle $T_{\mathbb{P}_1}$ have metrics, both of which are called the Fubini-Study metric. I have found that this fact causes confusion in discussions; Riemannian geometers tend to think of a metric as a "metric for the tangent bundle", while complex geometers also deal with metrics for line bundles, which often arise as weights in some $L^2$-theory (see, for example, Chapter 11). The relationship between these metrics is the following. The curvature of the Fubini-Study metric for $\mathbb{H} \to \mathbb{P}_1$ is a positive multiple of the metric form associated to the Fubini-Study metric for $T_{\mathbb{P}_1} \to \mathbb{P}_1$. ◇

**10.3.6. Existence of positive line bundles: Another approach.** Assuming the Uniformization Theorem, the work of the previous section almost gives a second proof of the existence, on any Riemann surface, of line bundles with metrics having strictly positive curvature.

THEOREM 10.3.8. *Every Riemann surface admits a positive holomorphic Hermitian line bundle.*

**Proof.** Indeed, if $X$ is covered by the disk then its canonical bundle $K_X$, being the dual of the holomorphic tangent bundle, is positively curved. For $\mathbb{C}$ and $\mathbb{C} - \{0\}$, the restriction of the Fubini-Study metric provides the desired metric. The only Riemann surfaces we have not handled are tori.

In fact, the canonical bundle of any complex torus $X$ is trivial, since one can use the form $d\pi$, where $\pi : \mathbb{C} \to X$ is the universal cover. (Recall that $\pi$ is multi-valued, but that $d\pi$ is a well-defined, nowhere-zero differential form on $X$.) It follows that a metric for $K_X$ is in $(1,1)$-correspondence with globally-defined functions of the form $e^{-\varphi}$, the correspondence being

$$e^{-\varphi} \mapsto \frac{e^{-\varphi}}{|d\pi|^2}.$$

Thus a metric is strictly positively curved on $X$ if and only if $\varphi$ is strictly subharmonic. Since $X$ is compact, and thus has no non-constant subharmonic functions, neither the canonical nor the anti-canonical bundle admits a metric of positive curvature. We have to go elsewhere for such a metric.

There are several ways to construct positive line bundles on complex tori, and we have already seen two such ways in Chapter 9. Here we take yet another approach. Recall that in Chapter 3 we showed, using $\theta$-functions, that on every complex torus $X$ there exists a non-constant meromorphic function, i.e., a holomorphic map $f : X \to \mathbb{P}_1$. The map $f$ certainly has branching, since otherwise it would

## 10.3. The Uniformization Theorem

define a cover of $\mathbb{P}_1$, and consequently an isomorphism of $X$ and $\mathbb{P}_1$, which is topologically impossible.

Let $D$ be the divisor of zeros of $f$, and $D'$ the divisor of poles of $f$. Then $D$ and $D'$ are effective divisors defining the same line bundle $L$ on $X$, and whose supports have empty intersection. Let $s$ and $s'$ be the global sections of $L$ whose zero divisors are $D$ and $D'$ respectively. These sections are determined up to a constant, and in fact $s/s'$ is a constant multiple of $f$.

We now define a Hermitian metric for $L$ as follows: if $v \in L_x$, then

$$h(v, \bar{v}) := \frac{|v|^2}{|s(x)|^2 + |s'(x)|^2}.$$

In other words, locally $h = e^{-\varphi}$ with $\varphi = \log(|s|^2 + |s'|^2)$. We then find that

$$\begin{aligned}
\sqrt{-1}\partial\bar{\partial}\varphi &= \sqrt{-1}\partial\left(\frac{s\overline{ds} + s'\overline{ds'}}{|s|^2 + |s'|^2}\right) \\
&= \sqrt{-1}\left(\frac{ds \wedge \overline{ds} + ds' \wedge \overline{ds'}}{|s|^2 + |s'|^2} + \frac{(\bar{s}ds + \bar{s}'ds') \wedge (s\overline{ds} + s'\overline{ds'})}{(|s|^2 + |s'|^2)^2}\right).
\end{aligned}$$

Letting $F = (s, s')$, we see that for a $(1, 0)$-tangent vector $\xi$,

$$\sqrt{-1}\partial\bar{\partial}\varphi(\xi, \bar{\xi}) = \frac{|F|^2 |dF(\xi)|^2 - |F \cdot \overline{dF(\xi)}|^2}{|F|^4}.$$

This shows that $\sqrt{-1}\partial\bar{\partial}\varphi$ is strictly positive away from the set

$$B := \{x \in X \; ; \; \text{for some } \xi \in T_{X,x}, dF(\xi) \text{ is proportional to } F\}.$$

Since $s/s' = cf$ for some constant $c$, we see that $B$ is exactly the branching locus of $f$. In particular, $B$ is non-empty.

REMARK. The above conclusion is not surprising: we leave it to the reader to check that $L$ is the pullback by $f$ of the hyperplane line bundle on $\mathbb{P}_1$ and $h$ is the metric on $X$ obtained by pulling back the Fubini-Study metric for the hyperplane line bundle on $\mathbb{P}_1$. ◇

To deal with the absence of strict positivity along $B$, we choose a second meromorphic function whose branching locus is different from $B$. We could appeal to $\theta$-functions to construct such a function, but instead we exploit the group property of tori: we compose our function $f$ with a translation $[z] \mapsto [z+\varepsilon]$ arbitrarily close to the identity, thereby obtaining a new function $f_\varepsilon$. The branching locus of this function is clearly $B_\varepsilon = \{b - \varepsilon \; ; \; b \in B\}$. If $\varepsilon$ is a sufficiently small complex number then $B \cap B_\varepsilon = \emptyset$.

We now get a new line bundle $L_\varepsilon$ associated with the meromorphic function $f_\varepsilon$ as above, and with metric $h_\varepsilon = e^{-\varphi_\varepsilon}$ whose curvature is strictly positive away from $B_\varepsilon$.

REMARK. We will see in Chapter 14 that, for small but non-zero $\varepsilon$, the line bundle $L_\varepsilon$ is not isomorphic to $L$ as a holomorphic line bundle, although it is isomorphic to $L$ as a complex line bundle. ◇

Now we take the line bundle
$$\Lambda := L \otimes L_\varepsilon$$
with the metric
$$e^{-(\varphi + \varphi_\varepsilon)}.$$
Evidently this metric has curvature
$$\sqrt{-1}\partial\bar{\partial}\varphi + \sqrt{-1}\partial\bar{\partial}\varphi_\varepsilon$$
which is everywhere positive. The proof of Theorem 10.3.8 is complete. □

## 10.4. Proof of the Uniformization Theorem

LEMMA 10.4.1. *Let $X$ be a Riemann surface and let $Y \subset\subset X$ be a connected, simply connected domain with non-empty smooth boundary. Then $Y$ is biholomorphic to the unit disk.*

**Proof.** Let $p \in Y$. Consider the line bundle $L$ associated to the divisor $1 \cdot p$. Then there is a holomorphic section $s$ of $L$ over $X$ with $\mathrm{Ord}(s) = 1 \cdot p$. By Theorem 8.2.1 there is a holomorphic section $t$ of $L$ over a small neighborhood of $\overline{Y}$ (which is an open Riemann surface by assumption) with no zeros. The function $f := s/t$ is then a holomorphic function on a neighborhood of $\overline{Y}$ with no zeros other than a simple zero at $p$. Moreover, $f$ is smooth up to $\partial Y$, where it also does not vanish.

Since $\partial Y$ is smooth and compact, we can solve the Dirichlet problem on $Y$. Let $u$ be a continuous function on $\overline{Y}$ such that $u$ is harmonic on $Y$ and
$$u = -\log|f| \quad \text{on} \quad \partial Y.$$
Since $Y$ is simply connected, there is a holomorphic function $g : Y \to \mathbb{C}$ such that $\mathrm{Re}\, g = u$. Let
$$F := fe^g.$$
Note that $f(z) = 0$ if and only if $z = p$. Moreover
$$|F| = |f|e^{-\log|f|} = 1 \quad \text{on} \quad \partial Y,$$
and thus for each $z \in Y$
$$|F(z)| < \sup_{\partial Y} |F| = 1.$$
Thus $F : Y \to \mathbb{D}$ and $F^{-1}(\partial \mathbb{D}) = \partial Y$. It follows that $F$ is proper. Indeed, if $K \subset\subset \mathbb{D}$ is a compact set, then $F^{-1}(K)$ is closed and contained in $Y \subset\subset X$. Therefore $F^{-1}(K)$ is compact.

Since $F$ is proper, the number $n(z)$ of preimages of a point $z$ in $\mathbb{D}$ is finite. Counting multiplicity, this number $n(z)$ is independent of $z$, as we explained in the

Remark following Theorem 3.3.14. Since $F^{-1}(0) = \{p\}$, $n(0) = 1$. It follows that $F$ is 1-1. □

**Conclusion of the proof of Theorem 10.3.1.** Let $X$ be a simply connected Riemann surface. If $X$ is compact, we already know that $X$ must be $\mathbb{P}_1$. Thus suppose $X$ is open. Let $Y_1 \subset\subset Y_2 \subset\subset ... \subset X$ be a normal exhaustion. By Lemmas 7.2.8 and 10.4.1, each $Y_j$ is simply connected and biholomorphic to the unit disk, say $F_n : Y_n \to \mathbb{D}$.

Now fix $p \in Y_1$ and a tangent vector $\xi \in T^{1,0}_{X,p}$. After composing from the left with a disk automorphism, we may assume that $F_n(p) = 0$ for all $n$. By scaling $F_n$ to $G_n := R_n \cdot F_n$ for some constant $R_n$, we may also assume that $dG_n(\xi) = 1$. Consider the maps $f_n : \mathbb{D}(0, R_n) \to \mathbb{D}(0, R_{n-1})$ defined by

$$f_n := G_{n-1} G_n^{-1}.$$

Then by the Schwarz Lemma, $1 > |f'_n(0)| =: R_{n-1}/R_n$. It follows that $R_n$ is increasing and thus converges to some $R \in (0, \infty]$. Let $\mathbb{D}(0, \infty) := \mathbb{C}$. We will show that, perhaps after passing to a subsequence, $G_n$ converges to a biholomorphic map $F : X \to \mathbb{D}(0, R)$.

To this end, consider the maps $g_n := G_n \circ F_1^{-1} : \mathbb{D} \to \mathbb{C}$. Then $g_n(0) = 0$ and $g'_n(0) = G'_n(p)\xi = 1$, provided we take $\xi = dF_1^{-1}(0)\frac{\partial}{\partial \zeta}$. By Köbe's Compactness Theorem, a subsequence of the $g_n$ converges, and thus so does a subsequence of the $G_n$ on $Y_1$. Replace $\{G_n\}$ by this subsequence, and now consider $Y_2$. Repeating the same argument, we obtain a subsequence that converges on $Y_2$. Continuing in this way, we may assume, after disposing of the unnecessary $G_i$ and $Y_i$, that $\{G_n|_{Y_m}\}_{n \geq m}$ converges uniformly. It follows that we have a holomorphic limit map $F : X \to \mathbb{D}(0, R)$ with $|F'(0)| = R$. An application of the Argument Principle shows that $F$ is injective.

We need only show that $F$ is surjective, and for surjectivity it suffices to show that $F$ is proper. To this end, let $\{p_k\} \subset X$ be a sequence such that for each compact set $K$, all but a finite number of the $p_k$ lie outside $K$. We wish to prove that for any $r < R$, $|F(p_k)| > r$ for all but finitely many $k$. But let $Y_m \supset K$. If $m$ is sufficiently large, then $G_n \to F$ on $Y_m$ uniformly, and thus $|F(p_k) - G_n(p_k)| < (R_m - r)/2$ for $n > m >> 0$. Since each $G_n$ is proper, for each $k$ sufficiently large there exists $n$ such that $|G_n(p_k)| > r + (R_m - r)/2$. But then $|F(p_k)| > |G_n(p_k)| - (R_m - r)/2 > r$, as claimed. The proof of Theorem 10.3.1 is complete. □

## 10.5. Exercises

10.1. Prove the curve lifting property for a covering space.

10.2. Prove that if $\Gamma \subset \mathbb{C}$ is a closed discrete subgroup then $\Gamma$ has at most two generators.

10.3. Let $Y_1$ and $Y_2$ be open Riemann surfaces with smooth 1-dimensional boundaries $\partial Y_1$ and $\partial Y_2$, respectively, such that $\overline{Y_i} := Y_i \cup \partial Y_1$ is a compact Riemann surface with boundary, $i = 1, 2$. Let $f : \overline{Y_1} \to \overline{Y_2}$ be a non-constant, continuous map whose restriction to $Y_1$ is holomorphic. Assume that $f(\partial Y_1) \subset Y_2$. Show that $f$ is proper (meaning the inverse images of compact sets are compact) and prove that, counting multiplicity, the cardinality of $f^{-1}(y)$ is finite independent of $y \in Y_2$.

10.4. What is the area, with respect to the area form $\sqrt{-1}\partial\bar{\partial}\varphi + \sqrt{-1}\partial\bar{\partial}\varphi_\varepsilon$, of the torus $X$ in the proof of Theorem 10.3.8?

10.5. Let $X$ be a compact Riemann surface of genus 2 without boundary, and let $L \to X$ be a holomorphic line bundle with metric $e^{-\varphi}$ whose curvature $\partial\bar{\partial}\varphi$ is assumed to be positive. Let $g_\varphi$ be a Hermitian metric for $X$ whose area form is $\sqrt{-1}\partial\bar{\partial}\varphi$. What is the integral over $X$ of the curvature of $g_\varphi$?

# Chapter 11

# Hörmander's Theorem

## 11.1. Hilbert spaces of sections

**11.1.1. The Hilbert space of sections.** Let $H \to X$ be a complex line bundle with Hermitian metric $e^{-\varphi}$ and let $\omega$ be an area form on $X$. Given two sections $s$ and $t$, written locally as $s = f \otimes \xi$ and $t = g \otimes \xi$ of $H$, we define

$$(s,t) := \int_X f\bar{g} e^{-\varphi} \omega$$

to be the $L^2$ inner product of $s$ and $t$. The space $\Gamma(X, H)$ of smooth sections, with the norm $\|\ \|$ induced by the inner product $(\ ,\ )$, is not complete. Its completion with respect to $\|\ \|$ is denoted

$$L^2(\varphi, \omega).$$

**11.1.2. The Hilbert space of line bundle-valued $(0,1)$-forms.** We will also need a Hilbert space of $H$-valued $(0,1)$-forms. Of course, $H$-valued $(0,1)$-forms are just sections of the line bundle $T_X^{*0,1} \otimes H$, so in some sense we have dealt with this situation; we can simply replace $H$ by $T_X^{*0,1} \otimes H$. However, we are then required to have a metric for the latter line bundle.

Instead of following this more general situation, we exploit the fact that $T_X^{*0,1}$ is a special bundle. Indeed, if $\alpha$ is a $(0,1)$-form, then locally $\alpha = fd\bar{z}$, from which it follows that the global $(1,1)$-form

$$\frac{\alpha \wedge \bar{\alpha}}{2\sqrt{-1}} = |f|^2 \frac{\sqrt{-1}}{2} dz \wedge d\bar{z}$$

is a smooth measure and can be integrated without reference to a metric for $T_X^{*0,1}$ and an area form for $X$. More generally, if $e^{-\varphi}$ is a metric for a holomorphic line

177

bundle $H$ and $\alpha, \beta \in \Gamma(X, T_X^{*0,1} \otimes H)$ are given locally by $\alpha = f\xi \otimes d\bar{z}$ and $\beta = g\xi \otimes d\bar{z}$, then

$$\frac{\alpha \wedge \bar{\beta}}{2\sqrt{-1}} e^{-\varphi} := f\bar{g} e^{-\varphi(\xi)} \frac{\sqrt{-1}}{2} dz \wedge d\bar{z}$$

is a global $(1,1)$-form. We define

(11.1) $$(\alpha, \beta) := \int_X \frac{\alpha \wedge \bar{\beta}}{2\sqrt{-1}} e^{-\varphi}.$$

The Hilbert space closure of the set $\Gamma(X, T_X^{*0,1} \otimes H)$ of smooth, $H$-valued $(0,1)$-forms with respect to the norm $||\ ||$ obtained from the inner product (11.1) is denoted

$$L_{0,1}^2(\varphi).$$

REMARK. There is an alternative way to explain what we have done. Instead of proceeding as above, we could also introduce a metric for the bundle $T_X^{*0,1}$, given in the local frame $d\bar{z}$ by, say, $e^{-\eta}$. In this setting, our $L^2$-norm would be

$$||\alpha||^2 := \frac{1}{2\sqrt{-1}} \int_X \alpha \wedge \bar{\alpha} e^{-(\varphi+\eta)} \omega.$$

However, a metric for $T_X^{*0,1}$ is the inverse of an area element. Thus the $(1,1)$-form $\Omega = e^\eta \sqrt{-1} dz \wedge d\bar{z}$ is a globally defined, nowhere-zero $(1,1)$-form. Therefore if, in the same local frame, $\omega = e^{-\psi}\sqrt{-1} dz \wedge d\bar{z}$, then $e^{\eta+\psi}$ is a globally defined function, and

$$||\alpha||^2 = \frac{1}{2\sqrt{-1}} \int_X e^{-(\eta+\psi)} \alpha \wedge \bar{\alpha} e^{-\varphi}.$$

In our setting, we have decided to take $\eta = -\psi$ for simplicity. ◇

**11.1.3. The $\bar{\partial}$ operator on $L^2$.** The operator $\bar{\partial} : \Gamma(X, H) \to \Gamma(X, T_X^{*0,1} \otimes H)$ can be defined on all of $L^2(\varphi, \omega)$, in the sense of distributions. However, to be most useful to us, we need the image of $\bar{\partial}$ to be contained in $L_{0,1}^2(\varphi)$, which is never the case.

EXAMPLE 11.1.1. Let $X = \mathbb{C}$ and let $H$ be the trivial line bundle with metric $e^{-|z|^2}$. Consider the function $f$ defined in polar coordinates by

$$f(r) := \begin{cases} 0, & r \in [0, 1) \\ 1, & r \geq 1 \end{cases}.$$

Then $f$ is clearly in $L^2$, but $\bar{\partial}f$, defined in the sense of distributions, is not represented by any $L^2$-form. Indeed, suppose $\bar{\partial}f = \alpha \in L^2$. A simple calculation shows that in polar coordinates,

$$\frac{\partial g}{\partial \bar{z}} = \frac{1}{2}\left(e^{\sqrt{-1}\theta} \frac{\partial g}{\partial r} + \sqrt{-1} \frac{e^{\sqrt{-1}\theta}}{r} \frac{\partial g}{\partial \theta}\right).$$

## 11.1. Hilbert spaces of sections

Let $h$ be a real-valued function with compact support on $(0, \infty)$, and let $\beta$ be the $(0,1)$-form
$$\beta(z) = e^{|z|^2} e^{-\sqrt{-1}\theta} h(|z|) d\bar{z}.$$
Since $h$ has compact support, $\beta$ is clearly in $L^2$. Thus
$$\begin{aligned}\frac{1}{2\sqrt{-1}} \int_{\mathbb{C}} (\bar{\partial} f) \wedge \overline{\beta} e^{-|z|^2} &= -\int_{\mathbb{C}} f(r) \frac{\partial}{\partial \bar{z}} \left( e^{-\sqrt{-1}\theta} h(r) \right) \frac{\sqrt{-1}}{2} dz \wedge d\bar{z} \\ &= -\int_0^{\infty} \int_0^{2\pi} f(r) \left( \frac{\partial h}{\partial r} + \frac{h(r)}{r} \right) d\theta r dr \\ &= -2\pi \int_1^{\infty} \frac{\partial}{\partial r} (rh(r)) \, dr = 2\pi h(1).\end{aligned}$$

If we take $h(r) = r^{-1} e^{-\frac{1}{a}(r-1)^2} \chi(r)$ for some smooth function $\chi : [0, \infty) \to [0,1]$ with compact support such that $\chi \equiv \frac{1}{2\pi}$ on $[1/2, 2]$, then on the one hand,
$$\frac{1}{2\sqrt{-1}} \int_{\mathbb{C}} (\bar{\partial} f) \wedge \overline{\beta} e^{-|z|^2} = 1.$$
On the other hand,
$$\frac{1}{2\sqrt{-1}} \int_{\mathbb{C}} (\bar{\partial} f) \wedge \overline{\beta} e^{-|z|^2} \leq \|\bar{\partial} f\| \cdot \|\beta\|,$$
and
$$\int_{\mathbb{C}} \frac{\beta \wedge \bar{\beta}}{2\sqrt{-1}} e^{-|z|^2} \lesssim \int_{\mathbb{C}} \frac{\beta \wedge \bar{\beta}}{2\sqrt{-1}} \lesssim \int_{1/2}^{2} e^{-2(r-1)^2/a} dr \to 0 \quad \text{as } a \to 0.$$
This contradiction shows that the current $\bar{\partial} f$ cannot be represented by an element of $L^2_{(0,1)}(e^{-|z|^2})$. ◇

Since we cannot define $\bar{\partial}$ on $L^2$, we settle for defining it on the largest possible subset of $L^2$.

DEFINITION 11.1.2. $\text{Domain}(\bar{\partial}) := \{f \in L^2(\varphi, \omega) \, ; \, \bar{\partial} f \in L^2_{0,1}(\varphi)\}$. ◇

We have seen that $\text{Domain}(\bar{\partial})$ is a proper subset of $L^2(\varphi, \omega)$. On the other hand, since it contains all of the smooth sections, $\text{Domain}(\bar{\partial})$ is dense in $L^2(\varphi, \omega)$ by the definition of $L^2(\varphi, \omega)$.

**11.1.4. The formal adjoint of $\bar{\partial}$.** If $\alpha = h\xi \otimes d\bar{z}$ is a smooth $H$-valued $(0,1)$-form with compact support then, for any smooth section $s = f\xi$ of $H$,
$$\langle \alpha, \bar{\partial} s \rangle = \int_X \overline{h \left( \frac{\partial f}{\partial \bar{z}} \right)} e^{-\varphi} \frac{\sqrt{-1}}{2} dz \wedge d\bar{z} = -\int_X e^{\psi} e^{\varphi} \frac{\partial}{\partial z} \left( e^{-\varphi} \overline{h} \right) \overline{f} e^{-\varphi} \omega,$$
where locally the area form $\omega$ is given by
$$\omega = e^{-\psi} \frac{\sqrt{-1}}{2} dz \wedge d\bar{z}.$$

DEFINITION 11.1.3. The operator $\bar{\partial}^*$, defined on smooth compactly supported $H$-valued $(0,1)$-forms $\alpha = h\xi \otimes d\bar{z}$ by

$$\bar{\partial}^*\alpha := -e^{\psi+\varphi} \frac{\partial}{\partial z}\left(e^{-\varphi}h\right)\xi, \tag{11.2}$$

is called the *formal adjoint* of $\bar{\partial}$. ◇

REMARK. We take this opportunity to note that, in view of formula (6.2),

$$(\bar{\partial}^*\alpha)\omega = -\nabla^{1,0}\alpha. \tag{11.3}$$

This observation is the first step toward a geometric interpretation that we shall make after we establish the Basic Identity. ◇

## 11.2. The Basic Identity

In this section we develop an identity that plays the central role in solving $\bar{\partial}$.

### 11.2.1. An integration-by-parts identity.
Let $X$ be a relatively compact open subset in a Riemann surface $M$. Let $\beta$ be a smooth, $H$-valued $(0,1)$-form with compact support, written locally as $\beta = f\xi \otimes d\bar{z}$. Then integration by parts yields no boundary terms, and we have

$$\begin{aligned}
&||\bar{\partial}^*\beta||^2 \\
&= (\bar{\partial}\bar{\partial}^*\beta, \beta) \\
&= -\int_X \left(\frac{\partial}{\partial \bar{z}}\left(e^{\psi}(f_z - \varphi_z f)\right)\right) \bar{f} e^{-\varphi} \\
&= -\int_X e^{\psi}\psi_{\bar{z}}(f_z - \varphi_z f)\bar{f} e^{-\varphi} - \int_X e^{\psi}(f_{z\bar{z}} - \varphi_z f_{\bar{z}})\bar{f} e^{-\varphi} + \int_X e^{\psi}\varphi_{z\bar{z}}|f|^2 e^{-\varphi} \\
&= -\int_X e^{\psi}\psi_{\bar{z}}(e^{-\varphi}f)_z \bar{f} - \int_X e^{\psi}(e^{-\varphi}f_{\bar{z}})_z \bar{f} + \int_X e^{\psi}\varphi_{z\bar{z}}|f|^2 e^{-\varphi} \\
&= \int_X e^{-\varphi}f(e^{\psi}\psi_{\bar{z}}\bar{f})_z + \int_X e^{-\varphi}f_{\bar{z}}(e^{\psi}\bar{f})_z + \int_X e^{\psi}\varphi_{z\bar{z}}|f|^2 e^{-\varphi} \\
&= \int_X e^{\psi}(\varphi + \psi)_{z\bar{z}}|f|^2 e^{-\varphi} + \int_X e^{\psi}\left(\psi_z \psi_{\bar{z}}|f|^2 + \psi_{\bar{z}} f \bar{f_z} + f_{\bar{z}}\overline{\psi_{\bar{z}} f} + f_{\bar{z}}\bar{f_{\bar{z}}}\right) e^{-\psi} \\
&= \int_X e^{\psi}(\varphi + \psi)_{z\bar{z}}|f|^2 e^{-\varphi} + \int_X e^{\psi}|f_{\bar{z}} + \psi_{\bar{z}} f|^2 e^{-\varphi}.
\end{aligned}$$

Let us define
$$\overline{\nabla}\beta := (f_{\bar{z}} + \psi_{\bar{z}} f) d\bar{z}^{\otimes 2},$$
$$|\beta|^2_\omega e^{-\varphi} := e^{\psi-\varphi}|f|^2, \quad \text{and} \quad |\overline{\nabla}\beta|^2_\omega e^{-\varphi} := e^{2\psi-\varphi}|f_{\bar{z}} + \psi_{\bar{z}} f|^2.$$

The reader can check that the first expression defines a section of the complex line bundle $H \otimes (T_X^{*0,1})^{\otimes 2}$, while the second and third expressions are well-defined functions on $X$, which are just the pointwise $L^2$-norms of the sections $\beta$ and $\overline{\nabla}\beta$ of the complex line bundles $H \otimes T_X^{*0,1}$ and $H \otimes (T_X^{*0,1})^{\otimes 2}$ with respect to the

## 11.2. The Basic Identity

metrics $e^{\psi-\varphi}$ and $e^{2\psi-\varphi}$, respectively. We have therefore proved the following theorem.

THEOREM 11.2.1 (The Basic Identity). *The identity*

$$(11.4) \qquad ||\bar{\partial}^*\beta||^2 = \frac{1}{2\sqrt{-1}} \int_X |\overline{\nabla}\beta|^2_\omega e^{-\varphi}\omega + \int_X |\beta|^2_\omega e^{-\varphi} \frac{\sqrt{-1}}{2} \partial\bar{\partial}(\varphi+\psi)$$

*holds for all smooth compactly supported $H$-valued $(0,1)$-forms $\beta$.*

REMARK. The Basic Identity is also called the Bochner-Kodaira Identity in the literature.  ◊

**11.2.2. A more geometric interpretation of the Basic Identity.** In this section, which is logically unnecessary for the proof of Hörmander's Theorem, we give an explanation of why curvature arises in the Basic Identity.

11.2.2.1. THE FORMAL ADJOINT OF $\overline{\nabla}$. The formal adjoint $\overline{\nabla}^*$ of $\overline{\nabla}$ is computed with respect to the inner product

$$(\Upsilon, \Xi) := \int_X \langle \Upsilon, \Xi \rangle_\omega e^{-\varphi}\omega = \int_X g\bar{h} e^{\psi-\varphi}$$

for sections of $H \otimes (T_X^{*0,1})^{\otimes 2}$, where $\Upsilon = g\xi \otimes d\bar{z}^{\otimes 2}$ and $\Xi = h\xi \otimes d\bar{z}^{\otimes 2}$ locally. If $\beta$ is an $H$-valued $(0,1)$-form given locally by $\beta = f\xi \otimes d\bar{z}$, we compute that

$$(\overline{\nabla}^*\Upsilon, \beta) = (\Upsilon, \overline{\nabla}\beta) = \int_X g e^{-\psi} \tfrac{\partial}{\partial \bar{z}}(e^\psi f) e^{\psi-\varphi} = -\int_X e^{\psi+\varphi} \tfrac{\partial}{\partial z}(e^{-\varphi}g) \bar{f} e^{-\varphi}.$$

We thus have the formula

$$(11.5) \qquad \overline{\nabla}^*(g\xi \otimes d\bar{z}^{\otimes 2}) = -e^{\psi+\varphi} \tfrac{\partial}{\partial z}(e^{-\varphi}g) \xi \otimes d\bar{z}.$$

11.2.2.2. FORMS AND VECTOR FIELDS. The metric form $\omega$ allows us to relate $(0,1)$-forms with $(1,0)$-vector fields in the following manner. To the $(0,1)$-form $\beta$ we associate the $(1,0)$-vector field $\xi_\beta$ as follows:

$$\beta(\bar{\eta}) = \omega(\xi_\beta, \bar{\eta}), \qquad \eta \in T_X^{1,0}.$$

We write $\xi_\beta =: \Phi_\omega(\beta)$. Locally, if we write $\beta = fd\bar{z}$, $\xi_\beta = F\tfrac{\partial}{\partial z}$, and $\eta = c\tfrac{\partial}{\partial z}$, then

$$\omega(\xi_\beta, \bar{\eta}) = e^{-\psi} F\bar{c} \quad \text{and} \quad \beta(\bar{\eta}) = f\bar{c},$$

and thus

$$F = e^\psi f.$$

Since the metric form is sesqui-linear, we can extend the definition of $\Phi_\omega$ to forms with values in a complex line bundle as follows. For an $L$-valued $(0,1)$-form $\beta$ given locally by $\beta = fd\bar{z} \otimes \xi$, we set

$$\Phi_\omega(\beta) = e^\psi f \tfrac{\partial}{\partial z} \otimes \xi.$$

REMARK. Of particular interest here is the following observation. While $H \otimes T_X^{*0,1}$ is not a holomorphic line bundle, the sections of this line bundle are mapped by $\Phi_\omega$ to sections of the holomorphic line bundle $T_X^{1,0} \otimes H$. There is also an induced action of $\Phi_\omega$ on forms with values in the line bundle $H \otimes T_X^{*0,1}$. With all this in mind, it is interesting to note that

(11.6) $$\overline{\nabla} = \Phi_\omega^{-1} \circ \bar{\partial} \circ \Phi_\omega.$$

This observation, together with the observation (11.3), opens the door to a geometric interpretation of the basic identity, as we shall soon see. ◇

11.2.2.3. COVARIANT FORMULAS. Let $\beta$ be an $H$-valued $(0,1)$-form given locally by $\beta = f\xi \otimes d\bar{z}$. Observe first that

$$-\nabla^{1,0}\Phi_\omega\beta = -e^{\varphi+\psi}\frac{\partial}{\partial z}(e^{-(\varphi+\psi)}e^\psi f)\xi \otimes \frac{\partial}{\partial z} \otimes dz = \bar{\partial}^*\beta \otimes \frac{\partial}{\partial z} \otimes dz,$$

and therefore

(11.7) $$\Phi_\omega^{-1} \circ (-\bar{\partial}\nabla^{1,0}) \circ \Phi_\omega \beta = \bar{\partial}\bar{\partial}^*\beta.$$

Next, formula (11.5) reads

(11.8) $$\overline{\nabla}^*\Upsilon = \Phi_\omega^{-1} \circ (-\nabla^{1,0}) \circ \Phi_\omega \Upsilon$$

for a section $\Upsilon \in \Gamma(X, H \otimes (T_X^{*0,1})^{\otimes 2})$, and using (11.6) we see that

(11.9) $$\Phi_\omega^{-1} \circ (\nabla^{1,0}\bar{\partial}) \circ \Phi_\omega\beta = -\overline{\nabla}^*\overline{\nabla}\beta.$$

We have therefore proved the following theorem.

THEOREM 11.2.2. *One has the identity*

$$\bar{\partial}\bar{\partial}^* - \overline{\nabla}^*\overline{\nabla} = \Phi_\omega^{-1} \circ (\nabla^{1,0}\bar{\partial} + \bar{\partial}\nabla^{1,0}) \circ \Phi_\omega.$$

REMARK. Some care should be taken with the notation here. In formula (11.7), $\nabla^{1,0}$ is the $(1,0)$-part of the Chern connection acting on the section $\Phi_\omega\beta$ of the holomorphic line bundle $H \otimes T_X^{1,0}$, while in formulas (11.8) and (11.9), $\nabla^{1,0}$ is the $(1,0)$-part of the Chern connection acting on the 1-forms $\Phi_\omega\Upsilon$ and $\bar{\partial}\Phi_\omega\beta$, with values in the holomorphic line bundle $H \otimes T_X^{1,0}$, respectively. Similarly, in formula (11.7) $\bar{\partial}$ acts on the 1-form $\nabla^{1,0}\Phi_\omega\beta$ with values in the holomorphic line bundle $H \otimes T_X^{1,0}$, while in formula (11.9) $\bar{\partial}$ acts on the section $\Phi_\omega\beta$ of the holomorphic line bundle $H \otimes T_X^{1,0}$. These interpretations are crucial, because the Chern connection is only defined for sections of holomorphic line bundles or for differential forms with values in a holomorphic line bundle. The latter also have an interpretation as sections of a complex line bundle which may not be holomorphic, but we do not take this second interpretation. ◇

### 11.2.2.4. The Basic Identity and Curvature.

Theorem 11.2.2 may be summarized as follows. After the appropriate identification, namely via equations (11.3) and (11.8), the operators $\bar{\partial}^*$ and $\overline{\nabla}^*$ agree with $-\nabla^{1,0}$. Therefore the difference $\bar{\partial}\bar{\partial}^* - \overline{\nabla}^*\overline{\nabla}$ is, after the aforementioned identification, just the graded commutator $[\nabla^{1,0}, \bar{\partial}]$. (It is a graded commutator because the operators map sections to 1-forms.)

Now, as the reader will recall, the action of the curvature of the Chern connection on a section $f$ of a holomorphic line bundle $L$ is

$$\nabla\nabla f = (\nabla^{1,0} + \bar{\partial})(\nabla^{1,0} + \bar{\partial}) = [\nabla^{1,0}, \bar{\partial}]f.$$

In the setting of Theorem 11.2.2, the holomorphic line bundle is $L = H \otimes T_X^{1,0}$, and its metric is $e^{-(\varphi+\psi)}$. Therefore the curvature of the Chern connection is given by the formula

$$\partial\bar{\partial}(\varphi + \psi).$$

(The reader was asked to show this in Exercise 6.4.) We thus arrive at our punchline: Theorem 11.2.2 recovers the unintegrated form of the Basic Identity, and also gives a geometric interpretation of the basic identity in terms of curvature.

## 11.3. Hörmander's Theorem

THEOREM 11.3.1 (Hörmander's Theorem). *Fix a smooth Hermitian metric $\omega = e^{-\psi}\frac{\sqrt{-1}}{2}dz \wedge d\bar{z}$ for $X$. Let $H \to X$ be a holomorphic line bundle with Hermitian metric $e^{-\varphi}$ whose curvature satisfies*

(11.10) $$\sqrt{-1}\partial\bar{\partial}(\varphi + \psi) \geq c\omega$$

*for some positive constant $c$. Then for every $\alpha \in L^2_{(0,1)}(\varphi)$ there exists $u \in L^2(\varphi, \omega)$ such that $\bar{\partial}u = \alpha$ in the sense of currents and*

$$\int_X |u|^2 e^{-\varphi}\omega \leq \frac{1}{c}\int_X \frac{1}{2\sqrt{-1}}\alpha \wedge \bar{\alpha}e^{-\varphi}.$$

**Proof.** Let

$$\Gamma_0(X, T_X^{*1,0} \otimes H) \subset L^2_{(0,1)}(\varphi, \omega)$$

denote the set of all smooth forms with compact support in $X$. By the basic identity, the hypothesis (11.10) implies the estimate $||\bar{\partial}^*\beta||^2 \geq c||\beta||^2$. It follows that for all $\beta \in \Gamma_0(X, T_X^{*1,0} \otimes H)$,

(11.11) $$|(\alpha, \beta)|^2 \leq ||\alpha||^2 ||\beta||^2 \leq \frac{1}{c}||\alpha||^2 ||\bar{\partial}^*\beta||^2.$$

Consider the anti-linear functional $\lambda : \bar{\partial}^*(\Gamma_0(X, H)) \to \mathbb{C}$ defined by

$$\lambda(\bar{\partial}^*\beta) := (\alpha, \beta).$$

The estimate (11.11) implies that $\lambda$ is well-defined and continuous on the subspace $\bar{\partial}^*(\Gamma_0(X, T_X^{*1,0} \otimes H))$ and has norm at most $c^{-1/2}||\alpha||$. By the Hahn-Banach Theorem, $\lambda$ extends to a continuous linear functional on $L^2(\omega, \varphi)$ with the same norm. By the Riesz Representation Theorem $\lambda$ is represented by inner product against some $u \in L^2(\varphi, \omega)$, i.e., $\lambda(f) = (u, f)$, such that $||u|| = ||\lambda|| \leq c^{-1/2}||\alpha||$. It follows that

$$(u, \bar{\partial}^* \beta) = (\alpha, \beta) \quad \text{for all } \beta \in \Gamma_0(X, T_X^{*1,0} \otimes H).$$

But the latter says precisely that $\bar{\partial} u = f$ in the weak sense. The proof is finished. $\square$

## 11.4. Proof of the Korn-Lichtenstein Theorem

In Chapter 2 we used the following result to prove that every orientable Riemann surface supports a complex structure.

THEOREM 2.2.4. *Let $g$ be a $\mathscr{C}^2$-smooth Riemannian metric in a neighborhood of $0$ in $\mathbb{R}^2$. There exists a positively oriented change of coordinates $z = u(x, y) + \sqrt{-1} v(x, y)$ fixing the origin, such that*

$$g = a^{(z)} dz d\bar{z}.$$

In this section we finally give a proof of Theorem 2.2.4.

**11.4.1. Derivation of the PDE.** Since the result is local, we allow ourselves to shrink neighborhoods when necessary. Let us choose an orthonormal basis of 1-forms in $\mathbb{R}^2$, i.e., $\alpha$ and $\beta$ are 1-forms such that $g = \alpha^2 + \beta^2$. (Such a choice can be made by choosing any two independent 1-forms and then applying the Gram-Schmidt Algorithm from linear algebra.) Let $\omega = \alpha + \sqrt{-1}\beta$. Observe that, with $\cdot$ denoting the complexification of the symmetric product,

$$\omega \cdot \bar{\omega} = \alpha^2 + \beta^2 = g.$$

Thus if we can find function $F : \mathbb{R}^2 \to \mathbb{C} - \{0\}$ and $z : \mathbb{R}^2 \to \mathbb{C}$ such that $\omega = F dz$, then we have

$$g = |F|^2 |dz|^2,$$

and thus the function $z$ gives us the change of coordinates we seek. In turn, $\omega = F dz$ if and only if we can find a non-vanishing function $G$ such that

(11.12) $$d \log G \wedge \omega = -d\omega.$$

Indeed, the equation $0 = dG \wedge \omega + G d\omega = d(G\omega)$ means that $G\omega = dz$ by the Poincaré Lemma, and then we take $F = 1/G$.

Equation (11.12) is a partial differential equation analogous to the $\bar{\partial}$-equation. To see how this is so, choose tangent vectors $\xi$ and $\eta$ such that

(11.13) $$\alpha\xi = \beta\eta = 1 \quad \text{and} \quad \alpha\eta = \beta\xi = 0.$$

Letting
$$\xi_\omega = \frac{\xi - \sqrt{-1}\eta}{2} \quad \text{and} \quad \xi_{\bar{\omega}} = \overline{\xi_\omega} = \frac{\xi + \sqrt{-1}\eta}{2},$$
we find that
$$\omega\xi_\omega = 1, \quad \bar{\omega}\xi_{\bar{\omega}} = 1, \quad \omega\xi_{\bar{\omega}} = 0, \quad \text{and} \quad \bar{\omega}\xi_\omega = 0.$$

It follows that
$$df = df(\xi_\omega)\omega + df(\xi_{\bar{\omega}})\bar{\omega},$$
and thus, with $\phi$ defined by $d\omega = \phi\omega \wedge \bar{\omega}$,
$$d\log G \wedge \omega = -d\omega \iff d\log G(\xi_{\bar{\omega}}) = \phi.$$

In other words, if we want to find our function $G$, we must solve the PDE

(11.14) $$\xi_{\bar{\omega}}\gamma = \phi$$

and then take $G = e^\gamma$.

REMARK. Note that if our original metric $g$ were the Euclidean metric and we had chosen the standard basis as our orthogonal vectors, then (11.14) is exactly the Cauchy-Riemann equations. ◇

**11.4.2. The operator $\bar{\partial}_\omega$.** To simplify the situation with the weights, it is convenient to work with differential forms of the form $f\bar{\omega}$, rather than with functions. Accordingly, we define the operator $\bar{\partial}_\omega$ on a function $f$ by
$$\bar{\partial}_\omega f = (\xi_{\bar{\omega}}f)\bar{\omega}.$$

**11.4.3. Hilbert space structures.**

REMARK. We note that
$$\tfrac{\sqrt{-1}}{2}\omega \wedge \bar{\omega} = \alpha \wedge \beta$$
is the area element associated to the metric $g$. ◇

Let us begin with the Hilbert space of functions. An inner product of functions $f$ and $h$ on $U$ with respect to a weight $\psi$ is defined by
$$(f, h)_0 := \int_U f\bar{h}e^{-\psi}\tfrac{\sqrt{-1}}{2}\omega \wedge \bar{\omega}.$$

We then take $H_0(U, e^{-\psi})$ to be the Hilbert space completion of the set of smooth functions with finite norm.

Next we define the inner product on smooth multiples of $\bar{\omega}$ by
$$(f\bar{\omega}, h\bar{\omega})_1 := \frac{1}{2\sqrt{-1}}\int_U f\bar{\omega} \wedge \overline{h\bar{\omega}}e^{-\psi},$$

and let $H_1(U, e^{-\psi})$ be the Hilbert space completion of the space of all such smooth multiples of $\bar\omega$.

**11.4.4. The formal adjoint of $\bar\partial_\omega$.** Suppose $f\bar\omega$ and $h$ are compactly supported on $U$. Then

$$\begin{aligned}
(f\bar\omega, \bar\partial_\omega h)_1 &= \frac{\sqrt{-1}}{2} \int_U e^{-\psi} f(d\bar h) \wedge \bar\omega \\
&= \frac{\sqrt{-1}}{2} \int_U -e^{\psi} d(e^{-\psi} f\bar\omega) \bar h e^{-\psi} \\
&= \frac{\sqrt{-1}}{2} \int_U (-e^{\psi} \partial_\omega(e^{-\psi} f) \omega \wedge \bar\omega + f d\bar\omega) \bar h e^{-\psi} \\
&= \frac{\sqrt{-1}}{2} \int_U (-\xi_\omega f + (\xi_\omega \psi - \bar\phi) f) \bar h \omega \wedge \bar\omega e^{-\psi}
\end{aligned}$$

where, as before, $d\omega = \phi \omega \wedge \bar\omega$. Thus we see that the formal adjoint of $\bar\partial_\omega$ is

$$\bar\partial_\omega^*(f\bar\omega) = -\xi_\omega f + (\xi_\omega \psi - \bar\phi) f.$$

**11.4.5. The basic estimate.** Guided by the derivation of the basic identity for the $\bar\partial$-operator, we now establish an *a priori* estimate suitable to solving $\bar\partial_\omega$. First,

$$\bar\partial_\omega \bar\partial_\omega^*(f\bar\omega) = -(\xi_{\bar\omega} \xi_\omega f)\bar\omega + \xi_{\bar\omega}(\xi_\omega \psi - \bar\phi) f\bar\omega + (\xi_\omega \psi - \bar\phi)(\xi_{\bar\omega} f)\bar\omega.$$

Assuming $f$ has compact support, we compute that

$$\begin{aligned}
&(\bar\partial_\omega \bar\partial_\omega^* f\bar\omega, f\bar\omega)_0 \\
&= \int_U -(\xi_{\bar\omega} \xi_\omega f) \bar f e^{-\psi} \frac{\bar\omega \wedge \omega}{2\sqrt{-1}} + \int_U \xi_{\bar\omega}(\xi_\omega \psi - \bar\phi)|f|^2 e^{-\psi} \frac{\bar\omega \wedge \omega}{2\sqrt{-1}} \\
&\quad + \int_U (\xi_\omega \psi - \bar\phi)(\xi_{\bar\omega} f) \bar f e^{-\psi} \frac{\bar\omega \wedge \omega}{2\sqrt{-1}} \\
&= \int_U -(\xi_\omega \xi_{\bar\omega} f) \bar f e^{-\psi} \frac{\bar\omega \wedge \omega}{2\sqrt{-1}} + \int_U -([\xi_{\bar\omega}, \xi_\omega] f) \bar f e^{-\psi} \frac{\bar\omega \wedge \omega}{2\sqrt{-1}} \\
&\quad + \int_U \xi_{\bar\omega}(\xi_\omega \psi - \bar\phi)|f|^2 e^{-\psi} \frac{\bar\omega \wedge \omega}{2\sqrt{-1}} \\
&\quad + \int_U (\xi_\omega \psi - \bar\phi)(\xi_{\bar\omega} f) \bar f e^{-\psi} \frac{\bar\omega \wedge \omega}{2\sqrt{-1}} \\
&= \int_U |\xi_{\bar\omega} f|^2 e^{-\psi} \frac{\bar\omega \wedge \omega}{2\sqrt{-1}} + \int_U -([\xi_{\bar\omega}, \xi_\omega] f) \bar f e^{-\psi} \frac{\bar\omega \wedge \omega}{2\sqrt{-1}} \\
&\quad + \int_U \xi_{\bar\omega}(\xi_\omega \psi - \bar\phi)|f|^2 e^{-\psi} \frac{\bar\omega \wedge \omega}{2\sqrt{-1}}.
\end{aligned}$$

Now, by Cauchy-Schwartz one has the estimate

$$\left| \int_U ([\xi_{\bar\omega}, \xi_\omega] f) \bar f e^{-\psi} \frac{\bar\omega \wedge \omega}{2\sqrt{-1}} \right| \leq \varepsilon \int_U |[\xi_{\bar\omega}, \xi_\omega] f|^2 e^{-\psi} \frac{\bar\omega \wedge \omega}{2\sqrt{-1}} + \frac{C}{\varepsilon} \int_U |f|^2 e^{-\psi} \frac{\bar\omega \wedge \omega}{2\sqrt{-1}}.$$

## 11.4. Proof of the Korn-Lichtenstein Theorem

Moreover, we have

$$\int_U |[\xi_{\bar\omega}, \xi_\omega] f|^2 e^{-\psi} \frac{\bar\omega \wedge \omega}{2\sqrt{-1}} \leq C \int_U |\bar\partial^*_\omega (f\bar\omega)|^2 \frac{\bar\omega \wedge \omega}{2\sqrt{-1}},$$

because $[\xi_\omega, \xi_{\bar\omega}]$ is a first-order differential operator, and any first-order differential operator can be estimated in $L^2$ by the differential operator $\bar\partial^*_\omega$, as the reader is asked to verify in an exercise. Combining with the identity above, we have the basic estimate:

$$(11.15) \qquad \|\bar\partial^*_\omega(f\bar\omega)\|_0^2 \geq c \int_U \xi_{\bar\omega}(\xi_\omega \psi - \bar\phi) |f|^2 e^{-\psi} \frac{\bar\omega \wedge \omega}{2\sqrt{-1}},$$

where $c > 0$ is a (possibly quite small) constant that is independent of $f$.

**11.4.6. Positively curved weights.** Finally, we need to know that there are functions $\psi$ such that the term $\xi_\omega(\xi_{\bar\omega}\psi - \bar\phi)$ is positive in our neighborhood $U$. To this end we use the following trick. We note that at least at one point, we can see our metric $g$ as Euclidean in some chosen coordinate system. Assume the point is the origin and the coordinates are $(u, v)$. Write $\zeta = u + \sqrt{-1}v$. Then the operator $\xi_\omega$ is given by

$$(11.16) \qquad \xi_\omega = (1 + A(\zeta))\frac{\partial}{\partial \zeta} + B(\zeta)\frac{\partial}{\partial \bar\zeta} \quad \text{and} \quad \xi_{\bar\omega} = \overline{\xi_\omega},$$

where $A(\zeta)$ and $B(\zeta)$ are $o(1)$ as $\zeta \to 0$. It follows that

$$\xi_\omega \xi_{\bar\omega} = \frac{\partial^2}{\partial \zeta \partial \bar\zeta} + O(|\zeta|).$$

If we now let $\psi(\zeta) = C|\zeta|^2$ for some large constant $C > 0$, then

$$|\xi_\omega \xi_{\bar\omega} \psi - C| = O(|\zeta|).$$

It follows that for $C \gg 0$,

$$\xi_{\bar\omega}(\xi_\omega \psi - \bar\phi) \geq 1$$

in a small (but fixed) neighborhood of 0. We thus obtain from (11.15) the estimate

$$(11.17) \qquad \|\bar\partial^*_\omega(f\bar\omega)\|_0^2 \geq \|f\bar\omega\|_1^2, \quad f\bar\omega \in H_1(\omega, e^{-\psi}).$$

**11.4.7. Statement of the regularity theorem for $\bar\partial_\omega$.** We can apply the method of proof of Hörmander's Theorem 11.3.1 with the estimate (11.17) to conclude that the equation

$$\bar\partial_\omega(\gamma) = \phi\bar\omega$$

has a solution (with estimates, which we do not use here). However, the solution is at present just an $L^2$ weak solution.

REMARK. By saying that $\gamma$ is a weak solution of $\bar{\partial}_\omega \gamma = f\bar{\omega}$, we mean that for all $\chi \in \mathscr{C}_0^\infty(U)$,

$$\int_U \gamma \overline{\bar{\partial}_\omega^*(\chi\bar{\omega})} \frac{\sqrt{-1}}{2} \omega \wedge \bar{\omega} = \frac{1}{2\sqrt{-1}} \int_U f\bar{\omega} \wedge \overline{\chi\bar{\omega}}.$$

Here the formal adjoint $\bar{\partial}_\omega^*$ is computed relative to the unweighted case $\psi = 0$, so the formula

$$\bar{\partial}_\omega^*(\chi\bar{\omega}) = -\xi_\omega \chi - \bar{\phi}\chi$$

computes $\bar{\partial}_\omega^*$. ◇

Thus to prove Theorem 2.2.4 we must show that the solution is smooth. While the equation $\bar{\partial}_\omega \gamma = f\bar{\omega}$ resembles the $\bar{\partial}$-equation in many ways, establishing regularity for the former is a little more involved. The method we will use to establish this regularity is one that is typical in the theory of elliptic partial differential equations. The reader familiar with the method of elliptic regularity will follow our *ad hoc* argument with ease.

We are going to prove the following theorem.

THEOREM 11.4.1. *If $f$ is a smooth function on a domain $U$ in $\mathbb{C}$ and $\gamma$ is a weak solution to the equation $\bar{\partial}_\omega \gamma = f\bar{\omega}$, then $\gamma$ is a smooth function.*

For the reader in the know, Theorem 11.4.1 is a consequence of the fact that $\bar{\partial}_\omega$ is a (vector-valued) elliptic operator. The reader who is happy with this claim and its consequences can consider Theorem 2.2.4 proved and move on. The rest of the chapter will be dedicated to the proof of Theorem 11.4.1.

**11.4.8. The Fourier transform.** The Fourier transform is intimately connected to smoothness of functions. Smoothness of a function is manifested in the absence of large high-frequency components in the frequency spectrum of the said function, which means that the Fourier transform of the function in question decays more rapidly. Students often first encounter quantitative forms of this principle in the so-called *Riemann-Lebesgue Lemma*. More refined versions lead to Sobolev spaces, which we shall discuss in the next paragraph.

We begin with a more thorough treatment of the Fourier transform, which we first employed in our proof of Rellich's Compactness Theorem in Chapter 9. We work strictly in dimension 2, though the modifications to arbitrary dimension can easily be made.

Recall that for $F \in L^1(\mathbb{R}^2)$ the function

$$\hat{F}(t) := \frac{1}{2\pi} \int_{\mathbb{R}^2} e^{-\sqrt{-1}\langle t, x \rangle} F(x) dA(x), \qquad t \in \mathbb{R}^2,$$

## 11.4. Proof of the Korn-Lichtenstein Theorem

is called the Fourier transform of $F$. Sometimes we will also write $\hat{F} = \mathscr{F}(F)$ when it is convenient for phrasing certain statements. The inverse Fourier transform is written

$$\check{F}(t) := \frac{1}{2\pi} \int_{\mathbb{R}^2} e^{\sqrt{-1}\langle t, x\rangle} F(x) dA(x), \qquad t \in \mathbb{R}^2,$$

and we will also use the notation $\mathscr{F}^{-1} F := \check{F}$.

**LEMMA 11.4.2.** *For all $a > 0$,*

(11.18) $$\int_{\mathbb{R}^2} e^{\sqrt{-1}x \cdot t - a|x|^2} dA(x) = \frac{\pi}{a} e^{-\frac{|t|^2}{4a}}.$$

**LEMMA 11.4.3.** *For any continuous function $f : \mathbb{R}^n \to \mathbb{C}$,*

$$f(x) = \lim_{\varepsilon \to 0} \frac{1}{4\pi\varepsilon} \int_{\mathbb{R}^2} f(y) e^{-\frac{|x-y|^2}{4\varepsilon}} dA(y).$$

The proofs of both lemmas are left as an exercise to the reader.

**LEMMA 11.4.4 (Plancherel Formula).** *If $u, v \in L^1(\mathbb{R}^2)$ then $\hat{u}, \hat{v} \in L^\infty(\mathbb{R}^2)$ and*

$$(u, \hat{v}) = (\hat{u}, v).$$

**Proof.** That $\mathscr{F}(L^1(\mathbb{R}^2)) \subset L^\infty(\mathbb{R}^2)$ is clear. It follows that $(u, \hat{v})$ is well-defined, and by Fubini's Theorem

$$\begin{aligned}
(u, \hat{v}) &= \int_{\mathbb{R}^2} u(x) \hat{v}(x) dA(x) \\
&= \frac{1}{2\pi} \int_{\mathbb{R}^2} \int_{\mathbb{R}^2} u(x) v(t) e^{-\sqrt{-1} x \cdot t} dA(x) dA(t) \\
&= \int_{\mathbb{R}^2} \hat{u}(t) v(t) dA(t) = (\hat{u}, v),
\end{aligned}$$

as desired. $\square$

**THEOREM 11.4.5 (Plancherel Theorem).** *Let $u \in L^1(\mathbb{R}^2)$. If also $u \in L^2(\mathbb{R}^2)$ then $\hat{u}, \check{u} \in L^2(\mathbb{R}^n)$ and*

$$||\hat{u}||_{L^2(\mathbb{R}^2)} = ||\check{u}||_{L^2(\mathbb{R}^2)} = ||u||_{L^2(\mathbb{R}^2)}.$$

**Proof.** For $\varepsilon > 0$, let $w_\varepsilon(x) = e^{-\varepsilon|x|^2}$. Then, by (11.18), $\hat{w}_\varepsilon(\xi) = (2\varepsilon)^{-1} e^{-\frac{|\xi|^2}{4\varepsilon}}$. Therefore, for $v \in L^1(\mathbb{R}^2)$, the Plancherel Formula implies that

$$\int_{\mathbb{R}^2} \hat{v}(t) e^{-\varepsilon|t|^2} dA(t) = \frac{1}{4\pi\varepsilon} \int_{\mathbb{R}^2} v(x) e^{\frac{-|x|^2}{4\varepsilon}} dA(x).$$

We now apply the latter formula as follows. Let $u \in L^1(\mathbb{R}^2) \cap L^2(\mathbb{R}^2)$ and set $n(x) = \bar{u}(-x)$. Then $v := u * n \in L^1(\mathbb{R}^2) \cap \mathscr{C}(\mathbb{R}^2)$. Now

$$\hat{n}(t) = \frac{1}{2\pi} \int_{\mathbb{R}^2} e^{-\sqrt{-1} t \cdot x} \bar{u}(-x) dA(x) = \bar{\hat{u}}(t),$$

and as we will prove momentarily, if $w_1, w_2 \in L^1(\mathbb{R}^2)$ then
$$\mathscr{F}(w_1 * w_1) = 2\pi \mathscr{F}(w_1)\mathscr{F}(w_2) \in L^\infty(\mathbb{R}^2).$$
It follows that $\hat{v} = 2\pi |\hat{u}|^2$. From Lemma 11.4.3 and the continuity of $v$,
$$\begin{aligned}
\int_{\mathbb{R}^2} |\hat{u}|^2 dA &= \lim_{\varepsilon \to 0} \frac{1}{2\pi} \int_{\mathbb{R}^2} \hat{v}(t) e^{-\varepsilon |t|^2} dA(t) \\
&= \lim_{\varepsilon \to 0} \frac{1}{4\pi\varepsilon} \int_{\mathbb{R}^2} v(x) e^{-\frac{|x|^2}{4\varepsilon}} dA(x) \\
&= v(0) \\
&= \int_{\mathbb{R}^2} u(x) n(-x) dA(x) \\
&= \int_{\mathbb{R}^2} |u|^2 dA.
\end{aligned}$$
The proof for $\check{u}$ is analogous. $\square$

Plancherel's Theorem can be used to extend the Fourier transform to $L^2(\mathbb{R}^2)$. To this end, note that

(1) $L^1(\mathbb{R}^2) \cap L^2(\mathbb{R}^2)$ is dense in $L^2(\mathbb{R}^2)$ and
(2) the Fourier transform maps $L^1(\mathbb{R}^2) \cap L^2(\mathbb{R}^2)$ into $L^2(\mathbb{R}^2)$ isometrically.

Let $u \in L^2(\mathbb{R}^2)$. Take any sequence $\{u_j\} \subset L^1(\mathbb{R}^2) \cap L^2(\mathbb{R}^2)$ converging to $u$ in $L^2(\mathbb{R}^2)$. Then $\{u_j\}$ is a Cauchy sequence in $L^2(\mathbb{R}^2)$, and since
$$||\hat{u}_j - \hat{u}_k|| = ||u_j - u_k||,$$
so is $\{\hat{u}_j\}$. Since $L^2(\mathbb{R}^2)$ is a Hilbert space, $\{\hat{u}_j\}$ has a limit, and it is easy to see that the limit is independent of the choice of sequence $\{u_j\}$.

DEFINITION 11.4.6. *The Fourier transform $\hat{u}$ of $u$ is the limit of $\{\hat{u}_j\}$.* $\diamond$

The inverse Fourier transform is defined similarly.

THEOREM 11.4.7. *Let $u$ and $v$ be measurable functions on $\mathbb{R}^2$.*

(1) *If $u, v \in L^2(\mathbb{R}^2)$ then*
$$\int_{\mathbb{R}^2} u \bar{v} \, dV = \int_{\mathbb{R}^2} \hat{u} \bar{\hat{v}} \, dV.$$
(2) *If also $D^\alpha u \in L^2(\mathbb{R}^2)$ then $\mathscr{F}(D^\alpha u)(t) = (it)^\alpha \mathscr{F}(u)(t)$.*
(3) *If $u, v \in L^1(\mathbb{R}^2)$ then $\mathscr{F}(u * v) = (2\pi) \mathscr{F}(u) \mathscr{F}(v)$.*
(4) *(Fourier Inversion Theorem) $\mathscr{F}(\check{u}) = u$.*

**Proof.** (1) In view of the so-called parallelogram law
$$(t, \eta) = \tfrac{1}{4}\left( ||t+\eta||^2 - ||t-\eta||^2 + \sqrt{-1}\,||t+\sqrt{-1}\eta||^2 - \sqrt{-1}\,||t-\sqrt{-1}\eta||^2 \right),$$

part (1) follows from the Plancherel Theorem.

(2) By approximation it suffices to verify the result when $u$ is smooth with compact support, in which case the result follows repeated integration by parts.

(3) Using Fubini's Theorem,
$$\begin{aligned}
\mathscr{F}(u*v)(t) &= \frac{1}{2\pi}\int_{\mathbb{R}^2} e^{-\sqrt{-1}t\cdot x}\left(\int_{\mathbb{R}^2} u(y)v(x-y)dA(y)\right)dA(x)\\
&= \frac{1}{2\pi}\int_{\mathbb{R}^2}\int_{\mathbb{R}^2} u(y)v(x-y)e^{-\sqrt{-1}t\cdot(x-y)}e^{-\sqrt{-1}t\cdot y}dA(x)dA(y)\\
&= \frac{1}{2\pi}\int_{\mathbb{R}^2}\int_{\mathbb{R}^2} u(y)v(z)e^{-\sqrt{-1}t\cdot z}e^{-\sqrt{-1}t\cdot y}dA(z)dA(y)\\
&= 2\pi\hat{u}(t)\hat{v}(t).
\end{aligned}$$

(4) Observe that $\check{u} = \overline{\mathscr{F}(\bar{u})}$. Now,
$$\begin{aligned}
\int_{\mathbb{R}^2} \check{u}(t)v(t)dA(t) &= \int_{\mathbb{R}^2}\int_{\mathbb{R}^2} e^{\sqrt{-1}t\cdot x}u(x)v(t)dA(x)dA(t)\\
&= \int_{\mathbb{R}^2} u(x)\check{u}(x)dA(x).
\end{aligned}$$

Thus for all $w \in L^2(\mathbb{R}^2)$
$$\begin{aligned}
\int_{\mathbb{R}^2} \check{\hat{u}}(t)w(t)dA(t) &= \int_{\mathbb{R}^2} \hat{u}(x)\check{w}(x)dA(x) = \int_{\mathbb{R}^2} \hat{u}(x)\overline{\hat{\bar{w}}(x)}dA(x)\\
&= \int_{\mathbb{R}^2} u(x)\overline{\bar{w}(x)}dA(x) = \int_{\mathbb{R}^2} u(x)w(x)dA(x).
\end{aligned}$$

We conlcude that $\check{\hat{u}} = u$. $\square$

**11.4.9. Sobolev spaces.** Let $\mathscr{C}_0^\infty$ be the space of all $\mathscr{C}^\infty$-functions on $\mathbb{R}^2$ having compact support, and let $\mathscr{S}$ be the Schwartz space, i.e., the space of rapidly decreasing functions on $\mathbb{R}^2$. By definition, $f \in \mathscr{S}$ if and only if for all multiindices $\alpha$ and $\beta$,
$$\sup_{x\in\mathbb{R}^2} |x^\alpha D^\beta f(x)| < +\infty.$$
For any $s \in \mathbb{R}$, define the operator $\Lambda^s : \mathscr{S} \to \mathscr{S}$ by
$$\Lambda^s f = \mathscr{F}^{-1}(t \mapsto (1+|t|^2)^{s/2}\hat{f}(t)).$$
In particular $\Lambda^0 f = f$ and, if $k$ is a positive integer, then
$$\Lambda^{2k} f = (1-\Delta)^k f.$$
We define an inner product on $\mathscr{S}$ by
$$(f,g)_{H^s} := \int_{\mathbb{R}^2} \Lambda^s f \overline{\Lambda^s g}\, dA.$$

DEFINITION 11.4.8. The space $H^s$ is the completion of $(\mathscr{S}, (\cdot, \cdot)_{H^s})$. ◇

REMARK. The space $H^0$ is the usual Hilbert space $L^2(\mathbb{R}^2)$. ◇

If $k > 0$ is an integer, then using properties (1) and (2) of the Fourier transform proved in Theorem 11.4.7, we find that for any $s \in \mathbb{R}$,

$$||f||^2_{H^{s+k}} \sim \sum_{|\alpha| \leq k} ||D^\alpha f||^2_{H^s}.$$

Thus we can define a weak distributional derivative for functions in $H^s$: if $f \in H^s$ and $\{f_j\} \subset \mathscr{S}$ is a sequence converging to $f$, then we define $D^\alpha f \in H^{s-|\alpha|}$ by

$$D^\alpha f := \lim_{j \to \infty} D^\alpha f_j,$$

the limit being taken in $H^{s-|\alpha|}$. One can check that

(1) $D^\alpha f$ is independent of the sequence $f_j \to f$ in $H^s$ and
(2) $D^\alpha H^s \subset H^{s-|\alpha|}$.

If $f \in \mathscr{C}^k$, we define

$$||f||_{\mathscr{C}^k} := \sup_{\mathbb{R}^2} |f|_{\mathscr{C}^k} = \left( \sum_{j=0}^k \sup_{\mathbb{R}^2} |D^j f|^2 \right)^{1/2}.$$

The relation between weak and strong derivatives is given by the following result of Sobolev.

PROPOSITION 11.4.9 (Sobolev Embedding Theorem). *The relations $H^s \subset \mathscr{C}^k$ and $|| \ ||_{\mathscr{C}^k} \lesssim || \ ||_s$ hold if and only if $s > k + 1$. Moreover, in this case if $f \in H^s$, then after a modification of $f$ on a set of measure zero, the weak derivatives of $f$ up to order $k$ are ordinary derivatives.*

**Proof.** Observe that $||u||_{\mathscr{C}^k} \lesssim ||u||_s$ if and only if there is a constant $C > 0$ such that for every $|\alpha| \leq k$, $D^\alpha : \mathscr{S} \subset H_s \to L^\infty(\mathbb{R}^n)$ satisfies the norm bound

$$||D^\alpha|| \leq C.$$

We claim that such a norm bound holds if and only if

(11.19) $$e^{\sqrt{-1}\langle x,t \rangle} t^\alpha (1 + |t|^2)^{-\frac{s}{2}}$$

has finite $L^2$-norm for all $|\alpha| \leq k$. To see this claim, observe first that

$$(D^\alpha f)(x) = \int_{\mathbb{R}^2} e^{\sqrt{-1}\langle x,t \rangle} t^\alpha \hat{f}(t) dA(t)$$
$$= \int_{\mathbb{R}^2} e^{\sqrt{-1}\langle x,t \rangle} t^\alpha (1+|t|^2)^{-s/2} (1+|t|^2)^{s/2} \hat{f}(t) dA(t).$$

## 11.4. Proof of the Korn-Lichtenstein Theorem

Thus if (11.19) has finite $L^2$-norm $C$, the Cauchy-Schwartz inequality shows that

$$|D^\alpha f(x)|^2 \leq \|f\|_{H^s}^2 \cdot \int_{\mathbb{R}^n} \frac{|t^\alpha|^2}{(1+|t|^2)^s} dV(t) \leq C^2 \|f\|_{H^s}^2$$

and thus $\|D^\alpha\|$ has the desired bound. Conversely if $D^\alpha$ is bounded, then with

$$\hat{f}_o(t) = t^\alpha (1+|t|^2)^{-s} e^{-(\sqrt{-1}\langle x,t\rangle + \varepsilon|t|^2)} \in \mathscr{S} \subset H_s$$

we compute that

$$C \geq |D^\alpha f_o(x)| = \int_{\mathbb{R}^2} \frac{|t^\alpha|^2 e^{-\varepsilon|t|^2}}{(1+|t|^2)^s} dA(t).$$

By the Monotone Convergence Theorem, we may let $\varepsilon \to 0$, and thus we obtain that (11.19) lies in $L^2(\mathbb{R}^2)$.

Now, clearly the $L^2$-norm of (11.19) is independent of $x$, and a calculation in polar coordinates shows that this function is $L^2$ if and only if $s > |\alpha| + 1$.

Finally, if $s > k+1$ and $f \in H^s$, let $f_1, f_2, \ldots \in \mathscr{S}$ be a sequence of functions converging to $f$ in $H^s$. We claim that $D^\alpha f_j \to D^\alpha f$ uniformly for all $|\alpha| \leq k$. Indeed, by what we just proved,

$$|D^\alpha f_j(x) - D^\alpha f(x)| \leq C \|f - f_j\|_{H^s}.$$

Thus $D^\alpha f$ is continuous and is the derivative of $f$ in the ordinary sense. The proof is complete. □

**11.4.10. Proof of Theorem 11.4.1.** We are considering the equation

$$\bar{\partial}_\omega \gamma = f\bar{\omega}$$

in a domain $U$, where $f$ is a smooth function, and our goal is to show that $\gamma$ is also smooth. The domain in question may be taken to be a very small coordinate neighborhood in $\mathbb{R}^2$, with coordinates $(x,y)$.

Since we use the Fourier transform in our definition of Sobolev spaces, we are forced to work on all of $\mathbb{R}^2$. However, our equation takes place in a small domain, and we must adapt to this situation. The usual way to adapt is by use of a cutoff function. More precisely, if we mean to prove that the equation

$$\xi_{\bar{\omega}} h = f$$

has smooth solution $h$ whenever $f$ is smooth, then we might take a relatively compact open set $V \subset\subset U$ and replace $f$ by $\zeta f$, where $\zeta \in \mathscr{C}_0^\infty(U)$ satisfies $0 \leq \zeta \leq 1$ and $\zeta|_V \equiv 1$. To make sure the function $\zeta h$ is the solution on $V$ (where $\zeta h = h$), we alter the PDE at $0^{\text{th}}$-order: we set

$$Tu := \xi_{\bar{\omega}} u - (\xi_{\bar{\omega}} \eta) u.$$

Then

$$T(\zeta h) = \zeta \xi_{\bar{\omega}} h = \zeta f.$$

For ease of notation, set $A = -\xi_{\bar{\omega}}\zeta$. We remind the reader that $\xi_{\bar{\omega}} = \frac{1}{2}(\xi+\sqrt{-1}\eta)$. Then we have the following lemma.

LEMMA 11.4.10. *Let $U$ be a small neighborhood such that $\xi$ and $\eta$ are uniformly linearly independent in $U$. Then for each integer $s \geq 0$ there exists a positive constant $C_s$ such that for all smooth functions $h$ with compact support in $U$ we have*
$$||h||^2_{H^{s+1}} \leq C_s(||h||^2_{H^0} + ||Th||^2_{H^s}).$$

**Proof.** For $s = 0$ we proceed as follows.
$$\begin{aligned}
\int_U |Th|^2 &= \int_U |\xi_{\bar{\omega}}h + Ah|^2 \\
&= \int_U \left(|\xi_{\bar{\omega}}h|^2 + A^2|h|^2 + 2\operatorname{Re}(\xi_{\bar{\omega}}h)\overline{Ah}\right) \\
&\geq (1-\varepsilon)\int_U |\xi_{\bar{\omega}}h|^2 - C\int_U |h|^2.
\end{aligned}$$

Moreover, since $\xi$ and $\eta$ are uniformly independent, there is a constant $c$ such that $|\xi_{\bar{\omega}}u|^2 \geq c|Du|^2$ holds for all smooth functions $u$ in $U$. It follows that
$$\int_U (|Dh|^2 + |h|^2) \leq \left(\frac{C}{c(1-\varepsilon)} + 1\right)\left(\int_U |h|^2 + \int_U |Th|^2\right),$$
which establishes the case $s = 0$.

For $s \geq 1$, we proceed by induction on $s$. Note first that
$$D^s(Th) = T(D^s h) + (D^s T - TD^s)h.$$
The operator $D^s T - TD^s$ is a differential operator of order $s$, and therefore
$$||(D^s T - TD^s)h||_{H^0} \leq C||h||_{H^s}.$$
On the other hand, with the same calculation as was done for the case $s = 0$, we find that
$$\int_U |T(D^s h)|^2 \geq \delta \int_U |D(D^s h)|^2 - C'\int_U |D^s h|^2.$$
Putting these two estimates together and using the inductive assumption gives the desired inequality. □

We use the notation of Theorem 11.4.1 again. In view of Lemma 11.4.10 and the Sobolev Embedding Theorem, we see that, since $f$ is smooth and thus so is $\zeta f$, $\zeta \gamma$ is smooth. Therefore $\gamma$ is smooth in $U$. The proof of Theorem 11.4.1 is thus complete, and therefore so is the proof of Theorem 2.2.4. □

REMARK. The only property of $A$ used is smoothness. Thus we leave it to the reader to prove that for any smooth function $A$, the operator $T = \xi_{\bar{\omega}} + A$ is regular in the sense that if $Tu = F$ and $F$ is smooth in the interior of a domain, then $u$ is smooth in the interior as well. ◇

## 11.5. Exercises

**11.1.** Let $g$ be a Hermitian metric for a Riemann surface $X$ with associated area form $\omega$ locally written $\omega = e^{-\psi}\sqrt{-1}dz \wedge d\bar{z}$. Define a metric $h$ for $K_X$ by
$$h(dz, d\bar{z}) := e^{\psi}.$$
Show that $L^2_{0,1}(X, \Omega)$ and $L^2_{0,1}(-\psi, \omega)$ are the same Hilbert spaces (with the same inner product).

**11.2.** Suppose $H \to X$ is a holomorphic line bundle with metric $e^{-\varphi}$ of strictly positive curvature $\partial\bar{\partial}\varphi$.
  (a) Show that for any $\alpha \in \Gamma(X, T_X^{*0,1} \otimes K_X \otimes H)$, written locally as $\alpha = g d\bar{z} \otimes dz \otimes \xi$,
$$\frac{\alpha \wedge \bar{\alpha} e^{-\varphi}}{2\sqrt{-1}\Delta\varphi} := \frac{|g|^2 e^{-\varphi}}{\Delta\varphi} \frac{\sqrt{-1}}{2} dz \wedge d\bar{z}$$
  is a globally-defined, smooth measure, where $\partial\bar{\partial}\varphi = \Delta\varphi dz \wedge d\bar{z}$.
  (b) Show that if
$$\int_X \frac{\alpha \wedge \bar{\alpha} e^{-\varphi}}{2\sqrt{-1}\Delta\varphi} < +\infty$$
  then there exists $f \in \Gamma(X, K_X \otimes H)$ such that
$$\bar{\partial}f = \alpha \quad \text{and} \quad \int_X \frac{\sqrt{-1}}{2} f \wedge \bar{f} e^{-\varphi} \leq \int_X \frac{\alpha \wedge \bar{\alpha} e^{-\varphi}}{2\sqrt{-1}\Delta\varphi}.$$

**11.3.** Use Hörmander's Theorem to prove Theorem 8.1.1.

**11.4.** In this exercise we will directly prove the twisted Hodge Theorem in the case of positive curvature for the $\bar{\partial}$-Laplacian and deduce from it Hörmander's Theorem.

Fix a Riemann surface $X$ with smooth, possibly empty 1-dimensional boundary $\partial X$, a Hermitian metric $g = e^{-\psi}|dz|^2$, and a holomorphic line bundle $H \to X$ with smooth Hermitian metric $h = e^{-\varphi}$ such that
$$\sqrt{-1}\partial\bar{\partial}(\varphi + \psi) \geq c\omega$$
for some fixed $c > 0$.

As in Exercise 9.2, we define $\Box = \bar{\partial}\bar{\partial}^*$ and define the Hilbert space $\mathfrak{H}$ as the closure of the space of smooth, $H$-valued $(0,1)$-forms with respect to the inner product $(\theta, \eta)_{\mathfrak{H}} := (\bar{\partial}^*\theta, \bar{\partial}^*\eta)$.
  (a) Show that $\mathfrak{H}$ is a vector subspace of $L^2_{(0,1)}(\varphi, \omega)$.
  (b) For a fixed $\eta \in L^2_{(0,1)}(\varphi, \omega)$, show that the conjugate-linear functional $\lambda(\xi) := (\eta, \xi)$ is continuous on $\mathfrak{H}$, with norm at most $\frac{1}{c}$.
  (c) Conclude that there exists $u \in \mathfrak{H}$ such that $||u|| \leq \frac{1}{c}$ and for all $\xi$ smooth with compact support,
$$(u, \Box\xi) = (\eta, \xi),$$

i.e., $\Box u = \eta$ in the weak sense.

(d) Show that $f := \bar{\partial}^* u$ satisfies the equation $\bar{\partial} f = \eta$ in the weak sense, as well as the estimate
$$\int_X |f|^2 e^{-\varphi} \omega \leq \frac{1}{c} \int_X \frac{\eta \wedge \bar{\eta}}{2\sqrt{-1}} e^{-\varphi}.$$

11.5. In this exercise we will show that a sufficiently positive line bundle on a compact Riemann surface has a holomorphic section.

Let $X$ be a compact Riemann surface and let $\omega$ be an area form on $X$ with curvature $(1,1)$-form $\Theta$. (If $\omega = e^{-\psi} \frac{\sqrt{-1}}{2} dz \wedge d\bar{z}$ in local coordinates, then $\Theta = \partial \bar{\partial} \psi$.) Let $p \in X$ be a point and let $L_p$ and $s_p$ denote the line bundle and canonical section associated to the divisor $1 \cdot p$. Fix a Hermitian metric $e^{-\eta_p}$ for $L_p$. Let $H, F \to X$ be holomorphic line bundles with Hermitian metrics $e^{-\varphi}$ and $e^{-\theta}$, respectively. Assume that
$$\sqrt{-1} \partial \bar{\partial} \varphi \geq -(\sqrt{-1} \Theta + \sqrt{-1} \partial \bar{\partial} \theta - \sqrt{-1} \partial \bar{\partial} \eta_p) + \omega.$$

The goal of the exercise is to show that $H \otimes F$ has a holomorphic section that is not identically zero.

(a) Show that there is a smooth non-zero section $\sigma$ of $H \otimes F \to X$ with support in a disk centered at $p$, and that is holomorphic on a smaller disk centered at $p$.

(b) Show that $\alpha := \frac{1}{s_p} \bar{\partial} \sigma \in L^2_{0,1}(\varphi + \theta - \eta_p)$.

(c) Show that there is a section $u \in L^2(\varphi + \theta - \eta_p, \omega)$ such that $\bar{\partial} u = \alpha$.

(d) Show that $s := \sigma - s_p u \in \Gamma_\mathcal{O}(X, H \otimes F) - \{0\}$.

11.6. Formulate and prove a version of Exercise 11.5 for an open Riemann surface.

11.7. Assume the notation of Exercise 11.5. Let $\xi$ be a frame for $H \otimes F$ near $p$, let $z$ be a local coordinate vanishing at $p$, and let $f$ be a polynomial of degree $d$. Show that if
$$\sqrt{-1} \partial \bar{\partial} \varphi \geq -(\sqrt{-1} \Theta + \sqrt{-1} \partial \bar{\partial} \theta - (d+1) \sqrt{-1} \partial \bar{\partial} \eta_p) + \omega$$
then there is a section $s \in \Gamma_\mathcal{O}(X, H \otimes F)$ such that $s = f(z)\xi + O(|z|^{d+1})$ near $p$.

11.8. Assume the notation of Exercise 11.5, and let $q \in X - \{p\}$. Let $e^{-\eta_q}$ be a Hermitian metric for $L_q$. Show that if
$$\sqrt{-1} \partial \bar{\partial} \varphi \geq -(\sqrt{-1} \Theta + \sqrt{-1} \partial \bar{\partial} \theta - \sqrt{-1} \partial \bar{\partial} (\eta_p + \eta_q)) + \omega$$
then for any $v_p \in (H \otimes F)_p$ and any $v_q \in (H \otimes F)_q$ there exists a section $s \in \Gamma_\mathcal{O}(X, H \otimes F)$ such that $s(p) = v_p$ and $s(q) = v_q$.

11.9. Prove Lemmas 11.4.2 and 11.4.3.

11.10. Prove the claim made in the last remark of the chapter.

Chapter 12

# Embedding Riemann Surfaces

In this chapter, we aim to prove the following fundamental facts.

(1) Every holomorphic line bundle admits a non-trivial meromorphic section.
(2) Every Riemann surface admits a non-constant meromorphic function.
(3) (a) Every compact Riemann surface embeds in $\mathbb{P}_3$.
    (b) Every open Riemann surface embeds in $\mathbb{C}^3$

It should be remarked that, in (1), the non-triviality of the meromorphic section means that it has a non-zero order divisor, while in (3b), an embedding is a proper holomorphic immersion. Thus, in particular, the Riemann surface in question is realized as a closed subset of $\mathbb{C}^3$, and its intrinsic topology agrees with the relative topology inherited from the ambient Euclidean space.

All four of the above facts rely on the existence of many holomorphic sections of holomorphic line bundles admitting smooth metrics with sufficiently positive curvature. In Section 12.1 we will use Hörmander's Theorem to obtain sections of a sufficiently positive line bundle on a Riemann surface $X$, with control over the values of the sections and their derivatives at a finite number of points. In Section 12.4 we will use those sections to construct an embedding of a compact Riemann surface $X$ in $\mathbb{P}_N$ for some large $N$, after which a generic projection argument will show that $X$ can be biholomorphically projected to a 3-dimensional projective subspace. In Section 12.5 we will carry out a similar argument to embed open Riemann surfaces in $\mathbb{C}^3$, only this time, because of the absence of compactness, we will carry out the projection argument in conjunction with the proof of the Embedding Theorem, using the technique of exhausting $X$ by compact, smoothly bounded subsets.

## 12.1. Controlling the derivatives of sections

Let $X$ be a Riemann surface, not necessarily compact. By Proposition 9.4.1 or Theorem 9.4.6 there is a line bundle $H \to X$ and a smooth Hermitian metric $e^{-\varphi}$ whose curvature $\partial\bar{\partial}\varphi$ is a strictly positive $(1,1)$-form. We fix such a line bundle $H \to X$ and metric $e^{-\varphi}$, as well as a second holomorphic line bundle $E \to X$ with smooth Hermitian metric $e^{-\lambda}$ whose curvature we know nothing about.

Let us fix an area form $\omega$ on $X$. (For example, we could take $\omega = \sqrt{-1}\partial\bar{\partial}\varphi$, but it does not matter what area form we take.) Locally we can write

$$\omega = e^{-\psi} \frac{\sqrt{-1}}{2} dz \wedge d\bar{z}.$$

The form $\omega$ determines a metric for the tangent bundle and the curvature of this metric is $\partial\bar{\partial}\psi$.

Fix a point $x_o \in X$. Let $z$ be a local coordinate in a relatively compact neighborhood $U$ of $x_o$ such that $z(x_o) = 0$. Fix open sets $V \subset\subset W \subset\subset U$ with $x_o \in V$, and let $\chi : X \to [0,1]$ be a smooth function such that $\text{Support}(\chi) \subset U$ and $\chi|_W \equiv 1$. For each $x \in V$ consider the function

$$\eta = \eta_{N,\varepsilon,x} := \chi \cdot \log(|z - z(x)|^2 + \varepsilon^2)^N.$$

Then $e^{-\eta_{N,\varepsilon,x}}$ is a Hermitian metric for the trivial bundle on $X$. This metric is smooth with curvature

$$\sqrt{-1}\partial\bar{\partial}\eta_{N,\varepsilon,x} = \chi \frac{N\varepsilon^2 \sqrt{-1}dz \wedge d\bar{z}}{(|z-z(x)|^2 + \varepsilon^2)^2} + \theta \geq \theta,$$

where $\theta$ is a smooth $(1,1)$-from on all of $X$ that may be negative at some points but is supported in $U$. Most importantly, the form $\theta$ can be chosen independent of $\varepsilon$. Indeed, $\eta_{N,\varepsilon,x}$ is supported in $U$ and $\sqrt{-1}\partial\bar{\partial}\eta_{N,\varepsilon,x} > 0$ in $W$, since $\chi|_W \equiv 1$. Thus $\sqrt{-1}\partial\bar{\partial}\eta_{N,\varepsilon,x}$ cannot become arbitrarily negative at any point of $X$.

If $X$ is compact, then for some integer $k_o >> 0$, the curvature of the metric

(12.1) $$e^{-(k\varphi + \lambda + \eta_{N,\varepsilon,x})}$$

for $H^{\otimes k} \otimes E$ is greater than $\omega - \sqrt{-1}\partial\bar{\partial}\psi$ for all $k \geq k_o$. Moreover, compactness of $X$ implies that $k_o$ can be chosen independent of $x \in V$. If on the other hand $X$ is open, then we can replace $\varphi$ by $h \circ \varphi$ for some convex, rapidly increasing function $h$. Then

$$\partial\bar{\partial}(h \circ \varphi) = h'' \circ \varphi \partial\varphi \wedge \bar{\partial}\varphi + h' \circ \varphi \partial\bar{\partial}\varphi \geq h' \circ \varphi \partial\bar{\partial}\varphi.$$

Thus for sufficiently rapidly increasing $h$ we can guarantee that the curvature of the metric (12.1), with $k = 1$, is greater than $\omega - \sqrt{-1}\partial\bar{\partial}\psi$.

Now assume that the line bundles $H$ and $E$ (hence $H^{\otimes k} \otimes E$) are trivial on $U$. Fix a frame $\xi$ for $H$ and $\eta$ for $E$ on $U$. Let $\sigma = \chi \cdot p(z - z(x))\xi^{\otimes k} \otimes \eta$ where $p$ is a polynomial of degree at most $N$. Then $\sigma$ is a section of $H^{\otimes k} \otimes E$, and the Taylor

## 12.1. Controlling the derivatives of sections

polynomial of $\sigma$ at $x$ to order $N$ in the local coordinate $z$ and the frame $\xi^{\otimes k} \otimes \eta$ is precisely $p$. Unfortunately, $\sigma$ is not holomorphic on $X$, although it is holomorphic on the neighborhood $W$ of $x$.

Let $\beta = \bar\partial \sigma$. Then $\beta$ is a $\bar\partial$-closed $(0,1)$-form with values in $H^{\otimes k} \otimes E$, supported on $U$ but identically zero in $W$. It follows that

$$\frac{1}{2\sqrt{-1}} \int_X \beta \wedge \bar\beta e^{-(k\varphi + \lambda + \eta_{N,\varepsilon,x})} \leq C$$

for some constant $C > 0$ independent of $\varepsilon$. But since $\sqrt{-1}\partial\bar\partial(k\varphi + \lambda + \eta_{N,\varepsilon,x} + \psi) \geq \omega > 0$, Hörmander's Theorem 11.3.1 tells us that there is a section $u_\varepsilon$ of $H^{\otimes k} \otimes E$ such that

$$\bar\partial u_\varepsilon = \beta \quad \text{and} \quad \int_X |u_\varepsilon|^2 e^{-(k\varphi + \lambda + \eta_{N,\varepsilon,x})} \omega \leq C.$$

We would like to take the limit as $\varepsilon \to 0$, but we must be a little careful since our $L^2$-norms keep changing. Note, however, that the metrics increase as $\varepsilon \searrow 0$, so that we have

$$\int_X |u_\varepsilon|^2 e^{-(k\varphi + \lambda + \eta_{N,x,\varepsilon_o})} \omega \leq C.$$

It follows that $\{u_\varepsilon\}$ is a bounded family in a fixed Hilbert space, and therefore we can extract a weakly convergent subsequence, which we call $u_{\varepsilon_j}$. We denote its weak limit by $u$. It then follows that

$$\int_X |u|^2 e^{-(k\varphi + \lambda + \eta_{N,x,\varepsilon_o})} \omega \leq C$$

for all $\varepsilon_o > 0$, and therefore

$$\int_X |u|^2 e^{-(k\varphi + \lambda + \eta_{N,x,0})} \omega \leq C.$$

Moreover, since $u$ is a weak limit of $u_{\varepsilon_j}$ and $\bar\partial u_{\varepsilon_j} = \beta$ in the weak sense, $\bar\partial u = \beta$ in the weak sense. Of course, a weak solution of $\bar\partial u = \beta$ for smooth $\beta$ is also a strong solution, as we have pointed out before: the difference of two weak solutions is holomorphic and therefore smooth, and so it suffices to locally produce one smooth solution, which can be done for example with the Cauchy-Green Formula.

Now, the finiteness of the integral implies in particular that on $V$ the function

$$|u|^2 e^{-(k\varphi + \lambda + \eta_{N,x} + \psi)}|_V = \frac{|u|^2 e^{-(k\varphi + \lambda + \psi)}}{|z - z(x)|^{2N}}$$

is locally integrable. Since $u$, $\varphi$, $\lambda$, and $\psi$ are smooth functions, it must be the case that the Taylor polynomial of $u$ to order $N$ at $x$ is zero. Indeed, in local polar coordinates with center $x$, the area form is a multiple of $rdrd\theta$ so that if $u$ vanishes to order $\ell$ at $x$, it must be the case that $r^{2\ell + 1 - 2N} dr$ is integrable near the origin on the non-negative real line, and this is only so if $\ell \geq N$.

Now consider the section $s = \sigma - u$. Then $\bar{\partial}s = \beta - \bar{\partial}u = 0$, so that $s$ is holomorphic. Moreover, the Taylor polynomial of $s$ at $x$ is $p$. Finally, if $X$ is compact, we can cover $X$ by a finite number of neighborhoods $V_j$ such that $V_j \subset\subset W_j \subset\subset U_j$ are neighborhoods as above. Thus we can choose the number $k_o$ above independent of the point $x \in X$. We have proved the following theorem.

**THEOREM 12.1.1.** *Let $X$ be a Riemann surface, and let $p$ be a polynomial of a complex variable. Let $H \to X$ be a positive holomorphic line bundle (taken to be trivial if $X$ is open) and let $E \to X$ be any holomorphic line bundle. Then there is an integer $k_o$ with the following property. For any $x \in X$ and any $k \geq k_o$ there is a holomorphic section $s$ of $H^{\otimes k} \otimes E \to X$ whose Taylor polynomial in some local coordinate near $x$ is $p$. In particular, there is an integer $k_o$ such that for any $x \in X$ and any $k \geq k_o$ there is a holomorphic section $s$ of $H^{\otimes k} \otimes E$ whose derivative in any local coordinate does not vanish at $x$.*

By a similar but slightly more delicate argument, we get the following result.

**THEOREM 12.1.2.** *Let $H \to X$ be a holomorphic line bundle on a Riemann surface. Suppose there is a positively curved Hermitian metric for $H$. Then there is an integer $k_o > 0$ with the following property. For any integer $k \geq k_o$, any distinct points $x, y \in X$ and any $a \in H_x^{\otimes k}$ and $b \in H_y^{\otimes k}$ there is a holomorphic section $s$ of $H^{\otimes k} \to X$ such that $s(x) = a$ and $s(y) = b$.*

**Proof.** The argument of Theorem 12.1.1 does not immediately give the present result because we have to worry about what happens when the points $x$ and $y$ get close together. Thus we will distinguish this case by slightly modifying the argument.

We begin by choosing a locally finite cover by relatively compact coordinate neighborhoods $U_j$ on which $H$ is trivial, and relatively compact open subsets $V_j \subset\subset W_j \subset\subset U_j$, such that $\{V_j\}$ is again a locally finite subcover. We also fix coordinates $z_j$ on $U_j$ and smooth functions $\chi_j$ with support on $U_j$ such that $\chi_j \equiv 1$ on $W_j$. Now define for $x \in V_j$ and $y \in W_\ell$ the singular function

$$\eta_{x,y,\varepsilon} := \chi_j \log(|z_j - z_j(x)|^2 + \varepsilon^2) + \chi_\ell \log(|z_\ell - z_\ell(y)|^2 + \varepsilon^2).$$

Of course there is some choice involved, since a given point may be in more than one neighborhood, but since any point can lie in at most a finite number of the neighborhoods $U_j$, we will see that the argument can be made independent of the choice of $j$ and $\ell$.

Let $\varphi$ be a positively curved metric for $H$. Consider the metric $e^{-(k\varphi+\eta_{x,y,\varepsilon})}$. We calculate that for some positive constant $C$ that is independent of $x$ and $y$ and of the possible choices of $j$ and $\ell$,

$$\sqrt{-1}\partial\bar{\partial}\eta_{x,y,\varepsilon} \geq \frac{\varepsilon^2}{(|z - z(x)|^2 + \varepsilon^2)^2} + \frac{\varepsilon^2}{(|z - z(y)|^2 + \varepsilon^2)^2} - C\omega.$$

## 12.2. Meromorphic sections of line bundles

It follows that there is a constant $k_o$ such that for all $k \geq k_o$,

$$\sqrt{-1}\partial\bar{\partial}(k\varphi + \eta_{x,y,\varepsilon}) + \sqrt{-1}\partial\bar{\partial}\psi \geq c\omega > 0,$$

again independent of the points $x$ and $y$ and the possible choices of $j$ and $\ell$. (If $X$ is not compact, we replace $\varphi$ by $h \circ \varphi$, for some convex, rapidly increasing function $h$, and take $k_o = k = 1$.)

The rest of the argument resembles more closely the argument in the proof of Theorem 12.1.1. Since $H$ is trivial on $U_j$ and $U_\ell$, the sections $a\xi_j^{\otimes k} \otimes \eta_j$ and $b\xi_\ell^{\otimes k} \otimes \eta_\ell$ are well-defined on $U_j$ and $U_\ell$, respectively. We then define the global section

$$\sigma = \chi_j \cdot a\xi_j^{\otimes k} \otimes \eta_j + \chi_\ell \cdot b\xi_\ell^{\otimes k} \otimes \eta_\ell.$$

This section is smooth on $X$, supported on $U_j \cup U_\ell$, and holomorphic on $W_j \cup W_\ell$. It follows that $\bar{\partial}\sigma$ is supported on $(U_j - W_j) \cup (U_\ell - W_\ell)$. By Hörmander's Theorem 11.3.1 there is a section $u_\varepsilon$ of $H^{\otimes k}$ such that

$$\bar{\partial}u_\varepsilon = \bar{\partial}\sigma \quad \text{and} \quad \int_X |u_\varepsilon|^2 e^{-(k\varphi + \eta_{x,y,\varepsilon})}\omega < +\infty.$$

As in the proof of Theorem 12.1.1, we obtain an integrable section $u$, this time with respect to the metric $e^{-(k\varphi + \eta_{x,y,0})}$. The latter integrability implies, together with the smoothness of $u$, that $u(x) = 0$ and $u(y) = 0$. The section $s = \sigma - u$ therefore has the desired properties. $\square$

### 12.2. Meromorphic sections of line bundles

In Chapter 4 we discussed the correspondence between divisors and line bundles. At that point, we were able only to complete the correspondence between divisors on the one hand and holomorphic line bundles with meromorphic sections on the other. We stated (as Theorem 4.5.8) the fact that every holomorphic line bundle has a non-trivial meromorphic section. We can now complete the classification by proving Theorem 4.5.8.

**Proof of Theorem 4.5.8.** Let $L \to X$ be a holomorphic line bundle. Consider the line bundle $H$ associated to the divisor of a point $p \in X$, and the associated global holomorphic section $s_p$ of $H$ such that $\text{Ord}(s_p) = p$. (Recall the discussion in Section 4.5.) By Theorem 6.6.2, $c(H) = 1$. Therefore, by Proposition 9.4.2, the line bundle $H$ has a metric $e^{-\varphi_o}$ of strictly positive curvature.

By Theorem 12.1.1, for all $m >> 0$ the line bundle $E_m := H^{\otimes m} \otimes L$ has a non-trivial global holomorphic section, say $t_m$. It follows that

$$\sigma := \frac{t_m}{s_p^{\otimes m}} \in \Gamma_{\mathscr{M}}(X, L) - \{0\}.$$

The proof of Theorem 4.5.8 is complete. $\square$

## 12.3. Plenitude of meromorphic functions

In this section we demonstrate the existence of many meromorphic functions on compact Riemann surfaces.

DEFINITION 12.3.1. A family $\mathscr{F}$ of differentiable functions on a manifold $X$ is said to separate points if for each pair of distinct point $x, y \in X$ there exists $f \in \mathscr{F}$ such that $f(x) \neq f(y)$. We say $\mathscr{F}$ separates tangents if for each $x \in X$ there exists $f \in \mathscr{F}$ such that $dF(x) \neq 0$. ◊

Since every holomorphic line bundle on an open Riemann surface is trivial, Theorems 12.1.2 and 12.1.1 already imply the following result.

COROLLARY 12.3.2. *The family $\mathcal{O}(X)$ of holomorphic functions on an open Riemann surface separates points and tangents.*

For compact Riemann surfaces we have the following easy corollary of Theorems 12.1.2 and 12.1.1.

COROLLARY 12.3.3. *The family $\mathcal{M}(X)$ of all meromorphic functions on a compact Riemann surface $X$ separates points and tangents.*

**Proof.** In view of Theorem 12.1.2, there are sections $s$ and $\sigma$ of some line bundle $L_1$ such that $s(x)$, $\sigma(x)$, and $\sigma(y)$ are non-zero while $s(y) = 0$. Consider the meromorphic function
$$f = s/\sigma.$$
Then $f(x) \neq 0$ while $f(y) = 0$.

Next, we can find sections $s$ and $\sigma$ of some line bundle $L_2$ such that $s(x)$ and $\sigma(x)$ are non-zero, $s'(x)$ is non-zero, and $\sigma'(x) = 0$ (in some local trivialization and local coordinates). Consider the meromorphic function
$$f = s/\sigma.$$
Then
$$f'(x) = \frac{s'(x)}{\sigma(x)} - \frac{s(x)\sigma'(x)}{\sigma(x)^2} = \frac{s'(x)}{\sigma(x)}$$
is non-zero at $x$. The proof is complete. □

## 12.4. Kodaira's Embedding Theorem

**12.4.1. Embedding Riemann surfaces in $\mathbb{P}_N$.** We are now ready to prove the following result.

THEOREM 12.4.1. *Every compact Riemann surface $X$ can be holomorphically embedded in some $\mathbb{P}_N$.*

## 12.4. Kodaira's Embedding Theorem

**Proof.** Fix a positive line bundle $H \to X$, which exists, for example, by Proposition 9.4.1. Let $k$ be the larger of the integers $k_o$ given by Theorem 12.1.1 for the case of Taylor polynomial $p(z) = z$ and by Theorem 12.1.2. Now consider $\Gamma_\mathcal{O}(X, H^{\otimes k})$. Then, as in Section 4.7, we have the map

$$\phi_{|H^{\otimes k}|} : X \to \mathbb{P}(\Gamma_\mathcal{O}(X, H^{\otimes k})^*).$$

In terms of a basis $s_0, ..., s_\ell$ of $\Gamma_\mathcal{O}(X, H^{\otimes k})$,

$$\phi_{|H^{\otimes k}|} = [s_0, ..., s_\ell].$$

If all of the sections $s_i$ did not separate a given pair $x$ and $y$, then neither would any linear combination. Thus by Theorem 12.1.2, $\phi_{|H^{\otimes k}|}$ separates any pair of distinct points $x$ and $y$. Next, perhaps after renumbering, we can assume that $s_0(x) \neq 0$. If all of the functions $f_i = s_i/s_0$ had vanishing derivatives at a given point $x$, then so would every linear combination of these, and this would contradict Theorem 12.1.1. It follows that the map $\phi_{|H^{\otimes k}|}$ has full rank at each point of $X$. Thus we have proved that $\phi_{|H^{\otimes k}|}$ is an embedding, as desired. $\square$

REMARK. The fact that a compact Riemann surface can be embedded in projective space much precedes Kodaira's work. In fact, it was probably known in one form or another before Riemann surfaces were officially defined. We have chosen to refer to Kodaira because the approach we are using is philosophically closer to the higher-dimensional version of the Embedding Theorem, which is indeed due to Kodaira. The proof we use is well known to experts in the field of complex analytic geometry. $\diamond$

### 12.4.2. Embedding Riemann surfaces in $\mathbb{P}_3$.

THEOREM 12.4.2. *Every compact Riemann surface $X$ can be holomorphically embedded in $\mathbb{P}_3$.*

**Proof.** The result will be proved by descending induction. To this end, suppose $n \geq 4$ and we have an embedding $\phi : X \to \mathbb{P}_n$. For dimension reasons, $\phi(X) \neq \mathbb{P}_n$. Let $Q \in \mathbb{P}_n - \phi(X)$. Suppose we pick a projective hyperplane $\Pi \subset \mathbb{P}_n - \{Q\}$. Then we can define a map

$$\text{Proj}_{Q,\Pi} : \phi(X) \to \Pi$$

as follows. If $x \in \phi(X)$, then there is a unique projective line $L$ in $\mathbb{P}_n$ containing $x$ and $Q$. Since $L \not\subset \Pi$, $L \cap \Pi$ consists of a single point, which we call $\text{Proj}_{Q,\Pi}(x)$.

Because the dimension $n$ of the ambient space is at least 4, after small perturbations of $Q$ and $\Pi$ we can ensure that the map $\text{Proj}_{Q,\Pi} : X \to \Pi$ is actually an embedding. To see this, we can introduce the so-called *chordal set*. Let

$$I := \{(x, y, p) \in \mathbb{P}_n \times \mathbb{P}_n \times \mathbb{P}_n \; ; \; x, y \in \phi(X) \text{ and } x, y, \text{ and } p \text{ are collinear}\}.$$

Note that $x$ and $y$ can lie in $\phi(X)$ freely (apart from coalescing), which gives us two complex dimensions, and once $x$ and $y$ have been chosen, we have one more line (so one more complex dimension) for $p$. It follows that $I$ is 3-dimensional, and thus, with $\pi_3 : I \to \mathbb{P}_n$ the projection sending $(x,y,p)$ to $p$, $\pi_3(I)$ is 3-dimensional. Hence for $n \geq 4$, $\mathbb{P}_n - \overline{\pi_3(I)}$ has interior, so it is not empty. It follows from the definition that if we choose $Q \in \mathbb{P}_n - \pi_3(I)$, then $\mathrm{Proj}_{Q,\Pi} : \phi(X) \to \Pi$ is 1-1. Moreover, if we choose $Q \in \mathbb{P}_n - \overline{\pi_3(I)}$, then no vector tangent to $\phi(X)$ will be annihilated by $\mathrm{Proj}_{Q,\Pi}$, and thus $\mathrm{Proj}_{Q,\Pi} : \phi(X) \to \Pi \cong \mathbb{P}_{n-1}$ is an embedding. The proof is thus complete by descending induction. $\square$

## 12.5. Narasimhan's Embedding Theorem

**12.5.1. Statement and outline of the proof.** The goal of this section is to prove the following theorem.

THEOREM 12.5.1. *Let $X$ be a non-compact connected Riemann surface. Then there is a proper injective holomorphic map $\phi : X \to \mathbb{C}^3$.*

In the case of open Riemann surfaces, the absence of compactness complicates matters, and we cannot simply imitate the proof of Theorem 12.4.1.

In fact, the proof consists of three parts. In the first part, we being by exhausting $X$ by certain compact sets. On each of these sets we produce $N$ functions separating points and tangents, where $N$ may depend on the compact set we are on. We then employ the analogue of the projection method used in the proof of Theorem 12.4.2 to produce three functions that separate points and tangents on the compact set. Passing to the limit will give us a 1-1 map $F : X \to \mathbb{C}^3$ that is an immersion. Unfortunately, we cannot conclude that this map is proper.

The second and hardest step is to produce a map $G : X \to \mathbb{C}^3$ that is proper. This step will occupy the majority of the section. Finally a simple argument will combine these two maps to produce a proper injective immersion $\phi : X \to \mathbb{C}^3$.

**12.5.2. The injective immersion of $X$ in $\mathbb{C}^3$.** Let $\rho : X \to (-\infty, a)$ be a strictly subharmonic exhaustion function, where $a \in \mathbb{R} \cup \{\infty\}$. Fix a sequence of numbers $a_j \nearrow a$ and set $K_j := \rho^{-1}((-\infty, a_j])$.

First, we claim that we can find global holomorphic functions $f_1^{(1)}, ..., f_N^{(1)}$ that separate points and tangents on $K_1$. To see this, we argue as follows. First, let $x \in K_1$. Then there exists a function $f_1$, defined on all of $X$, whose derivative does not vanish at $x$. It follows that $\tilde{f}_x$ separates any pair of distinct points near $x$. Now, since $K_1$ is compact, we can cover it by a finite number of neighborhoods, and for each such neighborhood $U_j$ we get a function $\tilde{f}_j$ that separates the points of $U_j$. By taking $f_j = \tilde{f}_j + C_j$ for appropriately chosen constants $C_j$, we can guarantee that in fact the functions $f_1^{(1)}, ..., f_N^{(1)}$ thus obtained separate the points of $K_1$. We let $\phi^{(1)} : X \to \mathbb{C}^N$ be the map

$$\phi^{(1)} = (f_1^{(1)}, ..., f_N^{(1)}).$$

## 12.5. Narasimhan's Embedding Theorem

The map $\phi^{(1)}$ separates points and tangents on $K_1$.

Now consider the set
$$I := \{(x, y, x + ty) \in \mathbb{C}^N \times \mathbb{C}^N \times \mathbb{C}^N \; ; \; x, y \in \phi^{(1)}(X), t \in \mathbb{C}\}$$
and the map $\pi_3 : I \to \mathbb{C}^N$ sending $(x, y, p)$ to $p$. Since $I$ has dimension 3, the closure $\overline{\pi_3(I)}$ of its image $\pi_3(I)$ is at most 3-dimensional. If $N \geq 4$ then the projection from any point not on $\pi_3(I)$ onto any hyperplane $\Pi$ restricts to $\phi^{(1)}(X)$ injectively, which amounts to saying that if $N \geq 4$ then $N - 1$ generic linear combinations $F_1^{(1)}, ..., F_{N-1}^{(1)}$ of the functions $f_1^{(1)}, ..., f_N^{(1)}$ produce a map $\Phi_{N-1}^{(1)} : X \to \mathbb{C}^{N-1}$ that still separates points and tangents on $K_1$.

Now suppose we have found functions $F_1^{(j)}, F_2^{(j)}, F_3^{(j)}$ such that the map
$$\Phi^{(j)} := (F_1^{(j)}, F_2^{(j)}, F_3^{(j)}) : X \to \mathbb{C}^3$$
separates points and tangents of $K_j$. Let $L_j := \overline{K_{j+1} - K_j}$. Then $L_j$ is compact, and the same argument as above shows that there are functions $g_1^{(j)}, g_2^{(j)}, g_3^{(j)} : X \to \mathbb{C}^3$ that separate points and tangents on $L_j$. We can put these together to make a map $(\Phi^{(j)}, g_1^{(j)}, g_2^{(j)}, g_3^{(j)}) : X \to \mathbb{C}^6$ that separates points and tangents on $K_{j+1}$. By taking projections close to the identity, it follows that the map
$$\Phi^{(j+1)} := \Phi^{(j)} + \left( \sum_{i=1}^{3} \varepsilon_{1i}^{(j)} g_i^{(j)}, \sum_{i=1}^{3} \varepsilon_{2i}^{(j)} g_i^{(j)}, \sum_{i=1}^{3} \varepsilon_{3i}^{(j)} g_i^{(j)} \right) : X \to \mathbb{C}^3$$
separates points and tangents on $K_{j+1}$. If we take the matrices $\varepsilon^{(j)} \in M_{3\times 3}(\mathbb{C})$ sufficiently small, evidently the process converges to give us the desired injective immersion. Thus we have proved the following result.

THEOREM 12.5.2. *Let $X$ be an open Riemann surface. Then there is an injective immersion $X \to \mathbb{C}^3$.*

REMARK. In fact, the above argument can easily be modified to show that the set $\mathscr{S}$ of holomorphic maps $X \to \mathbb{C}^3$ that are not injective immersions is of the first category in the sense of Baire, i.e., $\mathscr{S}$ is a countable union of closed sets without interior.

Indeed, since $X$ can be exhausted by a discrete family of compact sets, it suffices to show that the generic map separates points and tangents on some fixed compact set $K$, i.e., the set $\mathscr{S}_K$ of maps that do not separate tangents and points of $K$ is closed and has no interior.

To see that $\mathscr{S}_K$ is closed, consider such a sequence $\Phi_j : X \to \mathbb{C}^3$ converging to $\Phi$. Then for each $j$, there is either a point $x_j \in K$ such that $d\Phi_j(x_j) = 0$, or a pair of points $x_j, y_j \in K$ such that $\Phi_j(x_j) = \Phi_j(y_j)$. In the former case we set $y_j := x_j$. After passing to convergent subsequences, we can assume $x_j \to x$ and $y_j \to y$. Evidently if $x \neq y$ then $\Phi(x) = \Phi(y)$, while if $x = y$ then $d\Phi(x) = 0$. Thus $\Phi \in \mathscr{S}_K$.

To see that $\mathscr{S}_X$ has no interior, let $\Psi \in \mathscr{S}_K$ and take any map $\Phi : X \to \mathbb{C}^3$ separating the points and tangents of $K$. (We have already constructed such a map.) Then $(\Psi, \Phi) : X \to \mathbb{C}^6$ is an injective immersion, and thus for almost any $\varepsilon \in M_{3\times 3}(\mathbb{C})$, $\Psi + (\sum_{i=1}^{3} \varepsilon_{1i}\Phi_i, \sum_{i=1}^{3} \varepsilon_{2i}\Phi_i, \sum_{i=1}^{3} \varepsilon_{3i}\Phi_i) : X \to \mathbb{C}^3$ is again an injective immersion, which shows that $\Psi$ is not in the interior of $\mathscr{S}_K$. ◇

### 12.5.3. Analytic polyhedra.

DEFINITION 12.5.3. Let $X$ be an open Riemann surface. A relatively compact open set $P \subset\subset X$ is called an analytic polyhedron of order $N$ if there are holomorphic functions $f_1, ..., f_N \in \mathcal{O}(X)$ such that $P$ is a union of components of the set $\{z \in X \;;\; |f_i(z)| < 1, i = 1, ..., N\}$. ◇

In this paragraph we prove two crucial lemmas on the construction of analytic polyhedra.

LEMMA 12.5.4. *Let $X$ be an open Riemann surface, let $K \subset\subset X$ be a holomorphically convex compact set, and let $U \supset K$ be an open subset of $X$. Then there is an analytic polyhedron $P$ such that*

$$K \subset P \subset\subset U.$$

**Proof.** We may assume that $U$ is relatively compact. Since $K$ is holomorphically convex, for each $p \in \partial U$ we can find a function $f_p \in \mathcal{O}(X)$ such that $|f_p| < 1$ on $K$ and $|f_p(p)| > 1$. The latter also holds in a neighborhood of $p$ by continuity. By the compactness of $\partial U$ we can find finitely many functions $f_1, ..., f_N$ such that the sets $V_j := \{|f_j| > 1\}$ cover $\partial U$ and miss $K$. It follows that the intersection $P$ of $U$ with $\{|f_j| < 1\}$ contains $K$ and does not meet $\partial U$. The proof is complete. □

LEMMA 12.5.5. *Let $K$ be a compact set in an open Riemann surface $X$ and let $P$ be an analytic polyhedron of order $N \geq 3$. Then there exists an analytic polyhedron $P'$ of order $N - 1$ such that $K \subset P' \subset P$.*

**Proof.** Suppose $P$ is a union of components of the set $\{|f_j| < 1 \;;\; 1 \leq j \leq N\}$. We fix positive numbers $\varepsilon_o < \varepsilon_1 < \varepsilon_2 < \varepsilon_3 < 1$ such that $|f_j(z)| < \varepsilon_o$ for all $z \in K$ and $1 \leq j \leq N$. By the methods used in the proof of Theorem 12.5.2 we can choose functions $f'_j$, $1 \leq j \leq N - 1$, such that (i) the map

$$(f'_1/f_N, ..., f'_{N-1}/f_N)$$

separates tangents at the points of $\overline{P} \cap \{|f_N| \geq \varepsilon_2\}$ and (ii) $f'_j$ is so close to $f_j$ that $|f'_j(z)| < \varepsilon_o$ for $z \in K$ and

$$U := P \cap \{|f'_j(z)| < \varepsilon_3, 1 \leq j \leq N\} \subset\subset P.$$

(Since we do not have to separate points, the projection argument used to prove point separation simplifies, and we can assume $N \geq 3$ rather than $N \geq 4$. We leave the details of this remark to the reader.)

## 12.5. Narasimhan's Embedding Theorem

Now consider the open set

$$\Delta_\nu := \{x \in X \; ; \; |f'_j(x)^\nu - f_N(x)^\nu| < \varepsilon_1^\nu, \; 1 \leq j \leq N-1\}$$

for some positive integer $\nu$ to be chosen. We will prove that for $\nu >> 0$, the components of $\Delta_\nu$ that meet $K$ form the desired analytic polyhedron. Let us denote by $P'_\nu$ the collection of all these components.

First, if $(\varepsilon_1/\varepsilon_0)^\nu > 2$ then for any $z \in K$,

$$|f'_j(z)^\nu - f_N(z)^\nu| \leq 2\varepsilon_o^\nu < \varepsilon_1^\nu.$$

It follows that for $\nu >> 0$, $K \subset \Delta_\nu$, and thus $K \subset P'_\nu$. We now show that $P'_\nu \cap \partial U = \emptyset$. If not, let $y \in P'_\nu \cap \partial U$. If $|f_N(y)| < \varepsilon_2$, then

$$|f'_j(y)|^\nu = |f'_j(y)^\nu - f_N(y)^\nu + f_N(y)^\nu| < \varepsilon_1^\nu + \varepsilon_2^\nu < \varepsilon_3^\nu$$

for $\nu >> 0$, which contradicts the assumption that $y \in \partial U$. It follows that $y$ is contained in the compact set

$$L := \partial U \cap \{|f_N| \geq \varepsilon_3\}.$$

We assert that no component of $\Delta_\nu$ that meets $L$ can also meet $K$. We do this in local coordinates: let $L_1$ be the intersection of $L$ with some coordinate patch containing $y$ and having local coordinate $z$. Then with $F_j = f'_j/f_N$, $1 \leq j \leq N-1$, we have

$$|F_j(y)^\nu - 1| < (\varepsilon_1/\varepsilon_2)^\nu.$$

We now show that for any $z$ with $|z - y| = \nu^{-2}$,

(12.2) $$\max_{1 \leq j \leq N-1} |f'_j(z)^\nu - f_N(z)^\nu| > \varepsilon_1^\nu,$$

from which our assertion clearly follows. To prove (12.2), note first that

$$|f'_j(z)^\nu - f_N(z)^\nu| = |F_j(z)^\nu - 1| \cdot |f_N(z)^\nu|.$$

Since $|f_N(z)| \geq \varepsilon_2(1 + O(\nu^{-2}))$,

$$|f_N(z)^\nu| \geq \varepsilon_2^\nu(1 + O(\nu^{-1})) > \varepsilon_2^\nu/2$$

for $\nu >> 0$. Now, by Taylor's Theorem,

$$F_j(z) = F_j(y)(1 + \ell_j(z - y) + O(\nu^{-4})), \quad 1 \leq j \leq N-1.$$

Since the map $(F_1, ..., F_{N-1})$ separates tangents, the numbers $\ell_1, ..., \ell_{N-1}$ do not all vanish. Thus $\max_j |\ell_j| > c > 0$. We have

$$(F_j(z)/F_j(y))^\nu = 1 + \nu \ell_j (z - y) + O(\nu^{-2}), \quad 1 \leq j \leq N-1.$$

Now write

$$F_j(z)^\nu - 1 = F_j(y)^\nu \left( \left( \frac{F_j(z)}{F_j(y)} \right)^\nu - 1 \right) + F_j(y)^\nu - 1.$$

It follows that

$$\max_j |f_j'(z)^\nu - f_N(z)^\nu|$$
$$> \max_j |F_j(z)^\nu - 1|\varepsilon_2^\nu/2$$
$$= \max_j \varepsilon_2^\nu \left(|F_j(y)|^\nu \left|\left(\frac{F_j(z)}{F_j(y)}\right)^\nu - 1\right| - |F_j(y)^\nu - 1|\right)$$
$$> \max_j \varepsilon_2^\nu \left(|F_j(y)|^\nu \left|\left(\frac{F_j(z)}{F_j(y)}\right)^\nu - 1\right|\right) - \varepsilon_1^\nu$$
$$> \frac{\varepsilon_2^\nu}{4}(c/\nu + O(\nu^{-2})) > \varepsilon_1^\nu,$$

again for $\nu >> 0$. By the compactness of $L$, the constant $c$ can be chosen uniformly, and thus the final $\nu$ we choose is independent of the point $y \in L$. The proof of the lemma is complete. □

**12.5.4. Conclusion of the proof of Theorem 12.5.1.** By Theorem 12.5.2 there is a holomorphic map $g : X \to \mathbb{C}^3$ that separates points and tangents. We are going to construct a map $f : X \to \mathbb{C}^3$ with the following property: for every $k \in \mathbb{N}$,

(12.3) $$\{x \in X \ ; \ |f(x)| \le k + |g(x)|\} \subset\subset X.$$

If we have such a map $f$, then the map $(f, g) : X \to \mathbb{C}^6$ is an injective immersion. Indeed, the projection argument used above then shows that there are arbitrarily small numbers $a_{ij}, 1 \le i, j \le 3$, such that

$$\sum_{j=1}^3 |a_{ij}| < 1, \quad i = 1, 2, 3,$$

and the map

$$\tilde{f} = \left(f_1 + \sum_{j=1}^3 a_{1j}g_j, f_2 + \sum_{j=1}^3 a_{2j}g_j, f_3 + \sum_{j=1}^3 a_{3j}g_j\right)$$

separates points and tangents on $X$. But then

$$\{x \in X \ ; \ |\tilde{f}(x)| \le k\} \subset \{x \in X \ ; \ |f(x)| \le k + |g(x)|\} \subset\subset X,$$

and Theorem 12.5.1 follows.

Let

$$K_j \subset\subset \text{Interior}(K_{j+1}) \subset K_{j+1}, j \ge 1$$

be a sequence of compact sets such that

$$\widehat{K}_j = K_j \quad \text{and} \quad \bigcup_{j \ge 1} K_j = X.$$

## 12.5. Narasimhan's Embedding Theorem

By Lemmas 12.5.4 and 12.5.5, there are analytic polyhedra $P_j$, $j \geq 1$, such that
$$K_j \subset P_j \subset K_{j+1}.$$
Let
$$M_j := \sup_{P_j} |g|.$$
Condition (12.3) is implied by the following condition: for all $k \geq N$,

(12.4) $\qquad |f| \geq k + M_{k+1}$ in $P_{k+1} - P_k$.

Indeed, (12.4) implies that $|f| + k \geq |g|$ in $P_{k+1} - P_k$, and thus by holomorphic convexity $|f| \geq k + |g|$ in $X - P_k$.

We begin by constructing $f_1, f_2 \in \mathcal{O}(X)$ such that for every $k \in \mathbb{N}$,

(12.5) $\qquad \max(|f_1(x)|, |f_2(x)|) > k + M_{k+1}$ on $\partial P_k$.

By the definition of analytic polyhedron of order 2, we can find $h_{1,k}, h_{2,k} \in \mathcal{O}(X)$ so that $\max_j |h_{j,k}| < 1$ on $\overline{P_{k-1}}$ but $\max_j |h_{j,k}| = 1$ on $\partial P_k$. Fix numbers $a_k$ slightly larger than 1 and large integers $m_k$, and let
$$f_{j,k} = (a_k h_{j,k})^{m_k}, \quad j = 1, 2, \quad \text{and} \quad f_j = \sum_{k=1}^{\infty} f_{j,k}, \quad j = 1, 2.$$
With appropriate choices, we can ensure that for $j = 1, 2$,
$$|f_{j,k}| \leq 2^{-k} \text{ in } P_{k-1}$$
and
$$|f_{j,k}| > m_{k+1} + k + 1 + \max_j \left| \sum_{\ell=1}^{k-1} f_{j,\ell} \right| \text{ in } \partial P_k.$$
These conditions imply that the series defining the $f_j$ converge to holomorphic functions on $X$ satisfying (12.5).

Finally we construct $f_3$. To this end, set
$$G_k := \{x \in P_{k+1} - P_k \; ; \; \max_j |f_j(x)| \leq k + M_{k+1}\}$$
and
$$H_k := \{x \in P_k \; ; \; \max_j |f_j(x)| \leq k + M_{k+1}\}.$$
By (12.5), these disjoint sets are compact. The $\mathcal{O}(X)$-hull $\widehat{G_k \cup H_k}$ of $G_k \cup H_k$ is contained in $K_{k+2}$, and inspection shows that
$$\widehat{G_k \cup H_k} = G_k \cup H_k \cup L_k$$
for some $L_k \in X - P_k$. (In fact one can deduce that $L_k = \emptyset$, but we will not use this fact.)

By using Runge's Theorem 7.2.9 to approximate the functions
$$\tilde{h}_k(x) := \begin{cases} 0, & x \in H_k \cup L_k \\ A_k, & x \in G_k \end{cases},$$
for sufficiently large constants $A_k$, we obtain functions $h_k \in \mathcal{O}(X)$ such that
$$|h_k| < 2^{-k} \text{ in } H_k \quad \text{and} \quad |h_k| \geq 1 + k + M_{k+1} + \left|\sum_{\ell<k} h_\ell\right| \text{ in } G_k, \quad k \geq 1.$$

Since $G_k \subset H_{k+1} \subset H_{k+2} \subset H_{k+3} \subset ...$, the function $f_3 := h_1 + h_2 + ...$ satisfies the estimate $|f_3| \geq k + M_{k+1}$ in $G_k$. Thus (12.4) holds. Theorem 12.5.1 is established. □

## 12.6. Exercises

12.1. Let $x_1, ..., x_N \in \mathbb{P}_k$ be $N$ distinct points, let $F_1, ..., F_N$ be homogeneous polynomials of degree $d_1, ..., d_N$, respectively, such that $F_j(x_j) = 0$ and $dF(x_j) \neq 0$, and let $k_1, ..., k_N$ be positive integers. Show that there exist an integer $M > 0$ and an embedding $\Phi : X \to \mathbb{P}_M$ such that $\Phi(X)$ is tangent to $[F_j = 0]$ to order $k_j$ at $x_j$, for all $1 \leq j \leq N$.

12.2. Let $\{x_j\}$ be a closed, discrete subset of an open Riemann surface $X$. Fix local coordinates $z_j$ vanishing at $x_j$, and polynomials $p_j$ of degree $d_j$, $j = 1, 2, ...$. Show that there is a function $f \in \mathcal{O}(X)$ such that
$$f_j - p_j(z_j) = O(|z_j|^{d_j+1}), \quad j = 1, 2, ....$$

12.3. Let $X$ be a compact Riemann surface, let $L \to X$ be a holomorphic line bundle, let $p_1, ..., p_N$ be distinct points of $X$, and let $m_1, ..., m_N$ be positive integers. Show that there is a meromorphic section $s$ of $L$ such that for each $i \in \{1, ..., N\}$, $s$ has a pole of order $m_i$ at $p_i$.

12.4. Show that the family $\mathscr{M}(X)$ of all meromorphic functions on a compact Riemann surface $X$ realizes all Laurent jets at any finite number of points. That is to say, given $p_1, ..., p_N \in X$, local coordinates $z_i$ centered at $p_i$, $1 \leq i \leq N$, and Laurent polynomials $\phi_1, ..., \phi_N$, there is a meromorphic function $f$ such that the Laurent series of $f$ at $p_j$ in terms of $z_j$ is $\phi_j(z_j)$.

12.5. Give proofs of Proposition 12.3.3 and Theorem 12.4.1 using the conclusions of Exercises 11.7 and 11.8 of the previous chapter in place of Theorems 12.1.1 and 12.1.2.

Chapter 13

# The Riemann-Roch Theorem

In this chapter, our goal is to compute the dimension of the space of sections of a holomorphic line bundle in terms of other data related to the line bundle. The result we prove is the celebrated theorem of Riemann-Roch. This theorem is a fundamental result in the theory of compact Riemann surfaces, and we will derive several of its many consequences.

## 13.1. The Riemann-Roch Theorem

### 13.1.1. Statement of the theorem.

THEOREM 13.1.1 (Riemann-Roch). *Let $D$ be a divisor on a compact Riemann surface $X$ of genus $g$. Then*

$$\dim_{\mathbb{C}} \Gamma_{\mathcal{O}}(X, L_D) - \dim_{\mathbb{C}} \Gamma_{\mathcal{O}}(X, K_X \otimes L_D^*) = \deg(D) + 1 - g.$$

Consider the quantity

$$\rho(D) := \dim_{\mathbb{C}} \Gamma_{\mathcal{O}}(X, L_D) - \dim_{\mathbb{C}} \Gamma_{\mathcal{O}}(X, K_X \otimes L_D^*) - \deg(D).$$

Since $\rho(0) = 1 - g$, Theorem 13.1.1 is a consequence of the following result.

THEOREM 13.1.2 (Riemann-Roch II). *The function $\rho : \mathrm{Div}(X) \to \mathbb{N}$ is constant.*

### 13.1.2. Mittag-Leffler Problems.

Our approach to the Riemann-Roch Theorem avoids the direct use of sheaves. In place of sheaves, we use a method to construct meromorphic functions that resembles what we have done to solve the Mittag-Leffler Problem.

**THEOREM 13.1.3.** *Let $X$ be a compact Riemann surface and let $E \in \mathrm{Div}(X)$. Let $\{U_j\}$ be an open cover of $X$, and let $\{f_j\}$ be a principal part, i.e., a collection of meromorphic functions $f_i \in \mathscr{M}(U_i)$ such that $f_i - f_j \in \mathcal{O}(U_i \cap U_j)$. Then there is a meromorphic function $f$ such that $f s_E - f_j \in \mathcal{O}(U_j)$ if and only if for every $\omega \in \Gamma_\mathcal{O}(X, K_X \otimes L_E^*)$,*
$$\deg \mathrm{Res}(\{f_j \omega\}) = 0.$$

The proof of Theorem 13.1.3 requires the following, highly non-trivial lemma.

**LEMMA 13.1.4.** *Let $X$ be a compact Riemann surface, $L \to X$ a holomorphic line bundle, and let $\alpha$ be a closed, $L$-valued $(0,1)$-form. Then $\alpha = \bar{\partial}\sigma$ for some global section $\sigma \in \Gamma(X, L)$ if and only if*
$$\int_X \alpha \wedge \omega = 0 \quad \text{for all } \omega \in \Gamma_\mathcal{O}(X, K_X \otimes L^*).$$

**Sketch of proof.** Even in the case that $L$ is the trivial bundle, the proof of Lemma 13.1.4 is not easy; it is essentially the Hodge Theorem 9.3.1. However, the general case is no harder. One simply has to go through the proof of the Hodge Theorem, replacing forms by $L$-valued forms. All of the mathematics is exactly the same. In fact, we have left the proof as a series of exercises at the end of Chapter 9. □

**Proof of Theorem 13.1.3.** First suppose $\{f_j\}$ is the principal part of $f s_E$ for some meromorphic function $f$. Then $f s_E \omega$ is a meromorphic 1-form, and thus by Proposition 5.7.2
$$\deg(\mathrm{Res}(\{f_j \omega\})) = \deg(\mathrm{Res}(f s_E \omega)) = 0.$$

For the converse, fix a partition of unity $\{\chi_i\}$ subordinate to $\{U_i\}$, and define the $L_E$-valued $(0,1)$-form
$$\alpha := \sum_i \bar{\partial}\chi_i (f_i - f_j).$$

By the same calculation appearing in the proof of Theorem 8.5.8, we have

(13.1) $$\frac{1}{2\pi\sqrt{-1}} \int_X \omega \wedge \alpha = \deg \mathrm{Res}(\{f_j \omega\}).$$

Therefore by the residue hypothesis and Lemma 13.1.4, the Dolbeault cohomology class represented by $\alpha$ is 0. In other words, $\alpha = \bar{\partial}\sigma$ for some section $\sigma \in \Gamma(X, L_E)$.

Imitating the proof of the Mittag-Leffler Theorem 8.5.7 for compact Riemann surfaces, we let
$$s_j := -\sigma + \sum_i \chi_i (f_i - f_j).$$

Then $\bar{\partial}s_j = 0$, and thus
$$s_j \in \Gamma_\mathcal{O}(U_j, L_E).$$

## 13.1. The Riemann-Roch Theorem

Moreover,

$$s_j - s_k = \sum_i \chi_i((f_i - f_j) - (f_i - f_k)) = (f_k - f_j)\sum_i \chi_i = f_k - f_j.$$

Therefore $s_j + f_j = s_k + f_k$ on $U_j \cap U_k$, so the section $s$ given by $s_j + f_j$ on $U_j$ is a globally defined, meromorphic section of $L_E$, and $\{f_j\}$ is the principle part of $s$. The function we seek is therefore $f = s/s_E$. The proof is complete. □

**13.1.3. Exact sequences of vector spaces.** To establish Theorem 13.1.2 we will use a little bit of homological algebra. Specifically, we will use the following ideas.

DEFINITION 13.1.5. A sequence $A_1 \xrightarrow{a_1} A_2 \xrightarrow{a_2} A_3 \xrightarrow{a_3} \ldots$ of finite-dimensional vector spaces $A_j$ and linear maps $a_j$ is said to be *exact* if $\mathrm{Image}(a_j) = \mathrm{Kernel}(a_{j+1})$ for all $j$. ◊

We leave the reader to check that if $A_1 \xrightarrow{a_1} A_2 \xrightarrow{a_2} \ldots \xrightarrow{a_{N-1}} A_N \xrightarrow{a_N} A_{N+1}$ is an exact sequence then the dual sequence $A_{N+1}^* \xrightarrow{a_N^*} A_N \xrightarrow{a_{N-1}^*} \ldots \xrightarrow{a_2^*} A_2 \xrightarrow{a_1^*} A_1$ is also exact.

LEMMA 13.1.6. *Let* $0 \to A_0 \xrightarrow{a_0} A_1 \xrightarrow{a_1} \ldots \xrightarrow{a_{N-1}} A_N \xrightarrow{a_N} 0$ *be an exact sequence. Then*

$$\sum_{j=0}^{N}(-1)^j \dim(A_j) = 0.$$

**Proof.** Let $I_k = \mathrm{Image}(a_k)$ and $K_k = \mathrm{Kernel}(a_k)$. By definition of exactness, $A_k = I_k \oplus K_k$ and $K_{k+1} = I_k$. Since $I_N = 0 = K_0$, we have

$$\sum_{j=0}^{N}(-1)^j \dim(A_j)$$
$$= \dim(I_0) + \sum_{j=1}^{N-1}(-1)^j(\dim(I_j) + \dim(K_j)) + (-1)^N \dim(K_N)$$
$$= \dim(I_0) + \sum_{j=1}^{N-1}(-1)^j(\dim(I_j) + \dim(I_{j-1})) + (-1)^N \dim(I_{N-1})$$
$$= \sum_{j=0}^{N-1}(-1)^j(\dim(I_j) - \dim(I_j))$$
$$= 0.$$

The proof is complete. □

### 13.1.4. Proof of Theorem 13.1.2.

**LEMMA 13.1.7.** *Let $E \in \mathrm{Div}(X)$ be any divisor and let $D \in \mathrm{Div}(X)$ be a divisor with $D(x) \in \{0,1\}$ for all $x \in X$. Then the sequences of vector spaces*

$$0 \to \Gamma_{\mathcal{O}}(X, L_E) \xrightarrow{\otimes s_D} \Gamma_{\mathcal{O}}(X, L_{D+E}) \xrightarrow{r_{X|D}} \bigoplus_{D(x)=1} (L_{D+E})_x$$

*and*

$$0 \to \Gamma_{\mathcal{O}}(X, K_X \otimes L^*_{D+E}) \xrightarrow{\otimes s_D} \Gamma_{\mathcal{O}}(X, K_X \otimes L^*_E) \xrightarrow{\mathrm{Res}_D} \bigoplus_{D(x)=1} (L^*_{D+E})_x$$

*are exact.*

In the first sequence the map $r_{X|D}$ is the restriction of sections of $L_{D+E}$ to the support of $D$, while in the second sequence the map $\mathrm{Res}_D$ is the map that assigns to an $L^*_E$-valued holomorphic 1-form $\omega$ the residues along $D$ of the $L^*_{D+E}$-valued meromorphic 1-form $\omega/s_D$.

**Proof of Lemma 13.1.7.** The injectivity of the maps $\otimes s_D$ is obvious.

Consider $s \in \Gamma_{\mathcal{O}}(X, L_{D+E})$. Suppose $r_{X|D}(s) = 0$. Then $s$ vanishes along $D$, and since $D$ has only multiplicities 1 or 0, $s$ must be divisible by $s_D$. Thus $s$ is in the image of $\otimes s_D$.

Consider $\omega \in \Gamma_{\mathcal{O}}(X, K_X \otimes L^*_E)$. Suppose the residues of $\omega/s_D$ along $D$ are all 0. Then $\omega$ must vanish along $D$, and since $D$ has only multiplicities 1 or 0, $\omega$ must be divisible by $s_D$. The proof is complete. $\square$

Our next goal is to splice, in the right way, the first sequence in the statement of Lemma 13.1.7 with the dual of the second sequence by somehow identifying

$$\left( \bigoplus_{D(x)=1} (L^*_{D+E})_x \right)^* \quad \text{and} \quad \bigoplus_{D(x)=1} (L_{D+E})_x.$$

The two spaces are clearly isomorphic, but the particular isomorphism we choose determines whether or not the spliced sequence is exact. To achieve exactness, we choose the pairing between $\bigoplus_{D(x)=1}(L_{D+E})_x$ and $\bigoplus_{D(x)=1}(L^*_{D+E})_x$ as follows.

(13.2) $$\langle (s(x))_{D(x)=1}, (\sigma(x))_{D(x)=1} \rangle := \sum_{D(x)=1} \langle s(x), \sigma(x) \rangle.$$

This pairing is clearly perfect. Using this pairing to make the identification, we have the following crucial lemma.

## 13.1. The Riemann-Roch Theorem

LEMMA 13.1.8. *Let the notation and hypotheses be as in Lemma 13.1.7. The sequence*

(13.3) $$0 \to \Gamma_{\mathcal{O}}(X, L_E) \to \Gamma_{\mathcal{O}}(X, L_{D+E}) \to \bigoplus_{D(x)=1} (L_{D+E})_x$$
$$\to (\Gamma_{\mathcal{O}}(X, K_X \otimes L_E^*))^* \to (\Gamma_{\mathcal{O}}(X, K_X \otimes L_{D+E}^*))^* \to 0,$$

*with the identification of* $\left(\bigoplus_{D(x)=1} (L_{D+E}^*)_x\right)^*$ *and* $\bigoplus_{D(x)=1} (L_{D+E})_x$ *via the pairing (13.2), is exact.*

**Proof.** The sequence (13.3) is already exact at all spaces except possibly at the space

$$\bigoplus_{D(x)=1} (L_{D+E})_x.$$

To verify exactness at $\bigoplus_{D(x)=1} (L_{D+E})_x$, we must show that the kernel of $\operatorname{Res}_D^*$ equals the image of $r_{X|D}$. Since by linear algebra the kernel of $\operatorname{Res}_D^*$ is the annihilator of the image of $\operatorname{Res}_D$, we need to show the image of $r_{X|D}$ is exactly the annihilator of the image of $\operatorname{Res}_D$.

First we show that the image of $r_{X|D}$ lies in the annihilator of $\operatorname{Res}_D$. To this end, for $s \in \Gamma_{\mathcal{O}}(X, L_{D+E})$,

$$\langle r_{X|D}(s), \operatorname{Res}_D(\omega) \rangle = \sum_{x \in X} \operatorname{Res}_x(s \otimes \omega/s_D) = \sum_{D(x) \neq 0} s(x) \otimes \operatorname{Res}_x(\omega/s_D).$$

Since $s \otimes \omega/s_D$ is a meromorphic 1-form, the sum of its residues must vanish by Proposition 5.7.2, and therefore the image of $r_{X|D}$ lies in the annihilator of the image of $\operatorname{Res}_D$.

Next we show that the annihilator of $\operatorname{Res}_D$ lies in the image of $r_{X|D}$. Let

$$\lambda = (\lambda_x)_{D(x)=1} \in \bigoplus_{D(x)=1} (L_{D+E})_x$$

be in the annihilator of the image of $\operatorname{Res}_D$. Fix an open cover $\{U_j\}$ of $X$ such that each $U_j$ contains at most one $x$ with $D(x) = 1$. We can define (constant) holomorphic functions $\lambda_j \in \mathcal{O}(U_j)$ by

$$\lambda_j := \begin{cases} \lambda_x, & x \in U_j \text{ and } D(x) = 1 \\ 0, & D(y) = 0 \text{ for all } y \in U_j \end{cases}.$$

Now consider the meromorphic functions

$$f_j := \frac{\lambda_j}{s_D}.$$

Clearly $f_j - f_k \in \mathcal{O}(U_j \cap U_k)$, and moreover for every $\omega \in \Gamma_{\mathcal{O}}(X, K_X \otimes L_E^*)$ we have
$$\deg \operatorname{Res}(\{f_j\omega\}) = \sum_{x \in X} \lambda_x \operatorname{Res}_x\left(\frac{\omega}{s_D}\right) = \langle \lambda, \operatorname{Res}_D(\omega)\rangle = 0.$$
Thus by Theorem 13.1.3 there exists a meromorphic function $f$ such that $fs_E - f_j \in \mathcal{O}(U_j)$ for all $j$. It follows that the section
$$s_\lambda := fs_E s_D$$
of $L_{D+E}$ is holomorphic. Indeed, since $g_j := fs_E - f_j$ is holomorphic,
$$fs_D s_E = s_D(g_j + f_j) = s_D g_j - \lambda_j$$
is also holomorphic on $U_j$. Finally, we have
$$s_\lambda(x) = \lambda_x \text{ for all } x \in D^{-1}(1).$$
In other words, $r_{X|D}(s_\lambda) = \lambda$. The proof is complete. □

**Proof of Theorem 13.1.2.** By Lemma 13.1.6 the alternating sum of the dimensions of the spaces in the sequence (13.3) is 0. With the notation
$$h^0(X, \Lambda) := \dim_{\mathbb{C}} \Gamma_{\mathcal{O}}(X, \Lambda) \quad \text{and} \quad h^1(X, \Lambda) := \dim_{\mathbb{C}} \Gamma_{\mathcal{O}}(X, K_X \otimes \Lambda^*)$$
for a general line bundle $\Lambda \to X$, we have
$$h^0(X, L_E) - h^0(X, L_E \otimes L_D) + \deg(D) - h^1(X, L_E) + h^1(X, L_D \otimes L_E) = 0.$$
Adding $-\deg(E)$ to both sides and rearranging gives
$$\begin{aligned}&h^0(X, L_E) - h^1(X, L_E) - \deg(E)\\(13.4)\qquad &= h^0(X, L_D \otimes L_E) - h^1(X, L_D \otimes L_E) - \deg(D+E)\end{aligned}$$
for any divisor $D$ whose multiplicities are 1 or 0.

Let $D_1, D_2$ be two divisors. Choose a divisor $D_o$ such that $D_i - D_o \geq 0$ for $i = 1, 2$. We will prove that $\rho(D_i) = \rho(D_o)$ for $i = 1, 2$.

First, let $E = D_i - D_o$. Then we can write
$$E_k := \sum_{E(x) \geq k} x, \quad k = 1, 2, ..., \max_x(E(x)).$$

We let $E_0 := D_o$. Now fix $k \geq 1$ and apply (13.4) to the divisors $E := \sum_{j=0}^{k-1} E_j$ and $D := E_k$ to obtain
$$\begin{aligned}&h^0(X, L_{\sum_{j=0}^k E_j}) - h^1(X, L_{\sum_{j=0}^k E_j})\\&= h^0(X, L_{\sum_{j=0}^{k-1} E_j}) - h^1(X, L_{\sum_{j=0}^{k-1} E_j}) - \deg(E_k).\end{aligned}$$

Summing over $k$ from $1$ to $\max_x E(x)$, we have

$$h^0(X, L_{D_i}) - h^1(X, L_{D_i})$$
$$= h^0(X, L_{D_o}) - h^1(X, L_{D_o}) - \sum_{k=1}^{\max_x E(x)} \deg(E_k)$$
$$= h^0(X, L_{D_o}) - h^1(X, L_{D_o}) - \deg(E)$$
$$= h^0(X, L_{D_o}) - h^1(X, L_{D_o}) - \deg(D_i) + \deg(D_o).$$

Thus $\rho(D_i) = \rho(D_o)$, and consequently $\rho(D_1) = \rho(D_2)$, as claimed. The proof of the Riemann-Roch Theorem is thus complete. □

## 13.2. Some corollaries

**13.2.1. Necessity of meromorphic plenitude.** The Riemann-Roch Theorem provides precise control of meromorphic functions on a compact Riemann surface $X$. Recall the existence and plenitude of these functions, as stated in Corollary 12.3.3. We shall now show that point and tangent separation is a direct consequence of the Riemann-Roch Theorem.

PROPOSITION 13.2.1. *If $X$ is a compact Riemann surface that satisfies the conclusions of the Riemann-Roch Theorem, then $\mathscr{M}(X)$ separates points and tangents.*

**Proof.** Let $p, q \in X$. Applying Riemann-Roch to the divisor $D = (g+1) \cdot p$, we have
$$h^0(X, L_D) - h^1(X, L_D) = (g+1) + 1 - g = 2,$$
and thus $h^0(X, L_D) \geq 2$. We claim there is a section $s$ that vanishes to order no more than $g$ at $p$. Indeed, if $\sigma \in H^0(X, L_D)$ vanishes to order at least $g+1$ at $p$, then $\sigma/s_D \in \mathcal{O}(X)$, and thus $\sigma = cs_D$. Since $h^0(X, L_D) \geq 2$, the said section $s$ must exist. Thus $f := s/s_D$ is a meromorphic function with pole only at $p$. (The order of the pole might be more than 1; in fact it will certainly be more than 1 if the genus of $X$ is positive.) In particular, $f(q) \neq \infty$, and thus the meromorphic function $f$ separates $p$ and $q$. (In fact $f$ at once separates $p$ from any other point.)

Using the divisor $D_n = n \cdot p$ for sufficiently large $n$, we see in the same way that there exist meromorphic functions $g_m \in \mathscr{M}(X)$ having a pole of order $m$ at $p$ and no other poles, provided $m$ is large enough. It follows that $f = g_m/g_{m+1}$ has a simple zero at $p$ (though of course it may have other zeros) and so $f$ separates tangents at $p$. Thus $\mathscr{M}(X)$ separates tangents. □

**13.2.2. Ampleness.** In Chapter 12 we showed that the sections of a sufficiently positive line bundle on a compact Riemann surface $X$ serve as projective components for an embedding of $X$ in $\mathbb{P}_N$. The Riemann-Roch Theorem allows us to be more precise about how much positivity we need.

DEFINITION 13.2.2. A divisor $D \in \text{Div}(X)$ on a compact Riemann surface is said to be very ample if the map
$$\phi_{|L_D|} : X \to \mathbb{P}(H^0(X, L_D)^*)$$
is an embedding. The divisor $D$ is said to be ample if $mD$ is very ample for some $m \in \mathbb{N}$.

PROPOSITION 13.2.3. *If $X$ is a compact Riemann surface of genus $g$, then any divisor $D \in \text{Div}(X)$ of degree at least $2g+1$ is very ample. In particular, every divisor of positive degree is ample.*

Before proving Proposition 13.2.3, we will establish the following elementary lemmas.

LEMMA 13.2.4. *The degree of the canonical bundle $K_X$ of a compact Riemann surface of genus $g$ is $2g - 2$.*

**Proof.** By Riemann-Roch,
$$\deg(K_X) = \dim_{\mathbb{C}}(\Gamma_{\mathcal{O}}(X, K_X)) - \dim_{\mathbb{C}}(\Gamma_{\mathcal{O}}(X, K_X \otimes K_X^*)) + g - 1 = g - 1 + g - 1,$$
as claimed. $\square$

LEMMA 13.2.5. *Let $D$ be a divisor and choose (possibly non-distinct) points $p, q \in X$. If*
$$\dim_{\mathbb{C}}(\Gamma_{\mathcal{O}}(X, L_D)) = \dim_{\mathbb{C}}(\Gamma_{\mathcal{O}}(X, L_{D-p-q})) + 2,$$
*then the sections of $L_D$ separate the points $p$ and $q$ or, if $p = q$, tangents at $p$.*

**Proof.** First we show that not all the sections of $L_D$ over $X$ vanish at $p$. If not, then the well-defined map
$$\Gamma_{\mathcal{O}}(X, L_D) \ni s \mapsto s/s_p \in \Gamma_{\mathcal{O}}(X, L_{D-p})$$
is an isomorphism. Now, the subspace $V_q := \{s \in \Gamma_{\mathcal{O}}(X, L_{D-p}) \; ; \; s(q) = 0\}$ has codimension at most 1 (since it is defined by at most one linear relation), and the map
$$V_q \ni \sigma \mapsto \sigma/s_q \in \Gamma_{\mathcal{O}}(X, L_{D-p-q})$$
is an isomorphism. Thus $\dim_{\mathbb{C}} \Gamma_{\mathcal{O}}(X, L_{D-p-q}) + 1 \geq \Gamma_{\mathcal{O}}(X, L_D)$, which is a contradiction.

Now consider the codimension-1 subspace $W_p \subset \Gamma_{\mathcal{O}}(X, L_D)$ of sections of $L_D$ vanishing at $p$. As before, the map
$$W_p \ni s \mapsto s/s_p \in \Gamma_{\mathcal{O}}(X, L_{D-p})$$
is an isomorphism. We claim that not every section in $W_p$ vanishes at $q$. If not, then the map
$$W_p \ni s \mapsto s/s_p \mapsto (s/s_p)/s_q \in \Gamma_{\mathcal{O}}(X, L_{D-p-q})$$

## 13.2. Some corollaries

is an isomorphism. But then

$$\dim_{\mathbb{C}} \Gamma_{\mathcal{O}}(X, L_{D-p-q}) = \dim_{\mathbb{C}} \Gamma_{\mathcal{O}}(X, L_{D-p}) = \dim_{\mathbb{C}} W_p = \Gamma_{\mathcal{O}}(X, L_D) + 1,$$

which is again a contradiction. □

**Proof of Proposition 13.2.3.** First let $E$ be a divisor of degree at least $2g - 1$. By Lemma 13.2.4, $K_X \otimes L_E^*$ has negative degree and thus has no holomorphic sections. Consider the divisor $E = D - p - q$, where $p$ and $q$ are two (possibly non-distinct) points of $X$. Then $D$ and $E$ both have degree at least $2g - 1$, and thus $\dim_{\mathbb{C}} \Gamma_{\mathcal{O}}(X, K_X \otimes L_D^*) = \dim_{\mathbb{C}} \Gamma_{\mathcal{O}}(X, K_X \otimes L_E^*) = 0$. By Riemann-Roch

$$\dim_{\mathbb{C}}(\Gamma_{\mathcal{O}}(X, L_D)) = \deg(D) + 1 - g = \deg(E) + 3 - g = \dim_{\mathbb{C}}(\Gamma_{\mathcal{O}}(X, L_E)) + 2.$$

The proof is completed with an application of Lemma 13.2.5. □

**13.2.3. Curves of genus 0, 1, and 2.** Let $X$ be a curve of genus zero. Applying Proposition 13.2.3, we see that for any point $p \in X$, $\Gamma_{\mathcal{O}}(X, L_p)$ gives an embedding $\phi_{|1 \cdot p|} : X \hookrightarrow \mathbb{P}(\Gamma_{\mathcal{O}}(X, L_p)^*)$. By Lemma 13.2.4 the degree of $K_X$ is $-2$, and thus by Riemann-Roch $h^0(X, L_p) = h^1(X, L_p) + \deg(1 \cdot p) + 1 - g = 0 + 1 + 1 - 0 = 2$. Since any non-constant map between Riemann surfaces is surjective, the map $\phi_{|1 \cdot p|}$ is an isomorphism, i.e., we recover the following fact, already proved in several ways.

$$\text{If } g(X) = 0 \text{ then } X \cong \mathbb{P}_1.$$

Next let $X$ be an algebraic curve of genus 1, and let $D$ be a divisor of degree $3 = 2g + 1$. We know $D$ is very ample, and since $\deg K_X = 0$ by Lemma 13.2.4, Riemann-Roch implies that $h^0(X, L_D) = 3$. It follows that $X$ embeds as a curve in $\mathbb{P}_2$. Moreover, since $L_D = \phi^*(\mathbb{H})$, we see that the divisor $D$ is obtained from intersecting $\phi_{|D|}(X)$ with a hyperplane, i.e., the restriction to $\phi_{|D|}(X)$ of a homogeneous linear function on $\mathbb{P}_2$ has exactly 3 zeros, counting multiplicity. Thus $\phi_{|D|}(X)$ is a cubic curve in $\mathbb{P}_2$.

We would like to show that a curve $X$ of genus 1 is actually a complex torus. We know that it is topologically a torus, and thus if $\pi : Y \to X$ is the covering space, then $Y = \mathbb{R}^2$ as a topological space. To prove that $X$ is a complex torus, it suffices to show that $Y = \mathbb{C}$ as a complex space. While this fact can be deduced from uniformization, it is actually more elementary.

By the equivalence of arithmetic and geometric genus, we know there is a holomorphic 1-form $\omega$ on $X$. Let $K = \text{Ord}(\omega)$ be a canonical divisor on $X$. Then $\deg(K) = 0$ by Lemma 13.2.4, and thus $\omega$ has no zeros. It follows that the canonical bundle of $X$ is trivial. Thus its dual, the holomorphic tangent bundle, is also trivial, and we have a holomorphic vector field $\xi$ on $X$ with no zeros. Since $\omega(\xi)$ is a holomorphic function with no zeros, it is a non-zero constant. Thus by rescaling $\xi$ or $\omega$, we can even assume that $\omega(\xi) = 1$.

By the compactness of $X$, $\xi$ is complete. The integral curve
$$\psi : z \mapsto \varphi_\xi^z(p_o)$$
through some point $p_o \in X$ defines a local diffeomorphism $\psi : \mathbb{C} \to X$. Since $\pi : Y \to X$ is the universal cover, we have a lift $\tilde{\psi} : \mathbb{C} \to Y$. Since $\psi$ is an integral curve, $\tilde{\psi}$ is surjective by the Existence and Uniqueness Theorem for ODE.

On the other hand, consider the map $\varphi : Y \to \mathbb{C}$ defined by
$$\varphi(p) := \int_{c_o}^{p} \pi^*\omega,$$
where the integral is over a curve connecting $p$ to some point $c_o \in Y$. Since $\pi^*\omega$ is a holomorphic 1-form on $Y$ and $Y$ is simply connected, the map $\varphi$ is well-defined and holomorphic. (To see that the map varies holomorphically with $p$, we note that for small changes in $p$, we may compute the integral on $X$ by using $\pi$ as a change of variable.)

By definition,
$$\frac{\partial}{\partial z}\varphi \circ \tilde{\psi}(z) = \left\langle d\varphi, \tilde{\psi}' \right\rangle = \pi^*\omega(\psi') = \omega(d\pi(\tilde{\psi}')) = \omega(\xi) = 1$$
so that $\varphi \circ \tilde{\psi}(z) = z + c$. Hence $\tilde{\psi}$ is injective. Thus $Y \cong \mathbb{C}$ and we have proved the following result.

THEOREM 13.2.6. *Every genus-1 Riemann surface can be embedded as a smooth plane cubic curve in $\mathbb{P}_2$ and is of the form $\mathbb{C}/\Lambda$.*

Finally, we make one remark about genus 2 curves. By the equivalence of arithmetic and geometric genus,
$$h^0(X, K_X) = g = 2.$$
It follows that there is a non-constant meromorphic function
$$f : X \to \mathbb{P}_1 = \mathbb{P}(\Gamma_\mathcal{O}(X, K_X)^*).$$
(Note that $f$ is well-defined even if, *a priori*[1], all global holomorphic sections of $K_X$ vanish at some points; we simply use the local Normal Forms Theorem for holomorphic functions to factor out the zeros.) We claim that $f$ has degree 2. Indeed, the degree of $f$ is just the degree of $K_X$, which by Lemma 13.2.4 is $2g - 2 = 2$.

It is possible to show from the existence of the degree-2 holomorphic map $f : X \to \mathbb{P}_1$ that $X$ can be identified with a Riemann surface given locally by equations of the form $w^2 = P(z)$. Thus every genus 2 curve is hyperelliptic.

REMARK. It is the case that the generic compact Riemann surface of genus $g \geq 3$ is not hyperelliptic. ◇

---

[1] In fact, we shall see that the canonical bundle of a Riemann surface of positive genus is basepoint-free.

## 13.2. Some corollaries

**13.2.4. The canonical bundle of a curve of positive genus is free.** Recall that a holomorphic line bundle $L$ is free (or basepoint-free) if at each point $x$ there is a global holomorphic section $s$ of $L$ with $s(x) \neq 0$.

If $g = 0$, then $\deg(K_X) = -2$ and thus $K_X$ has no sections. However, in the case of genus 1, we know that the canonical bundle is trivial and thus free. (It has a global, nowhere-zero section.) In this paragraph, we analyze all Riemann surfaces of positive genus at once.

We begin with the following lemma.

LEMMA 13.2.7. *Let $X$ be a compact Riemann surface of positive genus and let $p \in X$. Then*
$$\dim_{\mathbb{C}} \Gamma_{\mathcal{O}}(X, L_p) = 1.$$

**Proof.** If there exists $s \in \Gamma_{\mathcal{O}}(X, L_p)$ such that $s(p) \neq 0$ then $s/s_p \in \mathcal{M}(X)$ has only one simple pole, which means $X \cong \mathbb{P}_1$. Therefore each $s \in \Gamma_{\mathcal{O}}(X, L_p)$ vanishes at $p$. But then $s/s_p \in \mathcal{O}(X) \cong \mathbb{C}$, and thus $s = \lambda s_p$ for some $\lambda \in \mathbb{C}$. The proof is finished. □

PROPOSITION 13.2.8. *The canonical bundle of any Riemann surface of positive genus is free.*

**Proof.** Let $p \in X$. If all $s \in \Gamma_{\mathcal{O}}(X, K_X)$ vanish at $p$ then $s \mapsto s/s_p$ gives an isomorphism of $\Gamma_{\mathcal{O}}(X, K_X)$ with $\Gamma_{\mathcal{O}}(X, K_X \otimes L_p^*)$. On the other hand, by Riemann-Roch we have

$$\begin{aligned}\dim_{\mathbb{C}} \Gamma_{\mathcal{O}}(X, K_X \otimes L_p^*) &= \Gamma_{\mathcal{O}}(X, L_p) - \deg(1 \cdot p) + g - 1 \\ &= g - 1 = \dim_{\mathbb{C}} \Gamma_{\mathcal{O}}(X, K_X) - 1.\end{aligned}$$

This is the desired contradiction. □

REMARK (Hyperellipticity and the canonical map). It follows from Proposition 13.2.8 that if $X$ has positive genus then the canonical map
$$\phi_{|K_X|} : X \to \mathbb{P}(\Gamma_{\mathcal{O}}(X, K_X)^*) \cong \mathbb{P}_{g-1}$$
is a morphism. If $g = 2$, we obtain our degree-2 projection that induces the hyperelliptic involution. For $g \geq 3$, one can ask if the canonical map is an embedding. It turns out that in general $\phi_{|K_X|}$ is an embedding if and only if $X$ is not hyperelliptic. If $X$ is hyperelliptic, $\phi_{|K_X|}$ is 2-1 onto its image (which also holds for the case $g = 2$). Thus the hyperelliptic locus is precisely the set of compact Riemann surfaces of genus $g \geq 2$ for which the canonical map is not an embedding. ◇

# Chapter 14

# Abel's Theorem

We have learned that every holomorphic line bundle on an open Riemann surface is trivial. By contrast, compact Riemann surfaces admit many non-trivial line bundles; the line bundle associated to any divisor of non-zero degree is never trivial. The next natural step is to consider the set of line bundles of a fixed degree. Since the collection of holomorphic line bundles forms an Abelian group, it suffices to look at the degree-zero case. As we have pointed out repeatedly, on a curve of positive genus not every degree-zero line bundle is trivial. Abel's Theorem characterizes those divisors of degree zero that are the divisors of a meromorphic function or, equivalently, whose associated line bundles are trivial.

## 14.1. Indefinite integration of holomorphic forms

**14.1.1. The Jacobian of a curve.** Let $X$ be a compact Riemann surface and let $\alpha$ be a holomorphic 1-form on $X$. Fix a point $o \in X$, and for a given point $x \in X$, choose a real curve $\gamma \subset X$ whose initial point is $o$ and whose final point is $x$. Then we obtain a number

$$\varphi_\gamma^x \alpha := \int_\gamma \alpha.$$

If we choose another curve $\gamma'$ connecting $o$ to $x$, then there is a closed curve $\sigma$ based at $x$ such that $\gamma'$ is obtained by first following $\gamma$ from $o$ to $x$, and then following $\sigma$ from $x$ to $x$. Unless $X$ is simply connected, the resulting number $\phi_{\gamma'}^x \alpha$ will in general differ from $\phi_\gamma^x \alpha$. But the difference will only depend on the homotopy (and thus homology) class of $\sigma$. Indeed,

$$\phi_{\gamma'}^x \alpha - \phi_\gamma^x \alpha = \int_\sigma \alpha.$$

It follows that the numbers $\phi_\gamma^x \alpha$ agree modulo the discrete set of periods of $\alpha$

$$\Pi(X, \alpha) := \left\{ \int_\sigma \alpha \, ; \, [\sigma] \in H_1(X, \mathbb{Z}) \right\} \subset \mathbb{C}.$$

Suppose now that the arithmetic genus of $X$ is $g$. Consider the multi-valued map

$$A : X \to (\Gamma_{\mathcal{O}}(X, K_X))^*$$

defined by $A(x) := \varphi_\gamma^x$. To make $A$ a single-valued map, we note that $H_1(X, \mathbb{Z})$ can be mapped into $(\Gamma_{\mathcal{O}}(X, K_X))^*$ via the group homomorphism

$$\mathscr{I} : [\sigma] \mapsto \int_\sigma .$$

Moreover, if $\int_\sigma \alpha = 0$ for all $\alpha \in \Gamma_{\mathcal{O}}(X, K_X)$ then $[\sigma] = 0$, and thus by Lemma 5.8.4 and Theorems 9.2.1 and 9.2.2, $\mathscr{I}$ is injective. Since $H_1(X, \mathbb{Z}) \cong \mathbb{Z}^{2g}$, $\mathscr{I}(H_1(X, \mathbb{Z}))$ is a lattice of maximal rank in $(\Gamma_{\mathcal{O}}(X, K_X))^*$. It follows that the map

$$\mathscr{A} : X \to J(X) := (\Gamma_{\mathcal{O}}(X, K_X))^* / \mathscr{I}(H_1(X, \mathbb{Z}))$$

induced by $A$ is single-valued.

DEFINITION 14.1.1. The $g$-dimensional complex torus $J(X)$ is called the *Jacobian* of $X$, and the map

$$\mathscr{A} : X \to J(X)$$

is called the *Abel map*. ◇

**14.1.2. Statement of Abel's Theorem.** To state Abel's Theorem, we let a general divisor of degree 0 be written

$$D = \sum_{j=1}^N x_j - y_j,$$

where the points $x_1, ..., x_N, y_1, ..., y_N$ are not necessarily distinct, but we assume that $x_i \neq y_j$ for all $i, j$. We denote by $\text{Div}^0(X)$ the set of divisors of degree zero, and define $\mu : \text{Div}^0(X) \to J(X)$ by

$$\mu\left(\sum_j (x_j - y_j)\right) = \sum_j (\mathscr{A}(x_j) - \mathscr{A}(y_j)).$$

Evidently $\mu$ is a group homomorphism.

THEOREM 14.1.2 (Abel's Theorem). *A divisor $D = \sum_j x_j - y_j$ of degree zero on a compact Riemann surface $X$ of positive genus is the divisor of a meromorphic function if and only if*

(14.1) $$\mu(D) = 0.$$

## 14.2. Riemann's Bilinear Relations

REMARK. Note that for $D = \sum x_j - y_j$, $\mu(D) = 0$ if and only if

$$\sum_j \int_{x_j}^{y_j} \theta \in \Pi(X, \theta)$$

for all $\theta \in \Gamma_{\mathcal{O}}(X, K_X)$.  ◇

The proof of Abel's Theorem will be given in Section 14.4, after we establish the Riemann Bilinear Relations and the Reciprocity Theorem.

### 14.2. Riemann's Bilinear Relations

To state and prove the relations we require, we briefly recall the basic construction of a compact oriented surface of genus $g$. Consider a $4g$-gon $\Gamma$ in the plane, with boundary $\partial \Gamma$ consisting of $4g$ sides $s_1, t_1, s'_1, t'_1, ..., s_g, t_g, s'_g, t'_g$, oriented counter-clockwise. Identifying $s_i$ with $-s'_i$ (i.e., the side $s'_i$ with reversed orientation) and $t_i$ with $-t'_i$ produces a handle. (For example, if $g = 1$ this is a standard way to produce the torus.) Making all of the identifications $i = 1, ..., g$ produces a sphere with $g$ handles $X$.

All of the vertices of $\Gamma$ are identified to a single point $x_o \in X$. Each side of $\Gamma$ is identified to a closed curve that cannot be continuously contracted to a point in $X$. We write

$$a_j := s_j / \sim \quad \text{and} \quad b_j := t_j / \sim .$$

The fundamental group of $X$ is generated by the curves $a_1, ..., a_g, b_1, ..., b_g$, and

$$U := X - \left( \bigcup_{j=1}^{g} a_j \cup b_j \right) \cong \Gamma - \partial \Gamma$$

is the interior of a disk and thus is simply connected.

For a smooth 1-form $\alpha$, define

$$A_j(\alpha) := \int_{a_j} \alpha \quad \text{and} \quad B_k(\alpha) := \int_{b_k} \alpha.$$

Fix a point $p \in U$. Consider a smooth closed 1-form $\beta$. Since $U$ is simply connected, we can define a function $f : \overline{U} \to \mathbb{C}$ by

$$f_\beta(x) := \int_p^x \beta,$$

where the integral is over any continuous curve $\gamma$ in $\overline{U}$ connecting $p \in U$ to $x \in \overline{U}$ and such that $\gamma \subset \Gamma$ is still a continuous curve.

LEMMA 14.2.1. *Let $\beta$ be a smooth closed 1-form on $X$ and let $\alpha$ be a smooth 1-form on an open subset of $X$ containing $\partial U = \bigcup_{j=1}^{g}(a_j \cup b_j)$. Then*

$$\int_{\partial U} f_\beta \alpha = \sum_{j=1}^{g} A_j(\beta) B_j(\alpha) - B_j(\beta) A_j(\alpha).$$

**Proof.** Let $P \in t_j$ and denote by $P'$ the corresponding point in $t'_j$. Let $\gamma$ be a curve joining $P$ to $P'$ and such that $\gamma - \{P, P'\} \subset U$. Then

$$\int_\gamma \beta = f_\beta(P) - f_\beta(P').$$

But $\gamma/\sim$ is homologous to $-b_j$, and thus

$$f_\beta(P) - f_\beta(P') = -\int_{b_j} \beta.$$

Similarly if we take a point $Q \in s_j$ and the corresponding point $Q' \in s'_j$, we obtain

$$f_\beta(Q) - f_\beta(Q') = \int_{a_j} \beta.$$

Thus

$$\begin{aligned}
\int_{\partial U} f_\beta \alpha &= \sum_{j=1}^{g} \int_{s_j} f_\beta \alpha + \int_{t_j} f_\beta \alpha + \int_{s'_j} f_\beta \alpha + \int_{t'_j} f_\beta \alpha \\
&= \sum_{j=1}^{g} \int_{P \sim P' \in a_j} (f_\beta(P) - f_\beta(P'))\alpha + \int_{Q \sim Q' \in b_j} (f_\beta(Q) - f_\beta(Q'))\alpha \\
&= \sum_{j=1}^{g} \int_{a_j} \beta \int_{b_j} \alpha - \int_{b_j} \beta \int_{a_j} \alpha.
\end{aligned}$$

The proof is complete. $\square$

We have the following important consequence.

PROPOSITION 14.2.2. *Let $X$ be a compact Riemann surface of genus $g > 0$. If $\omega \in \Gamma_\mathcal{O}(X, K_X)$ is not identically zero, then*

$$\operatorname{Im} \sum_{j=1}^{g} A_j(\omega)\overline{B_j(\omega)} < 0.$$

*In particular, if all of the $a_j$-periods of $\omega$ are zero, then $\omega \equiv 0$.*

**Proof.** Applying Lemma 14.2.1 with $\alpha = \omega = \bar{\beta}$, we have

$$\int_X \frac{\sqrt{-1}}{2} \omega \wedge \bar{\omega} = -\frac{1}{2\sqrt{-1}} \int_{\partial U} f_\omega \bar{\omega} = -\operatorname{Im} \sum_{j=1}^{g} A_k(\omega)\overline{B_k(\omega)},$$

## 14.2. Riemann's Bilinear Relations

where the first equality follows from Stokes' Theorem. □

COROLLARY 14.2.3. *If $\omega_1, ..., \omega_g \in \Gamma_\mathcal{O}(X, K_X)$ is a basis, then with*

$$a_{ij} := \int_{a_i} \omega_j,$$

*the matrix $A = (a_{ij})$ is invertible. In particular, we can choose a basis $\omega_1, ..., \omega_g$ of $\Gamma_\mathcal{O}(X, K_X)$ such that*

$$\int_{a_i} \omega_j = \delta_{ij} := \begin{cases} 1, & i = j \\ 0, & i \neq j \end{cases}.$$

(We call such a basis a *normalized basis*.)

**Proof.** Let $c \in \mathbb{C}^g$ with $Ac = 0$. Then with $\omega = \sum c_j \omega_j$, we have $\int_{a_i} \omega = 0$ for all $i = 1, ..., g$. Thus by Proposition 14.2.2, $\omega = 0$. Since $\omega_1, ..., \omega_g$ is a basis, $c = 0$. □

THEOREM 14.2.4 (Riemann Bilinear Relations). *Let $X$ be a compact Riemann surface of positive genus $g$ and let $\omega_1, ..., \omega_g \in \Gamma_\mathcal{O}(X, K_X)$ be a normalized basis. Then with*

$$B_{jk} := \int_{b_k} \omega_j,$$

*the matrix $B = (B_{jk})$ is symmetric and $\mathrm{Im}\, B$ is positive definite.*

**Proof.** By Stokes' Theorem and Lemma 14.2.1,

$$\begin{aligned} 0 &= \int_X \omega_j \wedge \omega_k \\ &= \int_{\partial U} f_{\omega_j} \omega_k \\ &= \sum_{\ell=1}^g A_\ell(\omega_j) B_\ell(\omega_k) - B_\ell(\omega_j) A_\ell(\omega_k) \\ &= B_j(\omega_k) - B_k(\omega_j). \end{aligned}$$

Thus $B$ is symmetric.

Next let $c = (c_1, ..., c_g) \in \mathbb{R}^g - \{0\}$ and set $\omega = \sum_j c_j \omega_j$. Then $\omega \not\equiv 0$, so by Proposition 14.2.2

$$0 > \mathrm{Im} \sum_j A_j(\omega) B_j(\bar{\omega}) = \mathrm{Im} \sum_{j,k} c_j \overline{B_{jk}} c_k = -\mathrm{Im} \sum_{jk} B_{jk} c_j c_k.$$

The proof is complete. □

## 14.3. The Reciprocity Theorem

### 14.3.1. Abelian differentials of the second and third kind.
We begin by constructing a convenient set of generators for the set of meromorphic forms.

PROPOSITION 14.3.1. *Let $X$ be a compact Riemann surface, let $x$ and $y$ be distinct points of $X$, let $n \geq 1$ be an integer, and let $z$ be a local coordinate vanishing at $x$.*

(1) *There exists a unique meromorphic 1-form $\eta_{x,y}$ all of whose $a_k$-periods are 0 and that is holomorphic on $X - \{x, y\}$, such that $\mathrm{Res}_x(\eta_{x,y}) = 1$ and $\mathrm{Res}_y(\eta_{x,y}) = -1$.*

(2) *There exists a unique meromorphic 1-form $\eta_x^n$ all of whose $a_k$-periods are 0 and that is holomorphic on $X - \{x\}$, such that $\eta_x^n - z^{-(n+1)} dz$ is holomorphic near $x$.*

**Proof.** Fix an open cover $\{U_i\}$ of $X$ by coordinate charts. For part (1), choose meromorphic 1-forms $\eta_i$ in $U_i$ that are holomorphic on $U_i - \{x, y\}$, have simple poles at $x$ and $y$, and have residue 1 at $x$ if $x \in U_i$ and residue $-1$ at $y$ if $y \in U_i$. (For example, one could arrange that each $U_i$ contains at most one element of the set $\{x, y\}$, and if $U_i \ni x$ (resp. $U_i \ni y$), take $\eta_i = (z_i - z_i(x))^{-1} dz_i$ (resp. $\eta_i = -(z_i - z_i(x))^{-1} dz_i$).) For part (2), choose meromorphic 1-forms $\eta_i$ in $U_i$ that are holomorphic on $U_i - \{x\}$, such that if $x \in U_i$ then $\eta_i - (z_i - z_i(x))^{-(n+1)} dz_i$ is holomorphic. In both cases, $\deg(\mathrm{Res}(\{\eta_i\})) = 0$. (In the second case $\mathrm{Res}(\{\eta_i\}) = 0$.)

An application of Theorem 8.5.8 shows the existence of a meromorphic 1-form with the desired polar structure. Any two such meromorphic sections clearly differ by a holomorphic 1-form. In view of Corollary 14.2.3, we can annihilate the periods of our meromorphic 1-forms by adding an appropriate holomorphic 1-form. Proposition 14.2.2 then implies that the resulting 1-forms are uniquely determined. □

REMARK. Meromorphic 1-forms of the types appearing in (1) and (2) of Proposition 14.3.1 are classically called (normalized) Abelian differentials of the second and third kind, respectively. Abelian differentials of the first kind are just holomorphic 1-forms. It is an exercise (using Proposition 5.7.2) to show that any meromorphic 1-form can be expressed as a linear combination of Abelian differentials of the first, second, and third kind. ◇

### 14.3.2. The Reciprocity Theorem.
We retain the notation of the previous section.

THEOREM 14.3.2. *Fix a compact Riemann surface $X$ of genus $g$. Let $\omega_1, ..., \omega_g \in \Gamma_{\mathcal{O}}(X, K_X)$ be a normalized basis, and let $\eta_{x,y}$ and $\eta_x^n$ be normalized Abelian differentials of the second and third kind, respectively.*

(1) If $\gamma_{x,y}$ is a curve joining $x$ to $y$ in $X - \{a_1, ..., a_g, b_1, ..., b_g\}$ then
$$\int_{b_k} \eta_{x,y} = 2\pi\sqrt{-1} \int_{\gamma_{x,y}} \omega_k, \quad k = 1, ..., g.$$

(2) If $z$ is the local coordinate used to define $\eta_x^n$ and $\omega_k = f_k(z)dz$, then
$$\int_{b_k} \eta_x^n = \frac{2\pi\sqrt{-1}}{n!} f_k^{(n-1)}(z(x)), \quad k = 1, ..., g.$$

**Proof.** We identify $X - \{a_1, ..., a_g, b_1, ..., b_g\}$ with the interior $U$ of the polygon used to construct the topological surface $X$. Let
$$g_k(z) := \int_{x_o}^{z} \omega_k.$$
Then by Lemma 14.2.1, together with the normalization condition on our basis $\omega_1, ..., \omega_g$ and the fact that the $a_j$-periods of $\eta_{x,y}$ all vanish,
$$\int_{\partial U} g_k \eta_{x,y} = B_k(\eta_{x,y}) = \int_{b_k} \eta_{x,y}.$$
On the other hand, by the Residue Theorem we have
$$\int_{\partial U} g_k \eta_{x,y} = 2\pi\sqrt{-1}(g_k(x) - g_k(y)) = 2\pi\sqrt{-1} \int_{\gamma_{x,y}} \omega_k.$$
This proves (1). Establishing (2) is similar:
$$\int_{b_k} \eta_x^n = \int_{\partial U} g_k \eta_x^n = 2\pi\sqrt{-1}\mathrm{Res}_x(g_k \eta_x^n) = \frac{1}{n!}\frac{\partial^n}{\partial z^n}g_k = \frac{2\pi\sqrt{-1}}{n!} f_k^{(n-1)}(z(x)).$$
The proof is complete. $\square$

## 14.4. Proof of Abel's Theorem

Fix a normalized basis $\omega_1, ..., \omega_g \in \Gamma_\mathcal{O}(X, K_X)$. We begin with the following reduction.

LEMMA 14.4.1. *A divisor $D = \sum_j (x_j - y_j)$ is the divisor of a meromorphic function $f$ if and only if there exist constants $c_k \in \mathbb{C}$, $k = 1, ..., N$, such that the meromorphic 1-form*

(14.2) $$\alpha := \sum_{j=1}^{N} \eta_{x_j, y_j} + \sum_{k=1}^{g} c_k \omega_k$$

*has $2\pi\sqrt{-1}$-commensurate periods, i.e.,*
$$A_k(\alpha), B_k(\alpha) \in 2\pi\sqrt{-1}\mathbb{Z}, \quad k = 1, ..., g.$$

**Proof.** Suppose there is a meromorphic function $f$ with $\mathrm{Ord}(f) = D$. Then by the Argument Principle, $\mathrm{Ord}(f) = \mathrm{Res}(df/f)$, and thus

$$\frac{df}{f} = \sum_{j=1}^{N} \eta_{x_j, y_j} + \sum_{k=1}^{g} c_k \omega_k$$

for some constants $c_k$. Since for any $f \in \mathcal{M}(X)$ and any closed curve $\gamma \subset X$,

$$\int_\gamma \frac{df}{f} \in 2\pi\sqrt{-1}\mathbb{Z},$$

we have proved one direction.

Conversely, if there exist $c_1, ..., c_g$ such that, with $\alpha$ defined by (14.2),

$$\int_\gamma \alpha \in 2\pi\sqrt{-1}\mathbb{Z}$$

for all closed curves $\gamma$ not passing through the support of $D$, then the (well-defined) function $f$ given by

$$f(x) := e^{\int_{x_o}^{x} \alpha}$$

has the property that $\mathrm{Ord}(f) = D$. □

**Conclusion of the proof of Abel's Theorem.** Let $\alpha$ be the form given by (14.2). Then

$$A_k(\alpha) = c_k,$$

since by definition the $a_k$-periods of $\eta_{x,y}$ are all 0. By the Reciprocity Theorem,

$$B_k(\alpha) = \sum_{j=1}^{N} 2\pi\sqrt{-1} \int_{x_j}^{y_j} \omega_k + \sum_{\ell=1}^{g} c_\ell B_{\ell k},$$

where

$$B_{\ell k} = \int_{b_\ell} \omega_k.$$

Thus

$$\sum_{j=1}^{N} \int_{x_j}^{y_j} \omega_k = \frac{1}{2\pi\sqrt{-1}} \left( B_k(\alpha) - \sum_{\ell=1}^{g} B_{k\ell} A_\ell(\alpha) \right)$$

$$= \sum_{\ell=1}^{g} \left( \frac{1}{2\pi\sqrt{-1}} B_k(\alpha) \int_{a_\ell} \omega_k - \frac{1}{2\pi\sqrt{-1}} A_\ell(\alpha) \int_{b_\ell} \omega_k \right).$$

Now, Lemma 14.4.1 tells us that $D$ is the divisor of a meromorphic function if and only if

$$\frac{1}{2\pi\sqrt{-1}} A_k(\alpha), \frac{1}{2\pi\sqrt{-1}} B_k(\alpha) \in \mathbb{Z}.$$

It follows that $D$ is the divisor of a meromorphic function if and only if there are integers $n_{k\ell}$ and $m_{k\ell}$, $1 \leq k \leq N$, $1 \leq \ell \leq g$, such that

$$\sum_{j=1}^{N} \int_{x_j}^{y_j} \omega_k = \sum_{\ell=1}^{g} n_{k\ell} \int_{a_\ell} \omega_k + m_{k\ell} \int_{b_\ell} \omega_k, \qquad k = 1, ..., g.$$

But the latter holds if and only if for every holomorphic form $\theta$,

$$\sum_{j} \int_{x_j}^{y_j} \theta \in \Pi(X, \theta).$$

In view of the remark following the statement of Abel's Theorem, we are done. □

## 14.5. A discussion of Jacobi's Inversion Theorem

In Paragraph 14.1.2 we defined a map $\mathscr{A} : X \to J(X)$ from a compact Riemann surface $X$ to its Jacobian, and an associated map $\mu : \text{Div}^0(X) \to J(X)$ from the set of divisors of degree 0 on $X$ to the Jacobian $J(X)$. The map $\mu$ is clearly a group homomorphism. Abel's Theorem tells us that the kernel of this homomorphism is the set of linearly trivial divisors, i.e., divisors of a meromorphic function.

Recall that the quotient of the set of divisors $\text{Div}(X)$ by the set of divisors of meromorphic functions is $\text{Pic}(X)$, the group of all holomorphic line bundles on $X$. The subgroup of equivalence classes of degree-zero line bundles is called $\text{Pic}^0(X)$. In view of Abel's Theorem, $\text{Pic}^0(X) = \text{Div}^0(X)/\ker \mu$, and therefore the map $\mu$ induces an injective map

$$\tilde{\mu} : \text{Pic}^0(X) \to J(X)$$

called the *Abel-Jacobi map*.

The Jacobi Inversion Theorem can be stated in this context as follows.

THEOREM 14.5.1 (Jacobi Inversion). *The Abel-Jacobi map $\tilde{\mu} : \text{Pic}^0(X) \to J(X)$ is surjective.*

We will not give a complete proof of Jacobi's Inversion Theorem. Instead we content ourselves with an argument that we hope is convincing.

Let $X$ be a compact Riemann surface of genus $g$. Consider the quotient

$$X^{(g)} := X^g/\mathbb{S}_g$$

of the $g$-fold Cartesian product of $X$ by the group of permutations of $g$ elements. That is to say, $X^{(g)}$ is the set of unordered $g$-tuples of points of $X$.

It is not hard to see that at a $g$-tuple of distinct points, $X^{(g)}$ looks like a manifold. Amazingly, the points of $X^{(g)}$ with repetitions are also smooth. To establish smoothness of $X^{(g)}$ at "multiple points", the reader can work locally with a disk. The idea is then to think of the points of $\mathbb{D}^{(g)}$ as roots of a polynomial of degree $g$ in $\mathbb{D}$ and use the symmetric functions of the roots as local coordinates. (The roots of a polynomial do not vary nicely with the coefficients of that polynomial, but the symmetric functions of the roots of a polynomial are holomorphic as functions of the coefficients of the polynomial, and a monic polynomial of 1 variable is uniquely determined by its roots up to permutation.)

By fixing a point $z \in X$, we can identify $X^{(g)}$ with a subset of $\text{Div}^0(X)$ by sending $\{x_1, ..., x_g\}$ to $\sum_{i=1}^g (x_i - z)$. Now define $\tilde{\mu}^{(g)} : X^{(g)} \to J(X)$ by

$$\tilde{\mu}^{(g)}(\{x_1, ..., x_g\}) = \mu\left(\sum_{j=1}^g (x_j - z)\right) = \sum_{i=1}^g (\mathscr{A}(x_i) - \mathscr{A}(z)).$$

The goal is to show that $\tilde{\mu}^{(g)}(X^{(g)}) = J(X)$. This goal is achieved in two steps.

(1) First, one shows that the image of $\mu$ is open and dense, as follows. We choose local coordinates $z_i$ centered at $x_i$. We can also choose torus coordinates on $J(X)$ as follows. By fixing a basis $a_1, b_1, ..., a_g, b_g$ for $H_1(X, \mathbb{Z})$ and an associated normalized basis $\omega_1, ..., \omega_g$ of $\Gamma_\mathcal{O}(X, K_X)$, we identify $(\Gamma_\mathcal{O}(X, K_X))^*$ with $\mathbb{C}^g$, and identify $\mathscr{I}(H_1(X, \mathbb{Z}))$ with the lattice $\Lambda$ in $\mathbb{C}^g \cong \mathbb{R}^{2g}$ generated by the $2g$ real vectors

$$e_{2j-1}, \sum_{k=1}^g \left(\int_{b_k} \omega_j\right) e_{2k}, \quad j = 1, ..., g.$$

The quotient of $\mathbb{C}^g$ by this lattice is biholomorphic to the complex torus $J(X)$.

In the coordinates chosen above,

$$\tilde{\mu}^{(g)}\{z_1, ..., z_g\} = \left[\sum_{i=1}^g \int_z^{z_i} \omega_1, ..., \sum_{i=1}^g \int_z^{z_i} \omega_g\right].$$

(The square brackets indicate that the vector is taken modulo periods.) By carefully computing the derivative of this map, one can show that there is some point $\{z_1^o, ..., z_g^o\}$ such that $d\tilde{\mu}^{(g)}(\{z_1^o, ..., z_g^o\})$ has full rank $g$. But the points of maximal rank occupy an open dense set. Thus since $X^{(g)}$ and $J(X)$ have the same dimension, the Implicit Function Theorem shows that the image of $\tilde{\mu}^{(g)}$ is open and dense.

(2) The map $\tilde{\mu}^{(g)}$ is a holomorphic map of equidimensional connected compact complex manifolds, and its Jacobian is not identically zero. Hence $\tilde{\mu}^{(g)}$ is proper. In particular, its image is closed. Thus $\tilde{\mu}^{(g)}$ is surjective.

In conjunction with Abel's Theorem we deduce the following fact.

THEOREM 14.5.2 (Structure of the moduli space of degree-zero line bundles). *The collection* $\text{Pic}^0(X)$ *of holomorphic line bundles of degree zero on a compact Riemann surface $X$ of positive genus is naturally identified with the Jacobian $J(X)$. In particular,* $\text{Pic}^0(X)$ *has the structure (group structure included) of a compact complex torus of dimension equal to the genus of $X$.*

# Bibliography

[Ahlfors-1978] Ahlfors, L., *Complex Analysis. An Introduction to the Theory of Analytic Functions of One Complex Variable.* McGraw-Hill, New York, 1978.

[Ahlfors & Sario-1960] Ahlfors, L.; Sario, L., *Riemann Surfaces.* Princeton Math. Series, No. 26. Princeton University Press, Princeton, NJ, 1960.

[Chern-1955] Chern, S.-S., *An elementary proof of the existence of isothermal parameters on a surface.* Proc. Amer. Math. Soc. 6 (1955), 771–782.

[Farkas & Kra-1990] Farkas, H.; Kra, I., *Riemann Surfaces.* Second Edition. Springer GTM 71, Springer-Verlag, 1990.

[Forster-1991] Forster, O., *Lectures on Riemann Surfaces.* Springer GTM 81, Springer-Verlag, 1991.

[Griffiths & Harris-1978] Griffiths, P.; Harris, J., *Principles of Algebraic Geometry.* Wiley Interscience, 1978.

[Hörmander-2003] Hörmander, L., *The Analysis of Linear Partial Differential Operators. I. Distribution Theory and Fourier Analysis.* Reprint of the second (1990) edition. Classics in Mathematics. Springer-Verlag, Berlin, 2003.

[Hörmander-1990] Hörmander, L., *An Introduction to Complex Analysis in Several Variables.* Third Edition. North-Holland, 1990.

[Narasimhan-1992] Narasimhan, R., *Compact Riemann Surfaces.* Lectures in Mathematics ETH Zürich. Birkhäuser Verlag, Basel, 1992.

[Rudin-1991] Rudin, W., *Functional Analysis.* Second Edition. International Series in Pure and Applied Mathematics. McGraw-Hill, New York, 1991.

# Index

Abel map, 224
Abel's Theorem, xvi, 224
Abel-Jacobi map, 231
Abelian differentials, 228
Argument Principle, 8, 41, 96
Automorphism group
   of a compact Riemann surface, xvi
   of a complex torus, 53
   of the plane, sphere, and disk, 165

Basic Identity, 181
Behnke-Stein Runge Theorem, 128
Bjorn, Anders, xviii
Bjorn, Jana, xviii
Bochner-Kodaira Identity, 181
Buckingham, Robbie, xviii

Canonical bundle, 66
   basepoint free, 221
   degree of, 218
Cauchy-Green Formula, 1, 123
Cauchy-Riemann operator, 89
   for holomorphic line bundles, 105
Chern number, 109
Complete intersection, 34
Connection, 101
   $(1,0)$-, 105
   Chern, 106
      curvature of, 107, 183
      Levi-Čivita, and,, 113
      twisted Hodge, and,, 163
   curvature of, 102
   Hermitian, 105

   induced, 103

D'Angelo, John P., xvii
de Rham cohomology, 97
Degree
   of a covering map, 167
   of a divisor, 74
   of a holomorphic line bundle, 111
   of a holomorphic map, 48
   of the canonical bundle, 218
Demailly, Jean-Pierre, xvii
Dolbeault cohomology, 99
Dolbeault Lemma, 98

Euler characteristic, 50

Finite-dimensionality of $\Gamma_{\mathcal{O}}(X, L)$, 84, 150
Fubini-Study metric, 104
   curvature of, 108
Fundamental group, 96
   action on covers, 167

Genus, 23
   and Euler characteristic, 50
   arithmetic, 85
   arithmetic vs. geometric, 150
   curves of low, 219
   Dolbeault cohomology and, 99
Green's Function, 119, 141, 168
   Cauchy-Green-type formula, 123
   symmetry of, 122
Green's Theorem, 1, 12
   Cauchy-Green Formula, 1
   Stokes' Theorem, 94

Hahn-Banach Theorem, 128, 141, 184
Harmonic forms, 145
   regularity of, 148
Harmonic functions, 11, 57, 147
   conjugate, 11
   regularity of, 14
Harnack Principle, 59
Herron, David, xviii
Hodge decomposition, 151
Hodge star, 145
Hodge Theorem, 151
   positive curvature, 195
   twisted version, 161, 212
Hyperbolic
   Poincaré, 170
   potential-theoretically, 168

Implicit Function Theorem, 29
Isothermal coordinates, 24
   existence theorem for, 184

Jacobi Inversion Theorem, 231
Jacobian of a curve, 224
   and $\mathrm{Pic}^0$, 232

Köbe, 9
   Compactness Theorem, 9, 175
Kodaira's Embedding Theorem, 202
Korn-Lichtenstein Theorem, 24
   Proof of, 184

Laplace-Beltrami Operator
   for $\bar{\partial}$, 162, 195
   for $\nabla^{1,0}$, 163
Laplace-Beltrami operator, 147
Laplacian, 11, 57
Linear equivalence of divisors, 77
Liouville's Theorem, 4
   $L^2$ version, 86
Lorent, Andy, xviii

McNeal, Jeffery D., xvii
Mejia, Diego, xviii
Minda, David, xviii
Mittag-Leffler Problem, 136
   and Dolbeault cohomology, 138
   and Riemann-Roch, 212
   for meromorphic 1-forms, 139

Narasimhan's Embedding Theorem, 204
Normal coordinate system, 148
Normal Forms Theorem, 7, 47
Numerical equivalence of divisors, 77

Order
   of a holomorphic function, 4
   of a meromorphic function, 39
   of a meromorphic section, 66

Poincaré Lemma, 98
Poincaré metric, 170
Poisson
   Equation, xiii
      solution of, 142, 151
   Equation, and curvature, 157
   Formula, 13, 58
Poplicher, Mihaela, xviii
Projective
   curves, 31
   line, 23, 26
   map, 81
   plane, 31
   space, 34

Raich, Andy, xvii
Ramachandran, Mohan, xviii
Residue, 5
   and Serre Duality, 214
   of a meromorphic 1-form, 76
   of a principal part, 138
   Theorem, 5, 95
Riesz Representation Theorem, 128, 141, 184
Robles, Colleen, xvii
Runge open set, 126

Shanmugalingam, Nages, xviii
Singularities, 4, 39
   Riemann's Theorem, 6, 78
Siu, Yum-Tong, xvii
Spectral Theorem for Compact Self-adjoint Operators, 155
Stokes' Theorem, 94
Subharmonic
   exhaustion, strictly, 160, 204
   function, 14, 59
   functions, local integrability of, 17
   functions, Perron family of, 115
   functions, smoothing of, 18

Weierstrass, 6
   $\wp$-function, 40
   -Casorati Theorem, 6
   points, xvi
   Product Theorem, 135

Zelditch, Steve, xv, xvii

# Titles in This Series

125 **Dror Varolin,** Riemann surfaces by way of complex analytic geometry, 2011

124 **David A. Cox, John B. Little, and Henry K. Schenck,** Toric varieties, 2011

123 **Gregory Eskin,** Lectures on linear partial differential equations, 2011

122 **Teresa Crespo and Zbigniew Hajto,** Algebraic groups and differential Galois theory, 2011

121 **Tobias Holck Colding and William P. Minicozzi II,** A course in minimal surfaces, 2011

120 **Qing Han,** A basic course in partial differential equations, 2011

119 **Alexander Korostelev and Olga Korosteleva,** Mathematical statistics: asymptotic minimax theory, 2011

118 **Hal L. Smith and Horst R. Thieme,** Dynamical systems and population persistence, 2010

117 **Terence Tao,** An epsilon of room, I: pages from year three of a mathematical blog. A textbook on real analysis, 2010

116 **Joan Cerdà,** Linear functional analysis, 2010

115 **Julio González-Díaz, Ignacio García-Jurado, and M. Gloria Fiestras-Janeiro,** An introductory course on mathematical game theory, 2010

114 **Joseph J. Rotman,** Advanced modern algebra: Second edition, 2010

113 **Thomas M. Liggett,** Continuous time Markov processes: An introduction, 2010

112 **Fredi Tröltzsch,** Optimal control of partial differential equations: Theory, methods and applications, 2010

111 **Simon Brendle,** Ricci flow and the sphere theorem, 2010

110 **Matthias Kreck,** Differential algebraic topology: From stratifolds to exotic spheres, 2010

109 **John C. Neu,** Training manual on transport and fluids, 2010

108 **Enrique Outerelo and Jesús M. Ruiz,** Mapping degree theory, 2009

107 **Jeffrey M. Lee,** Manifolds and differential geometry, 2009

106 **Robert J. Daverman and Gerard A. Venema,** Embeddings in manifolds, 2009

105 **Giovanni Leoni,** A first course in Sobolev spaces, 2009

104 **Paolo Aluffi,** Algebra: Chapter 0, 2009

103 **Branko Grünbaum,** Configurations of points and lines, 2009

102 **Mark A. Pinsky,** Introduction to Fourier analysis and wavelets, 2009

101 **Ward Cheney and Will Light,** A course in approximation theory, 2009

100 **I. Martin Isaacs,** Algebra: A graduate course, 2009

99 **Gerald Teschl,** Mathematical methods in quantum mechanics: With applications to Schrödinger operators, 2009

98 **Alexander I. Bobenko and Yuri B. Suris,** Discrete differential geometry: Integrable structure, 2008

97 **David C. Ullrich,** Complex made simple, 2008

96 **N. V. Krylov,** Lectures on elliptic and parabolic equations in Sobolev spaces, 2008

95 **Leon A. Takhtajan,** Quantum mechanics for mathematicians, 2008

94 **James E. Humphreys,** Representations of semisimple Lie algebras in the BGG category $\mathcal{O}$, 2008

93 **Peter W. Michor,** Topics in differential geometry, 2008

92 **I. Martin Isaacs,** Finite group theory, 2008

91 **Louis Halle Rowen,** Graduate algebra: Noncommutative view, 2008

90 **Larry J. Gerstein,** Basic quadratic forms, 2008

89 **Anthony Bonato,** A course on the web graph, 2008

88 **Nathanial P. Brown and Narutaka Ozawa,** $C^*$-algebras and finite-dimensional approximations, 2008

# TITLES IN THIS SERIES

87 **Srikanth B. Iyengar, Graham J. Leuschke, Anton Leykin, Claudia Miller, Ezra Miller, Anurag K. Singh, and Uli Walther,** Twenty-four hours of local cohomology, 2007

86 **Yulij Ilyashenko and Sergei Yakovenko,** Lectures on analytic differential equations, 2007

85 **John M. Alongi and Gail S. Nelson,** Recurrence and topology, 2007

84 **Charalambos D. Aliprantis and Rabee Tourky,** Cones and duality, 2007

83 **Wolfgang Ebeling,** Functions of several complex variables and their singularities (translated by Philip G. Spain), 2007

82 **Serge Alinhac and Patrick Gérard,** Pseudo-differential operators and the Nash–Moser theorem (translated by Stephen S. Wilson), 2007

81 **V. V. Prasolov,** Elements of homology theory, 2007

80 **Davar Khoshnevisan,** Probability, 2007

79 **William Stein,** Modular forms, a computational approach (with an appendix by Paul E. Gunnells), 2007

78 **Harry Dym,** Linear algebra in action, 2007

77 **Bennett Chow, Peng Lu, and Lei Ni,** Hamilton's Ricci flow, 2006

76 **Michael E. Taylor,** Measure theory and integration, 2006

75 **Peter D. Miller,** Applied asymptotic analysis, 2006

74 **V. V. Prasolov,** Elements of combinatorial and differential topology, 2006

73 **Louis Halle Rowen,** Graduate algebra: Commutative view, 2006

72 **R. J. Williams,** Introduction the the mathematics of finance, 2006

71 **S. P. Novikov and I. A. Taimanov,** Modern geometric structures and fields, 2006

70 **Seán Dineen,** Probability theory in finance, 2005

69 **Sebastián Montiel and Antonio Ros,** Curves and surfaces, 2005

68 **Luis Caffarelli and Sandro Salsa,** A geometric approach to free boundary problems, 2005

67 **T.Y. Lam,** Introduction to quadratic forms over fields, 2004

66 **Yuli Eidelman, Vitali Milman, and Antonis Tsolomitis,** Functional analysis, An introduction, 2004

65 **S. Ramanan,** Global calculus, 2004

64 **A. A. Kirillov,** Lectures on the orbit method, 2004

63 **Steven Dale Cutkosky,** Resolution of singularities, 2004

62 **T. W. Körner,** A companion to analysis: A second first and first second course in analysis, 2004

61 **Thomas A. Ivey and J. M. Landsberg,** Cartan for beginners: Differential geometry via moving frames and exterior differential systems, 2003

60 **Alberto Candel and Lawrence Conlon,** Foliations II, 2003

59 **Steven H. Weintraub,** Representation theory of finite groups: algebra and arithmetic, 2003

58 **Cédric Villani,** Topics in optimal transportation, 2003

57 **Robert Plato,** Concise numerical mathematics, 2003

56 **E. B. Vinberg,** A course in algebra, 2003

55 **C. Herbert Clemens,** A scrapbook of complex curve theory, second edition, 2003

54 **Alexander Barvinok,** A course in convexity, 2002

53 **Henryk Iwaniec,** Spectral methods of automorphic forms, 2002

52 **Ilka Agricola and Thomas Friedrich,** Global analysis: Differential forms in analysis, geometry and physics, 2002

For a complete list of titles in this series, visit the
AMS Bookstore at **www.ams.org/bookstore/**.